HUMAN REPRODUCTION:

PHYSIOLOGY AND PATHOPHYSIOLOGY

contributors

Mohammad M. Ahmad, M.D., Ph.D.
David R. Bolling, M.D.
Charles E. Gibbs, M.D.
Ronald S. Gibbs, M.D.
Robert H. Hayashi, M.D.
Gabriel S. Khodr, M.D.
Joseph Seitchik, M.D.
Marguerite K. Shepard, M.D.
Theresa M. Siler-Khodr, Ph.D.
Paul C. Weinberg, M.D.

The University of Texas Health Science Center at San Antonio
Department of Obstetrics and Gynecology

HUMAN REPRODUCTION:
PHYSIOLOGY AND PATHOPHYSIOLOGY

Edited by

ROBERT W. HUFF, M.D.

CARL J. PAUERSTEIN, M.D.
The University of Texas
Health Science Center at San Antonio

A WILEY MEDICAL PUBLICATION
JOHN WILEY & SONS
New York · Chichester · Brisbane · Toronto

Library of Congress Cataloging in Publication Data:

Main entry under title:

Human reproduction: physiology and pathophysiology.

1. Gynecology. 2. Obstetrics. I. Huff, Robert W. II. Pauerstein, Carl J., 1932–
[DNLM: 1. Reproduction. 2. Pregnancy. 3. Pregnancy complications. 4. Gynecologic diseases—Physiopathology. WQ200 P582]

RG101.H785 618 78-24480
ISBN 0-471-03562-9

Printed in the United States of America

10 9 8 7 6 5 4 3 2 1

preface

For the last ten years a basic course in Human Reproductive Physiology and Pathophysiology has been taught to sophomore medical students at the University of Texas Medical School at San Antonio. This book contains material presented in the course. Techniques which have proven successful in classroom teaching have been retained. The reader will notice the presence of objectives at the beginning of each chapter. Although there are no references listed in the text, several general references are provided at the end of each chapter. A series of questions with answers at the end of each chapter assures the reader that salient points have not been overlooked.

No attempt has been made to write an all-encompassing textbook of Obstetrics and Gynecology. Operative techniques, histopathology and modalities of treatment have been omitted except where absolutely necessary. Despite these omissions, the material in this book has been useful daily to junior medical students during their clinical clerkship. Houseofficers in Obstetrics and Gynecology have found the book helpful for a quick review of basic material, and it is thought that the book will be of value to practicing physicians preparing for examinations.

The editors wish to thank members of the department who wrote chapters for the book and Dr. C. E. Gibbs who contributed greatly even though he is not an author. Jo Wiglesworth and Mary Gonzales typed most of the original drafts, Pat Schriver did the art work, and a special thanks goes to Gretta Small for her excellent final draft of the manuscript. Our thanks also go to Ruth Wreschner of John Wiley and Sons for her patience in guiding the book to completion.

Robert W. Huff
Carl J. Pauerstein

contents

HUMAN REPRODUCTION:
PHYSIOLOGY AND PATHOPHYSIOLOGY

part one

reproductive biology

chapter one
developmental and anatomic correlates of reproduction

OBJECTIVES

1. To review the stages of embryogenesis during which the pattern common to the male and female urogenital systems is developed.
2. Describe control mechanisms in differentiation of the gonads, internal genital organs, and external genital organs.
3. List four important facts concerning differentiation of the gonads.
4. Describe the processes by which the paramesonephric cords become the fully developed internal genital organs in the female.
5. Describe the body changes at puberty in boys and girls.
6. Identify disorders associated with confusion of sex assignment at birth.
7. Describe the steps in differential diagnosis of a newborn with sexual ambiguity.
8. Know basic principles of management of neonates with ambiguous genitalia.

NORMAL DEVELOPMENT

The genotype exerts the strongest influence on gonadal differentiation. The precise mechanisms by which genetic factors control sex differentiation remain unresolved. However, the consensus holds that the gonads, internal ducts, and external genitalia develop along female lines unless induced toward male differentiation by a Y-linked determinant, which may exert its influence directly or by modifying the activity of other genes.

Although the sex genotype is determined at fertilization, sex differentiation of the gonads and internal and external genitalia does not become evident until the seventh week and is not completed until 12 weeks in the female and 14 weeks in the male. During the intervening weeks, the primordia of the male and female gonads are established. Just as the genotype determines the direction in which the gonads will differentiate, so the gonads determine the differentiation of the internal ducts and the external genitalia.

Gonad—Undifferentiated Stage

Both ovaries and testes develop from an undifferentiated gonadal primordium. At about the fifth week of embryonic life, the genital ridge may be distinguished as a thickening of the epithelium covering the medial aspect of the mesonephros. In this region, the epithelium is not separated from the underlying mesenchyme by a basement membrane. The gonadal blastema elevates the medioventral surface of the urogenital ridge, forming a lateral mesonephric ridge and a medial genital ridge (Fig. 1–1). During subsequent differentiation, primordial germ cells and mesoblastic cells form the epithelial structures of the definitive gonad, whereas the mesenchymal cells contribute to the interstitial tissues of the gonads.

The germ cells arise in the yolk sac or gut endoderm, migrate to the root of the mesentery of the primitive gut, and become incorporated into the urogenital ridge. The indifferent gonad has two major components, the cortex and medulla. The medullary element is capable only of becoming a testis. The cortical element is capable only of development as an ovary. Initially, the gonadal ridge is contiguous with the intermediate three-fifths of the mesonephros. As the mesonephros regresses, the gonads are suspended by a "urogenital" mesentery plus the ligaments of the degenerated mesonephros. The cranial ligament disappears, and the caudal suspensory ligament anchors the gonad to the genital swelling and persists as the gonadoinguinal ligament (gubernaculum).

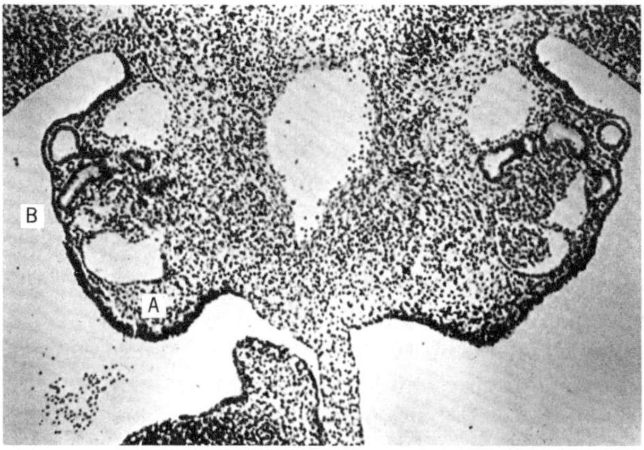

Figure 1–1. Section through an embryo at about the fifth week of embryonic life. The epithelium is thickened over the mesenchyme in the region of the developing gonad. (*A*). Laterally, mesonephric structures are seen in the region of the mesonephric ridge (*B*).

Gonad—Ovarian Differentiation

If so destined, the gonad differentiates into an ovary during the seventh week of embryonic life. One can then distinguish connective tissue and epithelial tissue, in the form of medullary cords. These cords contain primordial germ cells (oogonia) as well as primitive granulosa cells.

At this point, it is necessary to address a subject that causes confusion to all students of reproductive embryology: the nomenclature of the female gamete. The cells that migrate from the hindgut are called primordial germ cells. After reaching the genital ridge, they are designated oogonia. Early generations of oogonia enter into the formation of the medullary cords and degenerate. Later generations of oogonia remain in the cortical region. They then enter meiotic prophase and differentiate into oocytes. Meiosis is arrested in the diplotene phase of the first division and is only resumed in the mature ovary just prior to ovulation. The second meiotic division occurs after fertilization.

Late in embryonic life, the fetal ovary contains many primary follicles, each characterized by an oocyte surrounded by a single layer of flattened granulosa cells. Each of these follicles is surrounded by connective tissue. These changes allow ready recognition of the gonad as an ovary.

Gonad—Testicular Differentiation

The embryonal testis becomes distinguishable during the seventh week of embryonic life. The indifferent gonad is divided into cords by septae, formed of reticular fibers and fibroblasts. These cords are then segregated from the surface epithelium by a sheet of connective tissue, which will become the tunica albuginea. The primordial germ cells become incorporated into the testicular cords. The cords are lined by several layers of Sertoli cells. The incorporated germ cells are now called spermatogonia. At about 60 days, Leydig cells are seen in the interstitial tissues near the tunica albuginea. These cells reach their maximum number during the third and fourth months and begin to decrease during the fourth month. The fetal testes actively produce androgens, which influence the development of the internal ducts and the external genitalia.

Internal Ducts—Indifferent Stage

During the indifferent stage, the embryo possesses both mesonephric (Wolffian) and paramesonephric (Müllerian) ducts. The progenitor of the mesonephric duct, the pronephric duct, is completed by the seventh week, at which time it is called the mesonephric duct. The pronephric tubules arise from the lateral portion of the urogenital ridge to connect the pronephric duct by the fourth week of embryonic life. By the sixth week, the Müllerian cleft appears between the pronephric and gonadal components of the urogenital ridge. The cleft is the cranial end of the future Müllerian duct.

Just caudad to the cleft, the solid tip of the Müllerian cord grows downward, intimately associated with the Wolffian duct ventrally and reaches the urogenital sinus. Here the tips of the right and left cords elevate the dorsal wall of the urogenital sinus to form Müller's tubercle, which eventually opens into the urogenital sinus (Fig. 1–2). At the conclusion of the indifferent stage, each Müllerian cord has become a hollow tube, meeting in the midline. The Wolffian ducts are also present at this stage. They have also grown downward, as solid cords, to reach the region of the mesonephric blastema. They canalize and enter the caudal portion of the hindgut, which then becomes the cloaca. Mesonephric tubules differentiate next to the mesonephric duct, to degenerate after formation of the metanephros. Some of the mesonephric tubules will persist as vestigial structures.

The urogenital sinus initially appears during the fifth to sixth week.

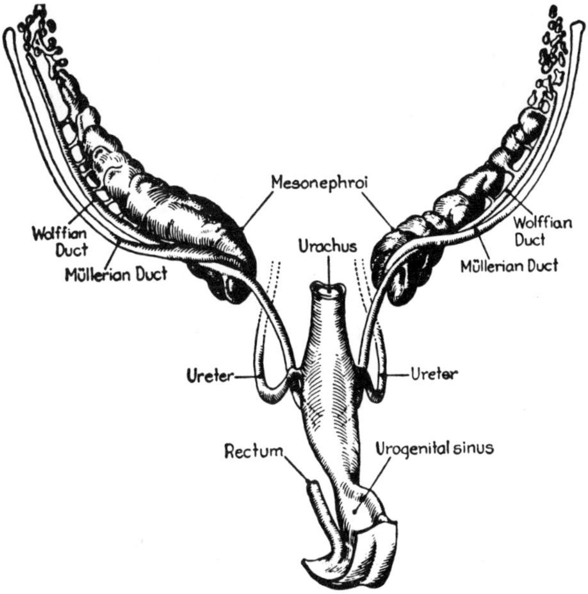

Figure 1–2. Reconstruction of the urogenital system during the indifferent stage. Note the relationships of the Wolffian and Müllerian ducts to each other and to the mesonephros. (From Hunter RH: Observations on the human female genital tract. *Contrib Embryol* 22:91–108, 1930. Courtesy of the Carnegie Institute of Washington, D.C.)

The transverse septum (urorectal septum) divides the cloaca into the rectum and a ventral segment that is the vesicourethral canal and the urogenital sinus. These regions give rise to structures of both the genital and urinary system. The most cranial portion gives rise to the urinary bladder. The midportion and caudal portion form the posterior urethra and membranous urethra. Subsequent urethral evaginations develop into the prostate.

Internal Ducts—Female Differentiation

If embryos are castrated at the indifferent stage, the mesonephric ducts regress, and the paramesonephric ducts develop. Local implantation of a testosterone crystal in a female fetus prevents mesonephric regression but does not induce paramesonephric regression. In contrast, local implantation of a testis both supports the mesonephric ducts and induces paramesonephric regression. The embryonic testis has two organizing functions, only one of which can be replaced by testosterone. In the absence of testes, the internal ducts always develop along female lines, whether or not ovaries are present.

By the end of the third month, the mesonephric ducts in the female have largely disappeared. The medial walls of the right and left Müllerian ducts fuse. Fusion usually starts at Müller's tubercle and progresses cranially, ending at the junction of each duct with the round ligament. This process results in a single tube with two lumens (Fig. 1–3). The medial septum then disappears, completing the creation of the uterus simplex. Canalization is complete by the 16th week. The

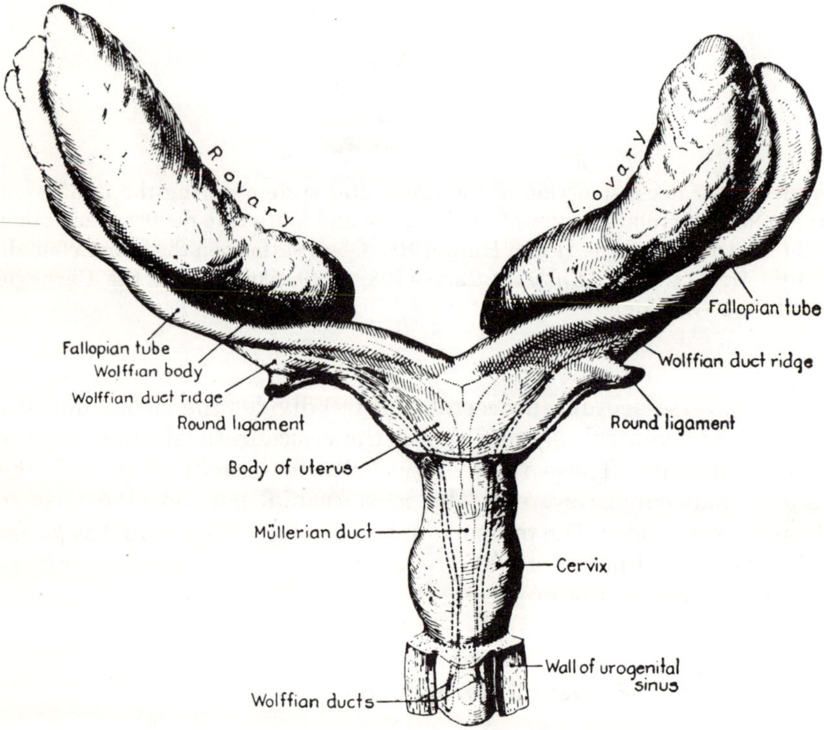

Figure 1–3. Internal ducts in embryo that has undergone female differentiation. The Wolffian ducts have regressed, and the Müllerian ducts have come together in the midline. (From Hunter RH: Observations on the human female genital tract. *Contrib Embryol* 22:91–108, 1930. Courtesy of the Carnegie Institute of Washington, D.C.)

muscular and connective tissue layers of the oviduct begin to appear as mesenchymal concentrations during the third month.

Internal Ducts—Male Differentiation

In male embryos, the paramesonephric ducts begin to regress during the ninth week. Paramesonephric regression begins in the region of the testes. The cranial end persists near the testes as the appendix testes, and the caudal end persists as the prostatic utricle. Subsequent male development involves the formation of the epididymis, vas, ejaculatory ducts, seminal vesicles, and prostate gland. Development of all these structures is androgen dependent.

External Genitalia—Indifferent Stage

During the undifferentiated stage, the external genitalia are bordered laterally by the genital swellings (or labioscrotal folds) and medially by the urethral groove. Just lateral to the urethral groove on each side, the urethral folds are interposed between the urethral groove and the genital swellings. The urethral groove is bounded anteriorly by the genital tubercle (Fig. 1–4B). The urogenital sinus opens into the urethral groove.

External Genitalia—Female Differentiation

At Müller's tubercle, the caudal extremity of the fused Müllerian duct ends against the urogenital sinus. The vaginal cord, which is derived from the blind ends of the Müllerian ducts, is interposed between the latter and the urogenital sinus (Fig. 1–5). As development proceeds, the urogenital sinus becomes progressively smaller as the vaginal cord progressively lengthens. Eventually, the vaginal plate acquires a cord and, at about 5 months, becomes the vagina. It is quite clear that the vagina is contributed to both by the Müllerian ducts and by the urogenital sinus. The precise site of this junction has been the subject of considerable debate.

In female differentiation, the urethral groove becomes the vestibule, the urethral folds become the labia minora, and the genital swellings form the labia majora. The genital tubercle becomes the clitoris (Fig. 1–4A).

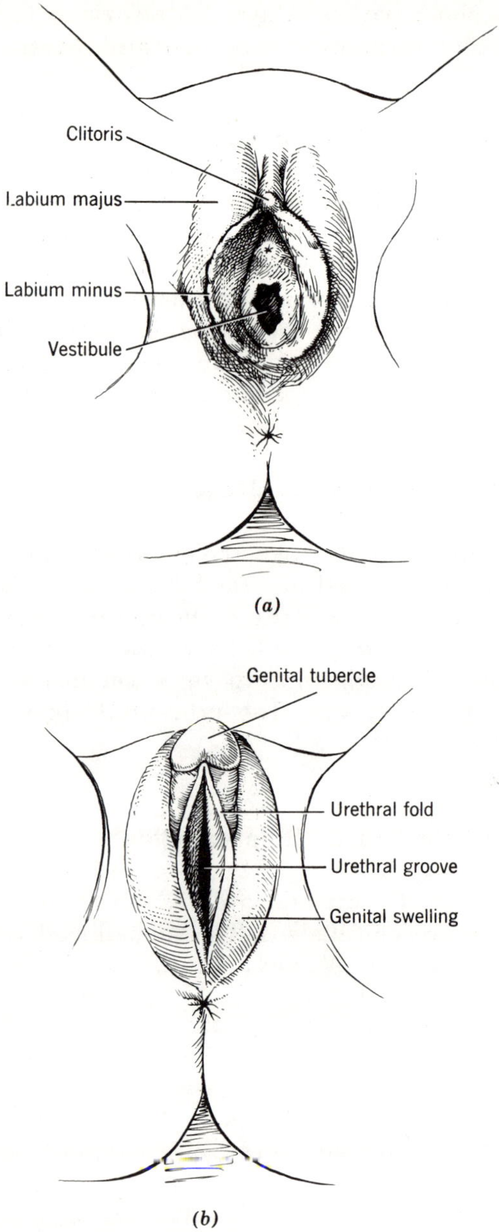

Figure 1–4. Diagrammatic representation of the development of the external genitalia. Panel *b* demonstrates the indifferent stage, Panel *a* female differentiation, and Panel *c* male differentiation. (See text for explanation.)

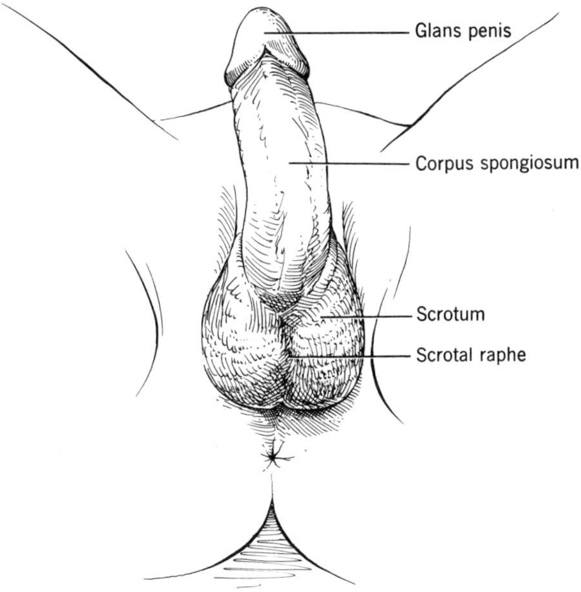

(c)

Figure 1–4. (continued)

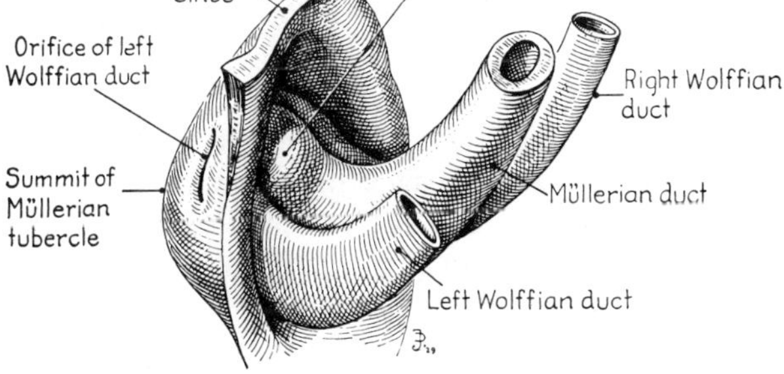

Figure 1–5. The development of the vagina. Note the relationships among the urogenital sinus, Müllerian tubercle, and vaginal cord. (From Koff AK: Development of the vagina in the human fetus. *Contrib Embryol* 24:61, 1933. Courtesy of Carnegie Institute of Washington, D.C.)

External Genitalia—Male Differentiation

Development of male external genitalia occurs in genetic males in response to androgen secreted by the embryonic testes (8th–13th week of gestation). Male genital development may also occur in genetic females exposed to androgens. Because testosterone is necessary for the development of male genitalia, genetic males with defective testes have female genitalia. The changes establishing the male genitalia involve expansion of the genital tubercle, genital swellings, urogenital folds, and fusion of the urogenital folds. The genital tubercle enlarges to form the glans penis. The dorsal portion of the penis is formed by mesodermal proliferation between the genital tubercle and each genital swelling, creating the corpora cavernosa penis. The urogenital folds fuse ventrally and convert the urogenital groove into the penile urethra. Proliferation of mesoderm creates the corpus spongiosum. The urogenital groove becomes continuous with the urethral groove distally and, with rupture of the urogenital membrane, is continuous with the urogenital sinus proximally. The urethral groove develops from an extension of the urethral plate, which appears at the distal end of the glans penis and extends ventrally into the genital tubercle in the midsagittal plane. The urogenital groove and its distal continuation, the urethral groove and urethral plate, are incorporated into the glans penis. As the urethral folds fuse, the urethral plate, groove, and sinus canalize to form the penis urethra. The genital swellings enlarge and move caudally. As the swellings enlarge, they override the caudal portion of the fused genital folds and fuse to become the scrotal swellings. The scrotal swellings are converted into a sac by an evagination of the peritoneal cavity (processus vaginalis) into the scrotal swellings. Connective tissue trapped between the fused vaginal processes forms the scrotal septum. The point of fusion creates the scrotal raphe (Fig. 1–4C).

The gubernaculum testes (caudal ligament of the mesonephros) is attached to the genital (scrotal) swelling. It participates passively (as an anchor) and actively (by contraction of its smooth muscle) in the movement of the testes from an intra-abdominal to a scrotal position. During this "descent," the testis carries the vas, nerves, and vessels with it. Normally, the communication of the vaginal processus with the peritoneal cavity is closed due to fusion of its walls—occasionally, it may persist in varying degrees to give rise to hydroceles or congenital hernias.

ABNORMAL DEVELOPMENT

Sometimes the sequence just described goes awry, and sexual development is abnormal. Abnormal sexual development may be due to faulty gonadal development or faulty endocrine environment. These disorders may come to attention at birth because the external genitalia are ambiguous or may be recognized at puberty because of failure to develop appropriate secondary sexual characteristics or found in reproductive life because of infertility.

Four disorders cause confusion in sex assignment at birth: 1) true hermaphroditism, 2) mixed gonadal dysgenesis, 3) dysgenetic male pseudohermaphroditism, and 4) female pseudohermaphroditism. They can be elucidated by history, physical examination, nuclear chromatin pattern, and measurement of 17-ketosteroids.

Physical examination of the newborn should include attention to the size of the phallus, to the location of the urethral opening, and to the presence and degree of labial fusion. If the child appears to be a male with hypospadias or incomplete scrotal fusion, the scrotum and groins should be palpated. A palpable gonad in the scrotum or in a hernia is always a testis. If the infant is bilaterally cryptorchid, a rectal exam should be done. If a uterus is felt, the probable diagnosis is adrenal hyperplasia, although true hermaphroditism or dysgenetic male pseudohermaphroditism should be considered.

In patients with ambiguous external genitalia, it is useful to obtain a buccal smear to search for the presence of sex chromatin (Barr bodies). The buccal smear allows the infant to be classified as either chromatin positive or chromatin negative. The chromatin-positive group includes true hermaphrodites and female pseudohermaphrodites. The term *hermaphrodite* means that some aspect of the patient's body sex is discordant with the others. Hermaphrodites are classified by their gonadal sex. Thus, the male hermaphrodite has testes, but either the internal ducts or external genitalia are female. Similarly, the female hermaphrodite has ovaries but is virilized. The true hermaphrodite has gonadal tissue of both sexes. This may be an ovotestis, a testis on one side and an ovary on the other, or some other combination. About 80% of true hermaphrodites are chromatin positive.

Female pseudohermaphroditism may be caused by congenital adrenal hyperplasia or by a virilizing tumor in the mother. These patients are genetic and gonadal females with varying degrees of external virilization. All have normal gonads and internal genitalia. The ambiguous external genitalia are due to exposure to androgens prior to the 16th week of gestation. They are capable of normal coitus and reproduction if properly treated. The adrenal hyperplasia subgroup is especially interesting because of the defective synthesis of hydrocortisone. The low cortisol levels cause increased ACTH production. Since the pathways for adrenal androgenesis are present, response to the increased ACTH causes virilization. The diagnosis is confirmed by elevated pregnanetriol and 17-ketosteroid levels.

The chromatin-negative group includes the other 20% of true hermaphrodites, patients with mixed gonadal dysgenesis and male pseudohermaphrodites.

Patients with unequivocal mixed gonadal dysgenesis are always chromatin negative. This syndrome is defined by the gonadal differentiation found. Usually, a streak gonad is found on one side and a testis on the contralateral side. Other combinations, such as a testis or streak on one side and a gonadal tumor on the other, are included in this classification. When sufficiently careful study of tissues is carried out, the usual cytogenetic finding in patients with mixed gonadal dysgenesis is mosaicism, including one 45, X stem line and another line with Y-containing cells. Cytogenetic findings and the syndrome do not correspond exactly in all cases. It seems likely that the 45, X stem has the dominant role in the development of the streak gonad and the Y-bearing stem in the appearance of a testis.

All patients with mixed gonadal dysgenesis have a uterus and at least one Fallopian tube. Most have bilateral tubes, but some display a vas deferens. The female internal genital development is accomplished by a complete spectrum of external virilization.

Dysgenetic male pseudohermaphrodites have two testes but incomplete virilization. They usually display genital development similar to that seen in mosaic gonadal dysgenesis. Some demonstrate 45, X/46,XY mosaicism. Because the abnormal testes fail to induce Müllerian suppression, the phenotype is ambiguous.

It is important to assign the correct sex to the newborn. The legal implications are obvious because the "nursery sex" appears on the birth certificate and thus becomes the "legal sex." Even more important is the impact that the assigned sex has on the child's experience in the family and society. Current evidence suggests that one's gender role is influenced to some degree by androgen exposure during critical periods of "imprinting" in utero. However, the sex of rearing seems a more important determinant of the development of a concept of sexual self. Thus, it is imperative that sexual anomalies be recognized at birth and that the sex be assigned that is consonant with the ability to function eventually as normally as possible regardless of phenotype or genotype.

MATURATION OF THE REPRODUCTIVE SYSTEM

At puberty, the gonads manufacture sex hormones in sufficient quantity to cause rapid growth of the genitalia, development of the secondary sex characteristics, and skeletal growth and maturation. The sequence and anatomic aspects of these maturational changes are detailed below.

Maturation—Female

The mean age of onset of pubertal changes in girls is eight years (range, 7–10 years). There is increased uterine and ovarian growth. Two years later, at approximately age 10, the first clinically discernible appearance of the pubertal sequence appears as an increase in diameter of the breast areola to form a small mound in approximately half of the girls. In the other half, growth of pubic hair is the first change noted. At first, lightly pigmented hair is present on the labia only. Estrogen output increases, the uterus begins to grow, and the ovaries enlarge. Approximately one year later, during the 11th year, other anatomic changes begin. The bony pelvis widens, and adipose tissue deposition begins on the hips, buttocks, and over the pubis. The pubic hair becomes coarser, and curlier, and some sparse growth begins over the mons pubis. The breasts enter a second stage of development. The breast bud emerges. The vaginal mucosa thickens. Uterine and ovarian size increases further. Gonadotrophins become detectable, and cyclic secretion of estrogen begins.

At about age 12, girls begin a premenarcheal growth spurt. The nipple and areola project to form the secondary breast mound. Axillary hair begins to appear (adrenarche). In most youngsters, menarche occurs within six months of the development of axillary hair.

Menarche is the single event that can be accurately documented. At the time of menarche, the pubic hair displays an adult pattern with a sharp superior border. The breasts demonstrate projection of the nipples and recession of the areola to the breast contour but no lateral pointing of the nipples. The uterus weighs 18 gms and the ovaries about 8 gms. Beyond puberty, the growth spurt will decelerate, and most women will grow only six inches over the next three years. The first menstrual episodes may be anovulatory. At the time of puberty (the first ovulation), pubic hair begins to grow toward the intercrural folds. The breasts display full adult contour and lateral pointing of the nipples. Some three to three and a half years beyond menarche, skeletal growth ceases. The ovaries are at the adult weight of 10 gms and the uterus weighs 30–40 gms.

Anatomy of the Female Pelvis

In the adult woman, the pelvic floor or pelvic diaphragm is formed by the levator ani and coccygeus muscles that meet in the midline. It is covered by the parietal layer of endopelvic fascia. Viewed from above, it appears as a broad hammock of muscle sweeping down from the pelvic brim, investing urethra, vagina, and rectum. There are two gaps in the muscular diaphragm. The anterior gap is bridged by the urogenital diaphragm and transmits the urethra and vagina. The posterior gap transmits the rectum and the anal canal.

The levator ani muscles are divided into three segments, which form the major part of the pelvic diaphragm. The coccygeus muscle completes the pelvic diaphragm posteriorly. The pubococcygeus muscle can be further divided into three parts: the pubourethralis, which skirts the urethra and blends with its intrinsic muscular coat; the pubovaginalis, which inserts into the lateral and posterior walls of the vagina; and the puborectalis, which forms a loop around the anorectal junction and is inserted into the lateral and posterior walls of the anal canal. The iliococcygeus and ischiococcygeus muscles complete the lateral parts of the pelvic floor (Fig. 1–6A, B).

The pelvic floor has inherent tone and reflexly responds to changes in the intra-abdominal pressure. The pubococcygeus is traversed by the urethra, vagina, and rectum and functions to support the vagina, the uterus (indirectly), and the bladder, urethra, and rectum. The pubovaginalis (bulbocavernosus) muscles also act as a sphincter for the vagina. Excessive spasm can cause a condition known as vaginismus. This spasm can be of such intensity that sexual intercourse may be painful or intromission impossible. The puborectalis muscle controls descent of fecal material and acts as an auxiliary sphincter for the anal canal.

The endopelvic fascia is reflected off the pelvic viscera. It is a continuation of the transversalis fascia of the anterior abdominal wall. Where these organs pierce the pelvic floor, tubular fibrous investments

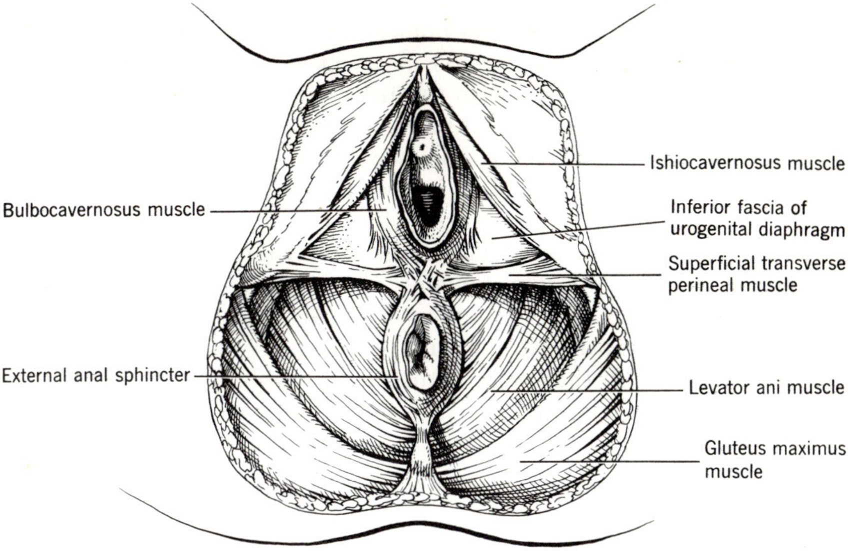

(a)

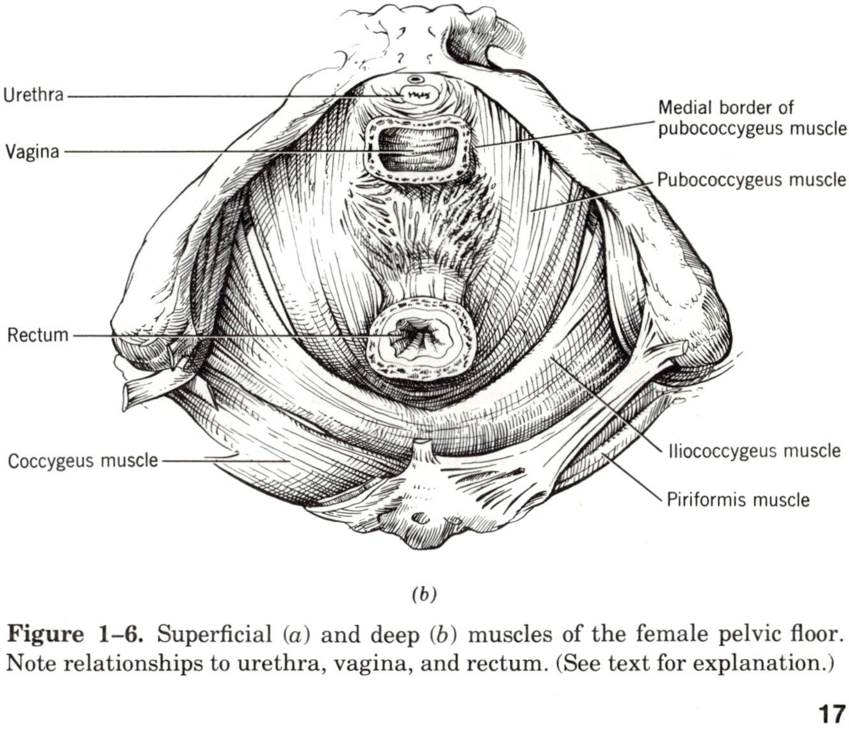

(b)

Figure 1–6. Superficial (*a*) and deep (*b*) muscles of the female pelvic floor. Note relationships to urethra, vagina, and rectum. (See text for explanation.)

are carried upward from the endopelvic fascia to form tightly fitting collars that blend with the outer muscular coats of the organs. A blend of endopelvic fascia, connective tissue, and the outer muscular coats of the various organs gives rise to the ligaments of support. The fascia between pubis and urethra is known as the pubourethral fascia; between pubis and cervix, the pubocervical fascia; between bladder and vagina, the vesicovaginal fascia; between rectum and vagina, the rectovaginal fascia. The condensations of tissue on the lateral aspects of the isthmus of the uterus extending to the lateral pelvic wall are called the cardinal ligaments. The condensations extending from the posterior aspect of the uterine isthmus to the sacrum are the uterosacral ligaments. The vesical, uterine, and rectal layers of endopelvic fascia are continuous with the fascia of the pelvic diaphragm, the obturator fascia, the iliac fascia, and the transversalis fascia.

Maturation—Male

The age of onset of puberty in boys is variable but occurs on the average at age 14. The secretion of androgens causes the penis to grow in width and length. The skin of the scrotum becomes pigmented and rugose. Changes in the internal genitals include enlargement of the seminal vesicles, the prostate gland, and the bulbourethral glands.

The general body conformation changes as the shoulders broaden and the muscles enlarge. General body hair increases, the male escutcheon appears in the pubic region, and hair grows in the axillae, on the chest, and around the anus. A beard appears, and some recession of the hair of the scalp is noted in the region of the temples, bilaterally. The skin may develop acne due to increased sebaceous gland secretion. Finally, the voice deepens as the larynx enlarges, and the vocal cords become larger and thicker.

Androgen production remains essentially unchanged until the seventh or eighth decade of life. Although there may be a reduction of androgen synthesis by the testes with age, there is a proportional decrease in metabolic clearance of male sex hormones. Thus, the aging process in the male is probably not directly related to diminished blood levels of male sex hormones.

Anatomy of the Male Pelvis

The penis is composed of three cylinders of erectile tissue, the paired corpora cavernosa and a single corpus spongiosum. The corpus spongiosum invests the urethra from the urogenital diaphragm to the meatus. All are enclosed within a fibrous capsule, the tunica albuginea, and covered by another fascial covering, Buck's fascia, which is attached to the urogenital diaphragm.

The body (corpora cavernosa) is attached to the pubic arch and to the urogenital diaphragm. The penis is fixed by the attachment of the crura of the corpora cavernosa to the medial aspect of the ischiopubic rami and the inferior rami of the pubis. The bulbous portion of the corpora spongiosum is attached to the urogenital diaphragm.

The blood supply is derived from branches of the internal pudendal artery. They pierce the suspensory ligament of the penis and run beneath Buck's fascia to terminate in anastomotic branches about the corona. Deep dorsal arteries arise from the internal pudendal artery and enter the bulb, ventral to the urethra. Superficial veins drain the integuments of the penis and return the blood through the superficial dorsal vein. The erectile tissue drains into the deep dorsal vein, which empties into the prostatovesical plexus.

The urethra extends the length of the penis, from the external urethral meatus to the internal orifice of the bladder. Three anatomic portions of the urethra are recognized: the penile urethra, which extends from the meatus to the distal leaf of the urogenital diaphragm; the membranous urethra, which lies within the urogenital diaphragm, and the prostatic urethra, which extends from the proximal aspect of the urogenital diaphragm to the vesical neck. This portion contains the ejaculatory duct and prostatic duct openings.

The prostate surrounds the proximal urethra. It is enclosed in a thick fibrous capsule derived from the pelvic fascia. The external surfaces are related to the rectum posteriorly, separated from the rectum by the reflection of the pelvic fascia. The apex of the prostate lies against the superior leaf of the urogenital diaphragm. In front of the prostate and between it and the symphysis pubis is a rich venous plexus into which the dorsal vein of the penis empties. All venous channels connect with the vesical plexus. Two fibrous thickenings of pelvic fascia extend from the anterior surface of the prostatic capsule and attach to the pubis—the puboprostatic ligaments.

The testis lies in a thin membranous sac, the tunica vaginalis, formed by the peritoneum during the descent of the testes. The tunica covers the testicle anteriorly but, posteriorly, the tunica vaginalis is lacking where the testis is capped by the epididymis. The outer layer of the tunica vaginalis is referred to as the parietal layer; the inner layer, which is closely applied to the testes, is the visceral layer. Each testis is covered by a thick fibroelastic capsule, the tunica albuginea. This tunica sends numerous septa through the organ that divide the testes into

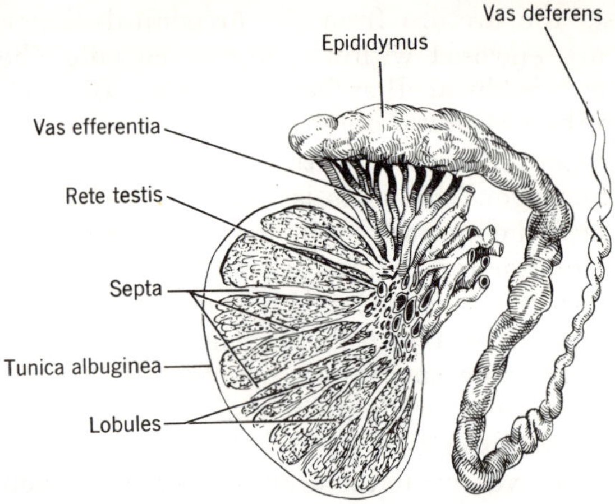

Figure 1–7. Diagrammatic representation of the adult testis and related ducts. (See text for explanation.)

pyramidal lobules that, in turn, contain the glandular elements of the testis. Each lobule is filled with convoluted seminiferous tubules. At the apex of the lobule, the tubules form straight tubules. These enter the mediastinum testes to form the rete testes. Near the upper pole of the testis, the tubules pierce the tunica albuginea and join the vasa efferentia. Thus, tubules and vasa efferentia unite to form a single convoluted tubule as the duct of the epididymis that joins the ductus deferens or vas deferens at the lower pole of the epididymis (Fig. 1–7).

At the upper pole of the testis, between it and the epididymis, is a small oval body called the appendix testes, or hydatid of Morgagni. It is a remnant of the Müllerian duct. From the rete testes, a collection of 15–20 ducts or vasa efferentia from the globus major of the epididymis form the body and tail of the epididymis. The ductus epididymis emerges from the tail of the epididymis or globus minor as the ductus deferens. The epididymis covers the posterior border of the testes, with the globus major at the upper pole and the globus minor at the lower pole. The tail of the epididymis is attached to the base of the scrotum by the genitoinguinal ligament.

The primary artery to the testes is the internal spermatic branch of the aorta, which arises just below the renal arteries. Two other blood supplies to the testes exist: the artery of the vas deferens, a branch of the inferior vesical artery, and the external spermatic or cremasteric artery, a branch of the deep epigastric artery.

Venous drainage from the testes and epididymis forms a plexus, the pampiniform plexus. This plexus consists of 8 or 10 veins that constitute a part of the spermatic cord. With cord, the veins enter the inguinal canal and near the internal inguinal ring form two main trunks. Higher up, these two trunks unite and form a single spermatic vein. The right spermatic vein drains into the inverior vena cava. The left spermatic vein enters the left renal vein.

The seminal vesicles are paired lobulated structures, 5–7 cms in length, that lie between the fundus of the bladder and the rectum. Their anterior surfaces are intimate to the wall of the bladder, and their posterior surfaces lie against the rectum, separated by the rectovesical fascia. At the fundus of the bladder, the ductus deferens becomes enlarged and tortuous to form the ampulla. Each seminal vesicle, at its anterior extremity, narrows into a straight duct that joins its corresponding duct to form the ejaculatory duct. The ejaculatory ducts end in tiny, slitlike openings in the prostatic urethra close to, or within, the prostatic utricle.

QUESTIONS

1. Match the organs listed in the first column with the normal control mechanisms listed in the second column:

A. Gonads	1. Fetal gonad
B. Internal genital organs	2. Hormones
C. External genital organs	3. Sex determinate chromosomes

2. Which of the following statements about gonadal differentiation are true?
   A. The gonad first appears at approximately 9 weeks gestation.
   B. The primordial germ cells originate from coelemic epithelium at the site of the future gonad in the urogenital ridge.
   C. Ovarian differentiation can be identified by the seventh week.
3. Vestigial remnants of the paramesonephric duct are identified in the male as prostatic utricle and appendix testes. TRUE FALSE
4. Which of the following processes occur during the maturation of the paramesonephric system into the fully developed internal female genital organ?
   A. Canalization
   B. Fusion
   C. Bifurcation
   D. Cord formation

5. Name the four disorders of sexual development that are associated with confusion concerning sex assignment at birth.

_______________ _______________

_______________ _______________

6. What specifics should be looked for at physical examination of a newborn with ambiguous external genitalia?

_______________ _______________

_______________ _______________

____

7. Which disorders causing ambiguous external genitalia are seen in chromatin-positive individuals?

_______________ _______________

8. Which are seen in chromatin-negative individuals?

_______________ _______________ _______________

9. Which of these conditions is compatible with normal reproduction if appropriately treated?_______________

10. The first clinical finding of puberal maturation in the female is often:
    A. Labial pubic hair growth
    B. Increase in areolar diameter
    C. Neither of the above
    D. Either of the above

11. Menarche is characterized by:
    A. Mean age, 12.5 years
    B. Occurs at an earlier age over the years
    C. Is the most accurate documented historical event
    D. Occurs coincident with axillary hair growth
    E. All of the above
    F. A, B, and C only

12. Coincident with the menarche:
    A. Peak height spurt has occurred
    B. Child can expect approximately 6% in increased statural growth
    C. Adrenarche has occurred approximately six months prior
    D. A and C only

E. A and B only
F. A, B, and C

13. In caring for a 16-year-old who has not yet menstruated, one finds, on examination, a definite breast and areolar development, pubic and *axillary* hair, and a normal pelvic examination with a patent hymen. Proper management would include:
    A. Further extensive endocrine evaluation
    B. Pituitary x-rays
    C. Reassurance that she is normal and menses will occur within six months
    D. Diagnostic curettage

ANSWERS

1. A/3, B/1, C/2
2. A. FALSE
   B. FALSE
   C. TRUE
3. TRUE
4. A, B, and D
5. true hermaphrodite, male pseudohermaphrodite, female pseudohermaphrodite, mixed gonadal dysgenesis
6. size of phallus, position of urethral orifice, labioscrotal fusion, presence of testes in scrotum, presence of uterus
7. true hermaphrodite (80%), female pseudohermaphrodite
8. true hermaphrodite (20%), mixed gonadal dysgenesis, male pseudohermaphrodite
9. female pseudohermaphrodite
10. D
11. F
12. F
13. C

chapter two
reproductive endocrinology

OBJECTIVES

1. Know the hormonal components of the hypothalamic-pituitary-ovarian axis and their sites of origin.
2. Know the major pathways of sex steroid synthesis.
3. Know the proposed mechanism and sites of action of the hypothalamic, pituitary, and gonadal hormones and understand the feedback relationships in both males and females.
4. Be familiar with the structure and function of the testes and ovary.
5. Be familiar with the changes in hypothalamic, pituitary, and gonadal function from fetal life to senescence.
6. Know the effect of sex steroids on nongender-related target tissues.
7. Know the patterns of metabolism and excretion of the gonadotrophins and steroids and common methods of measurement of the compounds and metabolites.

Normal reproductive endocrine function is characterized by tonic secretion of pituitary and gonadal hormones in males and cyclic secretion of these hormones in females.

The major components of the reproductive endocrine system are the hypothalamus, the pituitary, and the gonads, which are linked through a network of inhibitory and stimulatory feedback mechanisms. This axis is, in turn, affected by input from the pineal gland in the form of melatonin secretion, the limbic system via monoamine neurotransmitters, and higher cerebral centers via ill-defined mechanisms.

Hormonal Components of the Hypothalamic-Pituitary-Gonadal Axis

Feedback relationships are controlled by a number of hormones that are secreted by various components of the axis. The same hormones are synthesized by both male and female, but as would be expected, the patterns of secretion are different.

The hypothalamus secretes a releasing hormone for each of the pituitary trophic hormones except the gonadotrophins, which are under the control of a single releasing hormone. To date, the structures of only two of these releasing hormones, thyrotrophin releasing hormone (TRH or TRF) and gonadotrophin releasing hormone (GnRH, LHRH, or LRF) have been identified. Thyrotrophin releasing hormone is a tripeptide that causes the release of thyroid stimulating hormone (TSH) as well as release of prolactin (PRL). However, it is not thought to be the primary prolactin releasing hormone. Gonadotrophin releasing hormone is a decapeptide. Although it causes synthesis and secretion of both follicle stimulating hormone (FSH) and luteinizing hormone (LH), it is a much more potent stimulator of the latter than the former.

Hypothalamic releasing hormones are secreted and released by a group of a specialized neurosecretory cells in an area of hypothalamus at the base of the third ventricle. These cells are stimulated by synaptic transmission from neural cells secreting the monoamine neurotransmitters, norepinephrine and dopamine. Norepinephrine has a stimulatory effect on the synthesis and release of gonadotrophin releasing hormone. Dopamine has an inhibitory effect on the synthesis and release of prolactin, probably through synthesis of the as yet uncharacterized prolactin inhibiting factor (PIF).

The pituitary secretes three hormones that may have a trophic effect on the gonads: FSH, LH, and prolactin. FSH and LH are glycoprotein hormones composed of α and β subunits. They are closely related structurally and chemically to thyroid stimulating hormone (TSH) and human chorionic gonadotrophin (hCG). The α subunits of all four hormones are essentially identical, while the biologic and immunologic specificity of each hormone is conferred by the β subunit. FSH and LH are synthesized in the gonadotrophs, basophilic cells scattered throughout the anterior pituitary.

Prolactin is a protein hormone that is structurally and chemically related to both growth hormone (GH) and human chorionic somatomammotrophin (hCS). The protein hormones are composed of single polypeptide chains of differing length but with similar amino acid composition. The molecules have from two to three intramolecular disulfide bonds, with a characteristic number for each individual hormone.

Prolactin is synthesized and secreted by acidophilic cells in the lateral wings of the anterior pituitary. The primary role of prolactin in humans is preparation of the breast tissue for lactation and maintenance of lactation. In rodents, it has the additional function of maintaining the integrity of the corpus luteum in the postovulatory phase of the cycle and during early pregnancy. There is some evidence that prolactin may have a deleterious effect on luteal function in humans, but this has yet to be conclusively demonstrated.

The gonads secrete a series of steroid hormones including androstenedione (A), testosterone (T), dihydrotestosterone (DHT), and estradiol (E_2) in the male; and progesterone (P), androstenedione (A), testosterone (T), estrone (E_1), and estradiol (E_2) in the female. In addition, both testes and ovaries, but especially the former, appear to secrete an as yet incompletely identified protein hormone called inhibin, which may play an important role in negative feedback from the gonad to the hypothalamus.

Synthesis and Interconversion of Sex Steroids

Steroidogenesis is dependent on the conversion of cholesterol to Δ^5-pregnenolone. In the adrenal, ACTH is the major stimulating agent, while LH serves this function in the gonad. The end products of steroidogenesis in the adrenals, ovaries, and testes appear to be determined by the availability of critical enzymes in the respective target tissues, for example, absence of 21 hydroxylase blocks corticosteroid- and mineralocorticoid synthesis in the gonads. Similarly, relative abundance of aromatizing enzymes in the ovary and relative paucity in the testes favors estrogen synthesis in the former and androgen synthesis in the latter.

From Δ^5-pregnenolone, synthesis of the sex steroids may occur via either of two pathways, the Δ^5-3 β-hydroxyl pathway from pregnenolone to 17-hydroxypregnenolone to dehydroepiandrosterone (DHEA) or the Δ^4-3 β-ketone pathway from progesterone (P) to 17αhydroxyprogesterone (17αOHP) to androstenedione (A). Both pathways run a parallel course, and interconversions from the Δ^5 to the Δ^4 configuration may take place at any point along the way. In actuality, the major interconversion points in the gonad are from pregnenolone to progesterone or dehydroepiandrosterone to androstenedione (Fig. 2–1.)

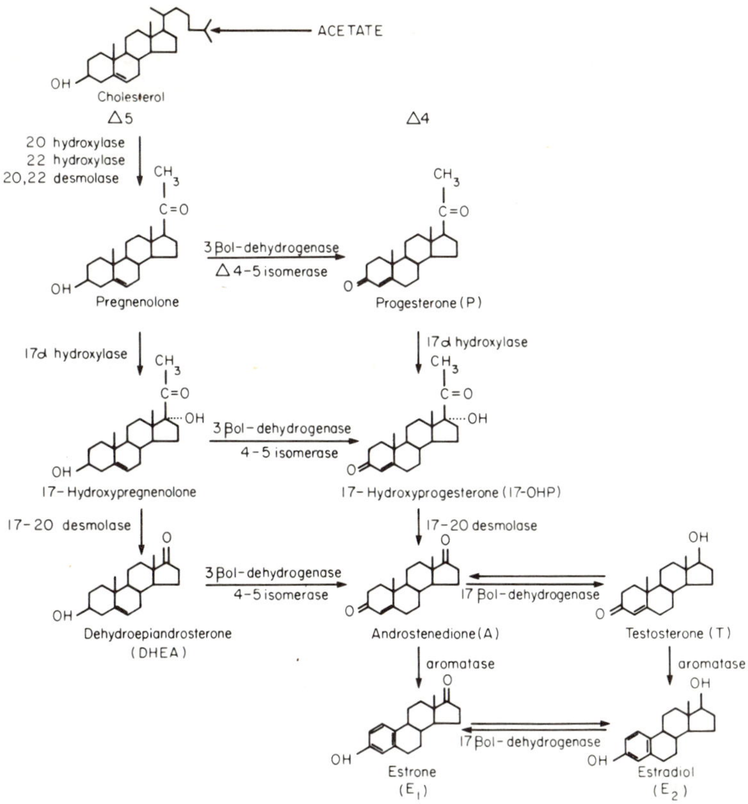

Figure 2–1. Synthetic pathways of sex steroids in the gonad. (See text for details.)

Interconversion from the Δ^5 to the Δ^4 pathway requires the activity of two enzymes, 3β-ol-dehydrogenase to convert the 3 hydroxyl group to a 3 ketone and $\Delta^{4\text{-}5}$ isomerase to transfer the double bond from the 5-6 to 4-5 position. The next important enzyme in the vertical pathway is 17α-hydroxylase, which introduces a hydroxyl group at the 17 position. This is a critical step, as the hydroxyl group is a necessary substrate for the action of 17-20 desmolase, the enzyme responsible for cleavage of the side chain and formation of the C-19 or androstane series from which the C-18 or estrane series is derived.

Synthesis of testosterone and estrone proceeds via Δ^4-androstenedione. The former is synthesized by the action of 17β-ol-dehydrogenase and the latter by aromatization,which involves loss of the 19 carbon and conversion of the 3β-ketone to a hydroxyl group. Estradiol can be synthesized from either estrone or testosterone by

17β-ol-dehydrogenase from the former or aromatization from the latter. In appropriate target tissues in which 5α-reductase levels are high, testosterone is preferentially converted to dihydrotestosterone, a high-potency androgen.

In the male, testosterone is synthesized by the Leydig cells preferentially via the Δ^4 pathway, that is, pregnenelone, progesterone, 17αOHP and A. Testosterone is then taken up by the Sertoli cells where it may be converted to dihydrotestosterone (DHT) or estradiol (E_2).

The work of Kase indicates that ovarian steroidogenesis may occur predominantly through the Δ^5-3β-hydroxyl pathway in the follicular phase of the cycle and through the Δ^4-3β-ketone pathway in the luteal phase. Androstenedione and dehydroepiandrosterone are the major secretory products of the stroma. The work of Ryan favors the hypothesis that the theca is the major secretory site of estradiol during the follicular phase, while the granulosa secretes progesterone. The virtual absence of progesterone in the follicular phase is explained by lack of vascularization of the granulosa cells until the formation of the corpus luteum occurs. The corpus luteum secretes both estradiol and progesterone.

Feedback Relationships of the Hypothalamic-Pituitary-Ovarian Axis and Hormone Action

Gonadotrophin releasing hormones synthesized by neurosecretory cells of the hypothalamus are carried via the hypophyseal portal system to the anterior pituitary where they stimulate synthesis and release of both FSH and LH.

The gonadotrophins, in turn, have two effects. There is recent evidence that they are transported in a reverse gradient fashion back to the hypothalamus, producing a short loop inhibitory feedback effect. Their primary effect, however, appears to be stimulation of appropriate target cells in the gonad. Although there may be some modest overlap in activity because of structural similarity, each gonadotrophin has its specific role in the economy of gonadal function. FSH stimulates follicular growth and maturation in the female and initiation of spermatogenesis in the male. In addition, there is compelling evidence that it catalyzes the aromatization step from testosterone to estradiol in the gonads of both sexes. LH stimulates steroidogenesis in the gonads of both sexes at the level of conversion of cholesterol to pregnenolone. The steroids elaborated by LH stimulation of both gonads are indispensable to the complete maturation of germ cells in both male and female gonads. However, steroidogenesis appears to proceed to a functional point with only modest levels of LH. FSH alone can produce neither complete germ cell maturation nor adequate steroidogenesis. Hence, it appears that LH in the presence of a paucity of FSH can induce sufficient steroidogenesis for normal sexual maturation, but both trophic hormones must be present in adequate amounts for normal fertility to exist.

The gonadal steroids have an effect on specialized target cells throughout the body and a feedback effect on the hypothalamus and pituitary. In the case of the male, feedback is always inhibitory. In the female, feedback is inhibitory except for a paradoxical stimulatory feedback that occurs at midcycle.

Gonadal Structure and Function

The testis is an organ in which medullary elements dominate. It consists of a series of 25–40 tubules that contain two specialized cell lines, Sertoli cells, which serve a nutritive and supportive function for spermatogenesis, and the germinal epithelium, which consists of germ cells in all stages of development. The seminiferous tubules are surrounded by interstitial cells called Leydig cells, which have as their primary function synthesis and release of testosterone. The seminiferous tubules converge into a secondary collecting tubular system called the rete testis and from there in a single, long, extremely convoluted tubule, the epididymis, which serves as an area for completion of maturation and then storage of spermatozoa. Testosterone secreted by the Leydig cells apparently diffuses directly to the Sertoli cells. It does not become absorbed into the blood stream until the area of the rete testis. A rather effective blood-testes barrier appears to exist up until this point.

Testosterone has two major roles in the male. It stimulates development of male secondary sex characteristics such as enlargement of the genitalia, increase in laryngeal size with concomitant deepening of the voice, increase in muscle mass, and growth of body hair. It also is indispensable to the maintenance of normal spermatogenesis.

FSH is necessary for the initiation of spermatogenesis. Whether this is by activity on the Sertoli cells alone or by activity on the Sertoli cells as well as directly on the germ cells has yet to be conclusively determined. Testosterone is definitely necessary for the later steps in spermatogenesis. One of the effects of both FSH and testosterone on the Sertoli cell is the elaboration of androgen binding protein (ABP). The exact role of this protein in the spermatogenic process is uncertain. The Sertoli cell also converts testosterone to both dihydrotestosterone and estradiol. Testosterone rather than dihydrotestosterone appears to be important in spermatogenesis. Estradiol is thought by some to be the source of feedback inhibition from the testis to the hypothalamus. However, there is strong evidence for the existence of a protein hormone called inhibin, which, acting either alone or in concert with estradiol, may be the real source of testicular negative feedback.

The spermatogenic cycle in man takes 74 ± 6 days. Unlike the situation in lower mammals, all stages of spermatogenesis can be observed simultaneously in the same seminiferous tubule. The process of spermatogenesis involves production of haploid germ cells (spermatids) from diploid cells (spermatogonia). Spermiogenesis is the process by which the spermatids mature into cells that are capable of independent movement and transport (spermatozoa).

The stem cells or spermatogonia are divided into roughly two populations: those that constantly divide to maintain the stem cell population and those that undergo growth preparatory to meioic division, producing two secondary spermatocytes; the secondary spermatocytes undergo further reduction division to produce two spermatids. From this point, specialized development of head, midpiece, and tail takes place, leading to the development of spermatozoa.

Spermatozoa that have been surrounded by the Sertoli cells up to this point are released into the lumen of the seminiferous tubules and are transported to the epididymis, where they undergo further maturation and storage. At the time of ejaculation, they are propelled as a bolus through the vas deferens. At the point of passage through the prostatic urethra, seminal plasma secreted by the seminal vesicles and prostate is added to the bolus of sperm, thus completing the composition of the ejaculate.

There has been much debate over what constitutes a normal ejaculate. Although many investigators have considered counts of 20,000,000/ml or less as effectively infertile, MacLeod has reported that at least 5% of fertile males tested have counts of 10,000,000/ml or less. Concentration of sperm is not the only criterion for a normal ejaculate. Volume, pH, immediate and prolonged motility, and percentage of normal forms are also important. Table 2–1 gives generally accepted mean values for the aforementioned components of the normal ejaculate.

The ovary is an endocrine organ in which cortical elements dominate. It consists of numerous follicles in varying stages of maturation surrounded by stromal cells. Follicular units consist of an ovum sur-

Table 2–1. Semen Characteristics of 2,400 Men with Proven Fertility

Study	*Number of Men*	*Volume (ml) Mean ± S.D.*	*Density (million per ml) ± S.D.*	*Percentage Motility Mean ± S.D.*	*Morphology (Percent Normal) Mean ± S.D.*
1	1000	3.38 ± 1.61	107 ± 74	58 ± 16	79 ± 13
2	100	2.85 ± .93	101 ± 45	61 ± 12	88
3	1300	3.2 ± 1.4	79 ± 57	65 ± 22	—

1. MacLeod J: Semen quality in one thousand men of known fertility and in eight hundred cases of infertile marriage. *Fertil Steril* 2:115, 1951.

2. Falk HC, Kaufman SA: What constitutes a normal semen? *Fertil Steril* 1:489, 1950.

3. Rehan NE, Sobrero AJ, Fertig JW: The semen of fertile men: statistical analysis of 1300 men. *Fertil Steril* 26:492, 1975.

rounded by a layer of granulosa cells and internal and external layers of thecal cells. The maximum number of follicles (millions) is achieved at approximately 20 weeks of intrauterine life. From this time, follicles become involved in an inexorable process of partial maturation and atresia that appears to be uninfluenced by pregnancy or hypothalamic inhibition and is the basic mechanism responsible for ovarian modulation of hypothalamic-pituitary output. In the average woman, only 400–500 ova develop to the stage of ovulation. The vast majority of follicles become atretic. The thecal component forms the bulk of the ovarian stroma and synthesizes steroids, predominantly androstenedione.

Although the components of the system have been identified, the mechanism of their interaction has yet to be elucidated. The following model appears most plausible (Fig. 2–2). In response to a nadir in the circulating level of estrogen, gonadotrophin releasing hormone is secreted by the hypothalamus. The pituitary, in turn, releases FSH, which stimulates growth of a crop of follicles. LH secreted by the pituitary in basal amounts stimulates synthesis of estradiol by the theca interna of the developing follicles. In response to increasing levels of FSH, increase in both follicular size and estradiol synthesis occurs. Because of asynchronous growth, one follicle develops slightly ahead of the remaining follicles in the cohort. This is destined to be the ovulatory follicle in that particular cycle. Estradiol synthesized by the follicle has a trophic effect on the follicle itself and an inhibitory effect on the release of FSH. In the face of declining FSH, the remaining follicles in the cohort begin to regress and become atretic. Meanwhile,

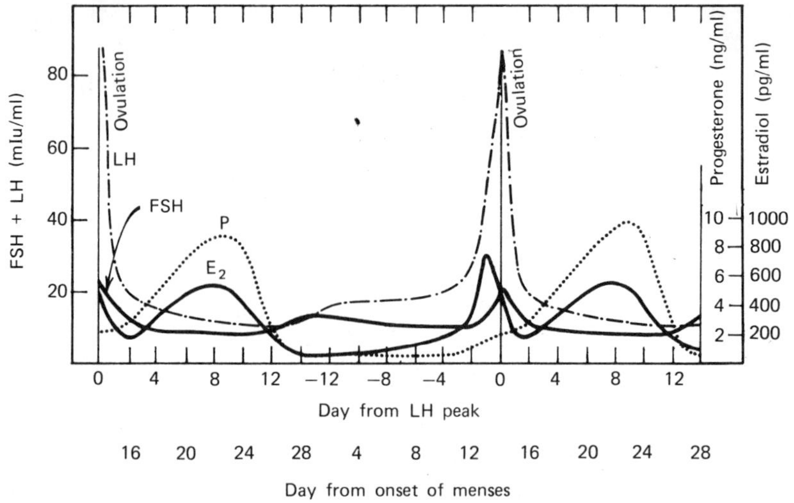

Figure 2–2. Levels of FSH, LH, estradiol (E_2), and progesterone (P) throughout one and a half menstrual cycles. See text for details. (From Speroff L, Vande Wiele RL: Regulation of the human menstrual cycle. *Am J Obstet Gynecol* 109:234, 1971.)

estradiol synthesis in the ovulatory follicle has increased geometrically. Peak levels of estradiol are reached approximately two days prior to ovulation. In response to the estradiol peak, a flood of LH is released by the pituitary. The flood appears to result from two factors: a sudden release of stored LH by the anterior pituitary and a marked increase in secretion of LH releasing hormone by the hypothalamus. A small increase in FSH snythesis is noted at the same time. Thus, estradiol has an inhibitory effect on the hypothalamus in the follicular phase but a stimulatory effect at midcycle.

Following ovulation, the granulosa cells become vascularized as the corpus luteum forms. Basal levels of LH stimulate secretion of both estradiol and progesterone by the corpus luteum. The combination of both steroids has a potent inhibitory effect on the synthesis of gonadotrophin releasing factors. In addition, progesterone probably has a local inhibitory effect on follicular maturation. In the absence of production of a luteotrophic factor (in the human, hCG), the corpus luteum ceases to function after 10–12 days. The cells of the corpus luteum show evidence of degeneration. Although the yellow color is maintained for months, the structure becomes smaller and is eventually replaced by hyaline tissue. The resultant structure is called a corpus albicans.

In response to the falling levels of estradiol and progesterone, an increase in FSH secretion begins, and a new crop of primed follicles begins to grow and synthesize estradiol, thus perpetuating the cyclic pattern. The nadir of steroid synthesis occurs approximately two days before the onset of menses. The rise in FSH begins shortly before this nadir is reached.

Variations in cycle length are generally considered to take place at the expense of the follicular phase. Thus, in normally ovulating women, the length of the postovulatory phase remains constant from cycle to cycle, while the length of the preovulatory phase may vary. These variations may be a reflection of very fine adjustments between FSH secretion and follicular response that take place until one follicle becomes large enough to assume major estradiol synthesis. Once this critical size has been achieved, only modest amounts of FSH and LH are required for continued growth of the follicle. Because of the intercycle variation in follicular length, Ross has stated that the onset of menstruation is a far better indication of the end of the previous cycle than of the beginning of the next cycle.

Function of the Hypothalamic-Pituitary-Gonadal Axis Throughout Life

Once the components of the reproductive endocrine system have been defined, it is possible to describe the method in which they are integrated throughout the life span of the individual from early embryonic life through senility. As would be expected, there are critical differences between male and female that appear to be determined primarily by responsiveness of the gonad rather than the hypothalamus or pituitary. Embryonic differentiation of the gonad has been treated elsewhere, so it will be discussed only briefly here.

The fetal testis differentiates earlier than the fetal ovary and has been noted to secrete testosterone and dihydrotestosterone at the end of the first trimester, the time at which differentiation of the external genitalia takes place. Early gonadal steroidogenesis is thought to occur in response to high levels of hCG secreted by the placenta. This is substantiated by the identification of Leydig cells in the fetal testis from the end of the first trimester. They revert to mesenchymal cells shortly after birth.

From birth until shortly before puberty, the testis shows little activity. The germinal epithelium, although present, displays no signs of active spermatogenesis. Although apparently inactive, the tissue is still quite sensitive to insult. Cryptorchid (intra-abdominal) testes that are not surgically released well before puberty rarely demonstrate competence of the germinal epithelium. Levels of both FSH and LH rise gradually but perceptibly as puberty approaches, and following early signs of increased testosterone synthesis, spermatogenesis ensues.

As males age, gonadotrophin levels, especially FSH, become gradually higher. Dihydrotestosterone levels increase, testosterone levels decrease, and spermatogenesis becomes less efficient. However, there is no evidence for sudden cessation of reproductive function in the male as in the female.

The fetal ovary does not demonstrate the early hormonal activity seen in the fetal testis. As mentioned in an earlier chapter, ovarian steroid synthesis is not necessary for differentiation of female external or internal genitalia. The fetal ovary is far from inactive. The full complement of ova is achieved by 20 weeks of intrauterine life, and the process of follicular maturation and regression is evident by the third trimester.

As in the male, gonadotrophin levels are quite low after birth and rise gradually as puberty approaches. In general, levels of FSH are slightly higher than those of LH, but the ratio is fairly close to unity. This same relationship of FSH to LH is noted in postmenopausal females. However, during the reproductive period of life, LH is always higher than FSH except in the very early follicular phase of the cycle, when growth of a new crop of follicles is stimulated.

As gonadotrophin levels gradually rise in late childhood, estrogen synthesis increases in the ovarian stroma with resultant maturation of the external genitalia and beginning growth of the breasts. Eventually, when there has been sufficient stimulation of the endometrium (uterine mucosa), some bleeding will occur. The first episode of bleeding is called menarche. This shedding, menstruation, occurs initially in response to a relative drop in circulating levels of estrogen. The first several menstrual episodes may occur at irregular intervals because ovulation has not yet occurred. Once an ovulatory pattern has been established, it normally occurs with a consistent periodicity for each individual woman. Toward the end of the reproductive age, ovulation and, hence, menstrual cycles again become irregular. When there are no further responsive units remaining in the ovary, menopause ensues, ovarian steroid synthesis decreases drastically, and gonadotrophin levels become markedly elevated. The reproductive life span of the average woman extends from the early teens to the late forties.

Mechanisms of Hormone Action at the Cellular Level

Gonadotrophins and steroids appear to select their appropriate target cells because of the presence of specific receptors on or in these cells, but the mechanism of hormone receptor interaction differs for each class of hormones.

FSH and LH appear to attach to receptors located on the cell membrane. The membrane receptor-hormone complex then activates cyclic AMP within the cell, and this reaction stimulates the synthesis of appropriate compounds by the cell. In the case of LH, steroidogenesis is initiated, and in the case of FSH, specific proteins are elaborated that presumably have an effect on germ cell growth and development. There is speculation that the hormone specific β subunit may be broken off and transported into the cell, but the evidence for this is not yet forthcoming.

Steroid hormones, on the other hand, have a different type of receptor interaction. These hormones react with receptors in the cytosol and are carried by these receptors to the nucleus of the cell where they interact with specific sequences of DNA to effect messenger RNA synthesis and eventually protein synthesis.

Transport of Steroids to Target Tissues

Steroids secreted by the gonads of either sex are transported to target tissues via the circulation in both free and bound forms. Testosterone and estradiol both bind to a specific binding globulin known as the TEBG (testosterone-estradiol binding globulin), or sex steroid binding globulin. Progesterone has no specific binding globulin of its own but appears to bind preferentially to CBG (corticosteroid binding globulin). Free steroids are rapidly conjugated to water-soluble products.

Response of Target Organs to Gonadal Steroids

In addition to their feedback relationship with the hypothalamus and pituitary, gonadal steroids affect a variety of other target organs, including the skin and its appendages, subcutaneous fat, muscle, breast, larynx, uterus, cervix, vaginal mucosa, external genitalia, and the gonad itself.

Androgens have a primary trophic effect on skin, muscle, laryngeal tissue, and male external genitalia. Androgens stimulate penile and testicular growth and are responsible for rugations of the scrotum. Androgens stimulate growth of body hair and secretory activity of sebaceous glands. Paradoxically, androgens are responsible for the temporal recession of the hair line seen in males. The active androgen at the peripheral level is dihydrotestosterone, which is usually derived by peripheral conversion of androstenedione and/or testosterone. Androgens are responsible for laryngeal enlargement, which produces the deep voice characteristic of males, and for the increase in muscle mass, which is characteristic of the male body habitus.

Estrogens have a trophic effect on a number of target tissues. Estrogen is responsible for growth in size of the breast and development of the ductile system. Estrogen is also responsible for the deposition of fat in the breasts, abdomen, hips, and thighs that produces the characteristic female body habitus. Estrogen also induces proliferative or hyperplastic growth of the endometrium, increased production of cervical mucus, and thickening and rugation of the vaginal mucosa. Estrogen induces growth of receptors for itself and for progesterone in appropriate target tissues, while progesterone eliminates these receptors.

Progesterone serves as a complementary agent to estrogen as well as an antagonist. In the breast, progesterone induces growth of the lobule-alveolar complexes after estrogen-induced growth of the ductile system has taken place. Progesterone halts proliferation of endometrial tissue and induces hypertrophic growth of the existing tissues. Estrogen causes increase in the water content of cervical mucus, resulting in a copious, clear mucoid secretion somewhat reminiscent of egg albumin; progesterone induces increase in viscosity, with production of an opaque gummy mucoid material. The vaginal mucosa also exhibits an increase in less mature cells under the influence of progesterone.

The endometrium is probably the single most important tissue that is responsive to the effects of both estrogen and progesterone. The complete effect of these two steroids is responsible for the cyclic bleeding referred to as "menstrual periods." This is covered in Chapter 10.

Effect on Nongender Related Target Tissues

Gonadal steroids have a potent effect on target tissues that are not related to sexual or reproductive function. Testosterone has an anabolic action that results in positive nitrogen balance and retention of potassium, phosphorus, and calcium. It also is a potent stimulus for hemoglobin synthesis. Both testosterone and estradiol are vasoactive, and abrupt depletion, as experienced in loss of gonadal function, produces vasomotor instability with accompanying hot flashes.

Estradiol increases levels of serum triglycerides and hormone binding globulins, including those for cortisol, sex steroids, and thyroxine. Estrogen also increases insulin response to carbohydrate and causes sodium retention through interaction with the renin-angiotensin system. Estrogen affects the ability of the liver to secrete substances such as bromosulfophthalein. Unlike estrogen, progesterone appears to have a natriuretic effect by competing with aldosterone.

Finally, all three major sex steroids have been related to changes in mood. Estradiol and testosterone are said to enhance libido and contribute to a general feeling of mood elevation, optimism, and well-being. Progesterone, on the other hand, is said to contribute to depression. Evidence for these effects is nebulous at best and deserves further study.

Measurement, Metabolism, and Excretion of Gonadotrophins and Steroids

Output and levels of gonadotrophins and steroids are conveniently measured in either blood or urine. The advantage of blood sampling lies in its convenience, simplicity, and accuracy because one is measuring the actual secretory product to be studied. The disadvantage of blood sampling lies in the fact that there may be both minute to minute and diurnal fluctuation in steroid and gonadotrophin secretion. This disadvantage can be overcome by measuring the urinary metabolite of the hormone under study in a 24-hour collection. However, 24-hour urine collections have the disadvantage of being inconvenient and cumbersome. More importantly, a single urinary metabolite does not represent all of the substrate from which it is derived. A good example of this phenomenon is progesterone. The urinary metabolite of progesterone is pregnanediol; however, only about 15–20% of progesterone secreted daily is actually metabolized as pregnanediol.

Gonadotrophins are secreted in a pulsatile pattern with a periodicity of approximately 90 minutes. The amplitude of the pulse is greater for LH than for FSH. In males, the pulsatile pattern is maintained in a relatively tonic fashion, while, in the female, the pulsatile pattern is superimposed on the change in baseline levels leading up to and away from the cyclic gonadotrophin peak seen just before ovulation. If a very accurate estimate of mean daily gonadotrophin level is required, this can be obtained by drawing three samples at 20-minute intervals, which are then pooled prior to assay.

The pattern of steroid secretion is somewhat different. Testosterone appears to be secreted in a pulsatile pattern in the male. In the female, there is evidence for a diurnal variation in steroid secretion with morning values being higher than late-afternoon values. However, the magnitude of this variation is nowhere near as great as that commonly seen with cortisol where evening values are generally half of morning values.

Mean gonadotrophin levels in the adult male range from 5–20 mIU/ml for both FSH and LH. In the female, mean gonadotrophin levels fall within the same range as those for the male except for the period of the midcycle peak when levels are markedly elevated. Mean values for serum gonadotrophin levels throughout the menstrual cycle are found in Figure 2–2.

The pattern of steroid secretion differs markedly between males and females as would be expected. Testosterone is the major secretory product of the testis, and the testis, in turn, is the major source of circulating testosterone. The average output is 8 mg in 24 hours. Average blood level is about 670 ng/dl of serum. The testis also secretes dehydroepiandrosterone (DHEA), androstenedione (A), and dihydrotestosterone (DHT) in much smaller amounts. Most of the circulating DHEA and A are secreted by the adrenal, while about 90% of circulating dihydrotestosterone (DHT) is derived from peripheral conversion of androstenedione and testosterone in androgen-sensitive target cells. DHEA and A are metabolized to 17-ketosteroids prior to conjugation and excretion, whereas only some testosterone is metabolized in this fashion. Testosterone itself is not a 17-ketosteroid. In the male, urinary 17-ketosteroids are primarily derived from adrenal steroid secretion, so they do not afford a very accurate reflection of testicular steroid biosynthesis.

The gonadotrophins are excreted in the urine with modest changes in amino acid composition. The steroids, on the other hand, undergo a number of metabolic transformations to make them both active and more water soluble. Much of this transformation takes place in the liver. Deactivation is usually accomplished by hydroxylation and reduction of the A ring. Water solubility is achieved by conjugation to a sulfate or glucuronide.

In the female, the pattern of ovarian steroid secretion varies throughout the menstrual cycle, and the variety of secretory products is much greater than in the male. The ovary and, more specifically, the ovulatory follicle and corpus luteum are the major sources of estradiol synthesis and secretion. Average daily secretion ranges from 0.07 mg/day in the early follicular phase to 0.8 mg/day at the time of the preovulatory peak. Mean values throughout the cycle are found in Figure 2–2. Estrone is also secreted by the ovary, but the majority of circulating estrone is derived by peripheral conversion of androstenedione. Androstenedione is also secreted by the ovary in a cyclic pattern that parallels that of estradiol and estrone. Approximately 50% of circulating androstenedione is of ovarian origin and 50% of adrenal origin in the normal female. Almost all dehydroepiandrosterone produced in the normal female is of adrenal origin. Mean testosterone levels in the normal female are about 40 ng/dl of serum. Approximately 50% is secreted by the ovary, and the remainder is derived from peripheral conversion of androstenedione.

About 50% of the estrone and estradiol secreted daily undergoes 16α hydroxylation to form estriol. As is the case with other steroids, estriol is then conjugated with sulfate or glucuronide to facilitate excretion via the urine. Androgen inactivation and excretion in the female follows the same pathways as those found in the male.

Progesterone and 17αOH progesterone are secreted predominantly by the corpus luteum during the luteal phase of the cycle. Low levels of progesterone found in the follicular phase of the cycle are predominantly of adrenal origin. Ovarian progesterone and 17αOH progesterone secretion become detectable in the periovulatory period concomitant with the LH surge. Peak secretion in the midluteal phase reaches a level of 24 mg/24 hours of progesterone and 4 mg/24 for 17αOH progesterone. Mean daily values throughout the cycle are found in Figure 2–2. While progesterone shows a gradual rise beginning in the periovulatory period, 17αOH progesterone shows a preovulatory spike and a secondary midluteal rise. Measurement of both steroids has been used to indicate the adequacy of corpus luteum function. In addition, 17αOH P has been used as a measure of ovarian contribution to steroidal maintenance of early gestation, as the placenta does not synthesize 17αOH P in detectable amounts in early pregnancy. Both progesterone and 17αOH P are metabolized in part to ring A reduced compounds, the former to pregnanediol and the latter to pregnanetriol. Urinary pregnanediol measurements have classically been used to measure adequacy of corpus luteum function, while pregnanetriol measurement is useful in suspected cases of 21 hydroxylase deficiency.

SUGGESTED READING

Baker HWG, Bremner WJ, Burger HG, et al: Testicular control of follicle-stimulating hormone secretion. *Recent Prog Horm Res* 32:429, 1976.

Hansson V, Calandra R, Purvis K, et al: Hormonal regulation of spermatogenesis. *Vitam Horm* 34: 187, 1976.

Means AR, Fakunding JL, Huckins C, et al: Follicle-stimulating hormone, the Sertoli cell, and spermatogenesis. *Recent Prog Horm Res* 32:477, 1976.

Speroff L, Glass RH, Kase NG: *Clinical Gynecologic Endocrinology and Infertility,* ed. 2. Baltimore, Williams & Wilkins, 1978.

Williams RH (ed): *Textbook of Endocrinology*, ed. 5. Philadelphia, WB Saunders Co, 1974.

Yuen BH, Keye WR, Jr., Jaffe RB: Human prolactin: secretion, regulation and pathophysiology. *Obstet Gynecol Surv* 28:527, 1973.

QUESTIONS

1. Which of the following are the major components of the hypothalamic-pituitary-gonadal axis?
   A. GnRH, FSH, TSH, thyroxine, cortisol, and testosterone
   B. TRH, FSH, LH, prolactin estradiol, progesterone, and androstenedione
   C. GnRH, FSH, LH, estradiol, testosterone, and progesterone
   D. GnRH TRH, TSH, LH, prolactin, testosterone, and estradiol
   E. GnRH, FSH, LH, estradiol, testosterone, and prolactin
2. Match each of the following with its correct biochemical classification.

A. TSH	a. tripeptide
B. TRH	b. decapeptide
C. prolactin	c. protein
D. GnRH	d. glycoprotein
E. FSH	e. steroid
F. LH	
G. hCG	
H. testosterone	
I. androstenedione	
J. progesterone	
K. estradiol	

3. Give the designation and name the components of the two major pathways of sex steroid synthesis.
4. Which of the following has the major negative feedback effect on FSH?
   A. testosterone
   B. estradiol
   C. progesterone
   D. androstenedione
   E. pregnenolone
5. Which cell type is the major site of testosterone synthesis in the male?
   A. thecal cell
   B. granulosa
   C. Leydig cell
   D. Sertoli cell
   E. germinal epithelium
6. What is the major effect of tonic LH secretion?
   A. follicular growth
   B. growth of the germinal epithelium
   C. sperm maturation
   D. steroidogenesis in target tissues
   E. luteinization of the endometrium
7. What is the length of the normal spermatogenic cycle?
   A. 14 days
   B. 28 days
   C. 42 days
   D. 74 days
   E. 120 days
8. The maximum number of follicles in the ovary is found at
   A. 20 weeks gestation
   B. birth
   C. puberty
   D. 20 years of age
   E. 25 years of age
9. The midcycle LH surge is caused by
   A. peak FSH secretion
   B. rising progesterone
   C. rising 17αOH progesterone
   D. peak progesterone
   E. peak estradiol

10. Peak progesterone secretion is found
    A. just prior to the LH surge
    B. just after ovulation
    C. five to nine days postovulation
    D. just prior to the onset of menses
    E. just after the cessation of menses
11. Peak estradiol secretion is found
    A. just pricr to the LH surge
    B. just after ovulation
    C. five to nine days postovulation
    D. just prior to the onset of menses
    E. just after the cessation of menses
12. Which of the following best describes the gonadotrophin pattern leading up to puberty?
    A. rapid rise in LH
    B. rapid rise in FSH
    C. rapid rise in both gonadotrophins
    D. gradual rise in both gonadotrophins
    E. gradual rise in FSH
13. Which of the following best describes the gonadotrophin pattern found at the menopause?
    A. rapid rise in LH
    B. rapid rise in FSH
    C. rapid rise in both gonadotrophins
    D. gradual rise in both gonadotrophins
    E. gradual rise in FSH
14. Target tissues for steroid and gonadotrophins are determined by
    A. presence of a specific receptor
    B. sex chromatin
    C. carrier proteins
    D. conjugation to water-soluble products
    E. attachment to specific sequences of DNA
15. Which of the following is the most active androgen in peripheral target tissues?
    A. androstenedione
    B. dehydroepiandrosterone
    C. testosterone
    D. dihydrotestosterone
    E. androstenediol
16. Which of the following is *NOT* a primary effect of estrogen?

A. deposition of fat in breasts, hips, abdomen, buttocks, and thighs
B. growth of the ductile system of the breast
C. proliferative growth of the endometrium
D. growth of the lobule-alveolar complexes of the breast
E. increase in water content of the cervical mucus

17. Which of the following is *NOT* a primary effect of progesterone?
A. hypertrophic growth of the endometrium
B. increase in water content of the cervical mucus
C. increase in less mature cells at the surface of the vaginal mucosa
D. growth in the lobule-alveolar complexes of the breast
E. elimination of receptors for estrogens and progesterone in target cells

18. Which of the following is *NOT* an effect of testosterone?
A. retention of potassium, phosphorus, calcium
B. stimulation of hemoglobin synthesis
C. anabolic weight gain
D. vasomotor stabilization
E. excretion of sodium

19. Which organ is the major site of metabolic transformation and deactivation of steroid hormones?
A. liver
B. adrenal
C. kidney
D. gonad
E. gall bladder

20. What is the major mechanism by which steroid hormones are deactivated and excreted?

ANSWERS

1. C
2. A/d
B/a
C/c
D/b

E/d
F/d
G/d
H/e
I/e
J/e
K/e

3. Δ^5 pathway-pregnenolone-17αOH pregnenolone-dehydroepiandrosterone
   Δ^4 pathway-progesterone-17αOH progesterone-androstenedione
4. B
5. C
6. D
7. D
8. A
9. E
10. C
11. A
12. D
13. C
14. A
15. D
16. D
17. B
18. E
19. A
20. Hydroxylation and reduction of the A ring followed by conjugation to a sulfate or glucuronide to achieve water solubility.

chapter three

human sexuality, sexual counseling, and rape

OBJECTIVES

1. Consider the determinants of the development of human sexuality.
2. Understand the biologic input to sexuality.
3. Understand the behavioral aspects of developing sexuality.
4. Appreciate how sociocultural imprints affect sexuality.
5. Consider the student's role in dealing with problems of sexuality.
6. Assess the student's comfort in this area.
7. Consider the student's level of knowledge and skill.
8. Know some successful modalities of sexual therapy.
9. Recognize the importance of proper referral.
10. Understand the resources necessary to give the rape victim the care that she needs.
11. Realize the health professional's role in objective history, physical examination, laboratory evaluation, and prophylaxis in treating the victim of sexual assault.
12. Understand proper immediate care and proper follow-up of rape victims.

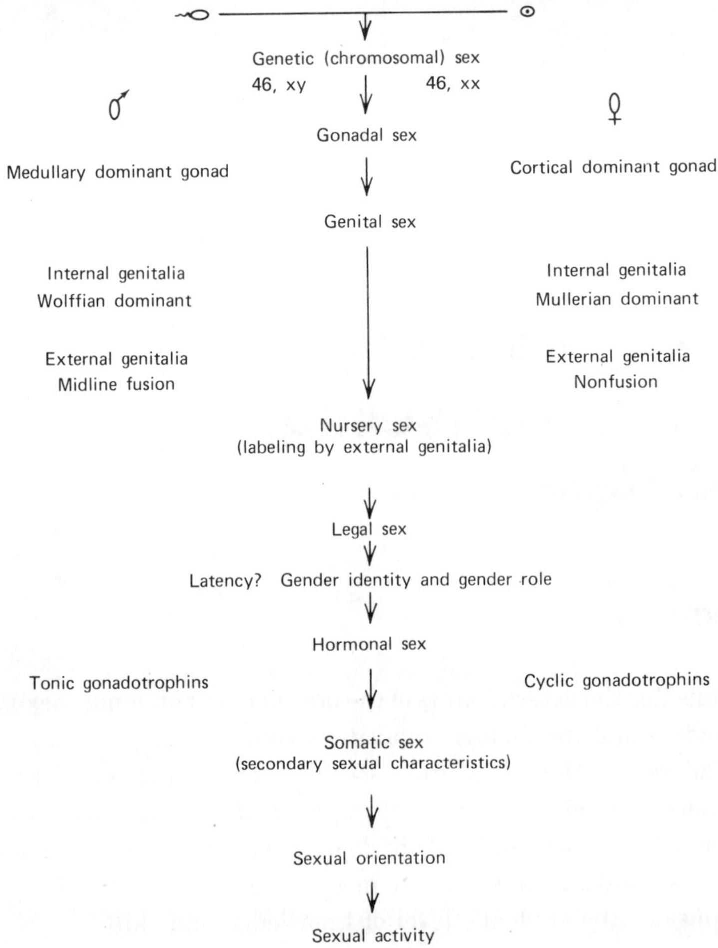

Figure 3–1. Levels of sexual development. (See text for details.)

SEXUALITY OVER THE LIFE CYCLE

The inputs into sexuality during an individual's life span are many, varied and complex. Sexual development can be considered on nine levels (Fig. 3–1). In this developmental context, each level flows from the complex inputs of prior levels.

At conception, chromosomal (genetic) sex is encoded. The genetic component of the sperm determines the male or female genotype of the developing fetus. If the sex chromosome component is an X, the genotype will be 46,XX, and the fetus will develop as a female. If the genetic sex chromosome contribution by the sperm is a Y chromosome, the fetus will be 46,XY and develop as a male. This is the first determinant of sexuality—genetic (chromosomal) sex.

The genotype of chromosomal sex influences the gonad to evolve with cortical dominance in the female or medullary dominance in the male. Beyond this level of sexual determination, the process becomes more complex and not as well understood. This is the *gonadal level* of sexual development.

The hormonal output of the gonad in concert with an unidentified organizing substance determines the structure of the internal genitalia. In the female, the Müllerian system is preserved, forming Fallopian tubes, uterus, and upper vagina. In the male, the Wolffian system develops, forming the seminal vesicles, vas deferens, epididymus, and the collecting system in the testicle itself.

The structure of the external genitalia is also determined by the gonadal hormonal milieu. Midline fusion takes place in the male under androgenic influence, forming the scrotum and penile shaft and its structures. The analogous structures in the female develop into the labia, clitoris, and vestibule. The formation of internal and external genitalia is completed by the 20th week. (See Chapter 1.)

Genetic, gonadal, and genital sexual determinants lead to recognition of the sex of an individual at birth. This recognition is primarily based on the appearance of the external genitalia. Thus, the external genital structure leads to a nursery identification and legal (birth certificate) labeling of the sex of the neonate (Fig. 3–1).

Crucial, but less understood, is the imprinting of the brain structures. The most important is hypothalamic imprinting. In males, this imprint leads to a tonic, acyclic output of gonadotrophic hormones, whereas in females this imprint leads to cyclic output of gonadotrophins. Gonadotrophins (FSH and LH) are the same in males and females. The gonadal steroids, however, differ according to sex, either androgens of testicular origin or estrogen and progesterone of ovarian origin. This imprinting is not fully expressed until secondary sexual development at puberty and will be discussed later. The presence or absence of androgen, primarily of gonadal origin, is also of importance at higher brain centers. Some aspects of behavior are determined by the gonadal hormonal milieu. The best-documented example is the tomboyish behavior of females exposed to androgens in utero.

As sexual development continues, the newborn has now been identified phenotypically as a "boy" or "girl." Next comes sociocultural and behavioral input into the individual's sexuality. The parents and others in the environment verbally and nonverbally program *gender identity*. Some obvious examples are dressing the infant in blue or pink, the way the child is handled, and the type of early toys. This child is conditioned to a gender identity that is quite firm by the age of approximately $2\frac{1}{2}$ years. This is of clinical importance in dealing with intersex problems since clinical experience has shown that beyond this age reassignment of gender identity causes serious emotional problems. In cases of ambiguous genitalia, a sex of rearing must be decided on prior to this age even though corrective surgery may not take place until later.

From the age of firm gender identity in the second year through approximately the ninth year, the child enters what psychoanalytic theory has called the "latent phase." Though the child appears physically undifferentiated, these years are actually dynamic rather than latent. There is continued verbal and nonverbal input from the societal environment of the child. This environment includes family, peers, teachers, and religious figures. Their attitudes suggest and at times dictate a *gender role*. The prior gender identity designated the child as male or female, whereas *gender role* dictates how the child is expected to behave within the context of that identity. The differences in clothing, athletic endeavor, hobbies, and roles assumed at play are examples. Parental expectations are also of prime importance in shaping gender role.

During the 9th or 10th year, the changes of the pubertal sequence begin. These changes are not physically apparent at first, but serial laboratory determinations show elevation of gonadotrophins common to both male and female (FSH and LH) and later the appearance of ovarian estrogens and testicular androgens. The first physical change noted is the beginning growth of pubic hair in the male, whereas, in the female, pubic hair growth or breast budding may be the first evidence of puberty. The development of secondary sexual characteristics follows a predictable pattern over the next four to five years. After a growth spurt in the early teens, full statural growth is achieved in the late teens. The gonadal hormones and growth hormone are crucial in the control of these somatic changes. (See Chapter 1.)

The male develops a muscular body. Body-hair growth increases with growth of a beard, chest hair, and a broader distribution of pubic hair. The larynx widens, and thus the voice deepens. Penile and testicular growth are noted, and ejaculatory capability is seen with masturbatory activity or nocturnal emission.

The female develops a more rounded body habitus, especially over the hips, buttocks, and mons pubis. Body hair is more localized to the pubic area. Adrenarche, the growth of axillary hair, is of clinical interest since menarche follows within six months. The average age of first menses is 12.5 years in the United States. The pattern of menstrual flow may be irregular at first, but usually within six months it becomes cyclic, denoting ovulation.

The difference in gonadotrophin and hypothalamic regulating hormone patterns is of crucial importance and may be the primary biologic difference between male and female. The male pattern of gonadotrophin output is tonic, whereas the female pattern is cyclic. The cyclic quality in the female continues through the reproductive years. The site of this difference is probably hypothalamic and is considered to be of genetic origin.

By midteenage, the person has achieved physical sexual maturity and is capable of sexual function and procreation. This has been a period of rapid physical change. Behaviorally, the adolescent has considered and acted on the emotional complexities of dependence versus independence, passivity versus aggression, markedly differing interpersonal relationships with peers and family, and the important aspect of his or her individuality and self image.

One of the most interesting models of emotional development is Erik Erikson's. He divides psychologic growth into stages of basic trust, autonomy, and initiative in childhood with identity in adolescence. He defines identity as that period where the individual differentiates from parental beliefs and value systems. This should not be confused with earlier gender identity. By the late teens, the majority of individuals have acted upon erotic impulses. This may take the form of masturbation or, in a substantial number, coitus (over 50% by age 19).

The last two levels of sexual developmental are sexual orientation and sexual activity (Fig. 3–1). In the late teens, individuals develop sexual orientation. The majority opt for an interpersonal sexual relationship and most choose heterosexuality. A significant minority (approximately 10%) develop a preference for homosexuality. Some relate both homosexually and heterosexually (bisexuality). Some choose to not relate overtly sexually—asexuality. Asexuality may be dictated by religion or be chosen because of other significant interests in life, such as a career.

Sexual relationships involve experimentation. Homosexual and heterosexual orientations evolve for the most part into dyadic pairing. Bisexuality is most frequently triadic. More open sexual relationships, such as the communes of the 1960s with group sexual relationships, have evolved, as have "swinging" and "key clubs." There is little sociobehavioral data as to the "success" of these relationships. They seem a part of the "sexual revolution," and only the future will shed light on the durability of such sexual choices.

One must be aware that sexual orientation is not a static choice. A given orientation may be followed throughout the life cycle; however, it is often dynamic and changing. The ability of many to change orientation over time while remaining functional supports the hypothesis that none of these orientations in themselves connote psychopathology.

During the life cycle, an abundance of experiences must be considered in the overview of sexuality. The majority of persons come to adulthood with a heterosexual orientation. They are then faced with the developmental "crises" of heterosexual orientation. Most choose marriage. Many then face the crisis of parenthood, shared by both male and female. The experiences of pregnancy, delivery, and breast-feeding are uniquely female. However, they are increasingly shared by male partners, at least to the extent of the male's presence and interest at being involved. These shared experiences have been enhanced by prepared childbirth and educational efforts concerning parenthood.

It is impossible to consider all of the complex aspects of sexuality over the life cycle, but one must consider sexuality at later ages. Young people have a tendency to "neuter" all adults, especially their parents. Recently, it has become apparent that the perception of old age as neuter is in error. Though the aging process may physiologically change sexual function, coital activity can and does continue into older age in a significant number of individuals who retain a meaningful dyadic relationship. Coital frequency may diminish, the male may ejaculate less frequently, and the duration of arousal may lengthen. With a clear understanding of physiologic changes of aging presented so that they do not impose a performance anxiety, many older couples find sexual experiences different but fulfilling.

Sexuality is a pervading force in our society. More precise biologic and behavior knowledge can only enhance the positive force that sexuality exerts on our civilization.

SEXUAL COUNSELING AND SOME THERAPY CONSIDERATIONS

The medical practitioner's role in dealing with patients with sexual concerns should be defined on an individual basis, based on the practitioner's comfort with his or her own sexuality, factual knowledge of human sexuality, counseling skills, and willingness to deal with such problems. At a minimum, the physician should be competent to arrive at a proper diagnosis and to make an appropriate referral. At a maximum, the physician can deal with these patients and refer only if psychopathology indicates need for intensive psychotherapy.

An excellent model leading to the understanding of sexual counseling is utilized by the group at the University of California, San Francisco, and is known as the "VEST" model. Patients with sexual concerns fit into this concept in the following fashion:

V = validation—persons in this group require only reassurance that their concern is not a serious one. This may be permission giving and would be exemplified by the person who is experiencing anxiety with some new sexual practice, such as oral-genital relationships or masturbation for the first time, and is concerned about the practice being unhealthy or pathologic.

E = education—in this group, the person may require only a presentation of factual data involving sexuality. Examples include the couple coming in for contraceptive advice or the person who has questions of orgasmic frequency, coital frequency, and the like. There is some overlap with the first group.

S = suggestion—in this group, the counselor deals with patients with a specific dysfunction such as premature ejaculation in the male or oligoanorgasmia in females. Here one might utilize the "squeeze technique" for the male and one of the many behavior modification programs for the female, such as the programs of Barbach and Lo Piccolo.

T = therapy—this group of patients needs more objective therapeutic intervention. An example is the anorgasmic female or the male with impotence.

As one can see, the problems become increasingly severe as one progresses down the VEST scale, and the professional must decide the level of his or her skills and interest in dealing with such patients.

Table 3–1 is a helpful catalog of sexual dysfunctions. Some problems are analogous male to female. In general, secondary problems have a better prognosis and also are more apt to have an organic component. Primary sexual dysfunction means that the problem has always been present, whereas secondary implies normal function at some time. Impotence, nonejaculation, and any of the dysfunctions involving pain require medical evaluation.

Primary problems of arousal are invariably psychophysiologically based. If secondary, there is a possibility of an organic basis. (Most authorities quote an approximate 10% incidence of organic basis.)

Table 3–1. Sexual Dysfunction (Dissatisfaction) Syndromes[a]

Male	*Female*
Impotence (erectile incompetence)	Lack of arousal
Premature ejaculation	
Oligoejaculation	Oligoanorgasmia
Retarded ejaculation	
Erectile or ejaculatory pain	Dyspareunia
	Vaginismus

[a]Each problem should be considered in the context of primary or secondary.

The following are some frequently successful therapeutic modalities for the various dysfunctions.

Impotence and Nonarousal

The cotherapy team approach (Masters and Johnson) has been often successful. Where the problem is associated with rejection or revulsion of sexual activity, long-term psychotherapy is the only modality that may be successful. "Frigidity" is a specific syndrome with *no* arousal and often conscious rejection or repulsion of sexual activity. This connotes serious psychopathology, is rare, and should be managed by a professional with excellent psychotherapeutic skills.

Premature Ejaculation

This can be dealt with by utilization of the "squeeze technique." There are several valuable demonstration films that can be used with patients. The problem can also be dealt with successfully in a cotherapy setting.

Oligoanorgasmia

Therapy programs of Lo Piccolo and Barbach have been quite successful. Adjunctive films and reading materials are available to go with these programs. Therapy groups of preorgasmic women have also had documented success.

Dyspareunia and Vaginismus

These can be dealt with effectively by counseling and self-dilatation. Surgery should be utilized only for precise organic pathology.

In general, one-to-one therapy is successful only with well-motivated patients who have no element of interrelationship or communication problems. Couple therapy (one to two) is necessary when there are communication problems, which is usually the case. The cotherapy (two-to-two) approach has been very successful with well-motivated couples. Referral to a skilled therapist is in order for problems with identifiable psychopathology.

The therapy approaches are diverse and relate to the therapist's training background. Dynamic, behavior modification, transactional analysis, gestalt, and various group-therapy modalities have proven successful in many instances.

Summary

Problems of sexuality can be competently dealt with by professionals who:

1. are comfortable with their own sexuality.
2. have a fund of knowledge that includes basic factual material of human sexuality.
3. have mastered basic interview and counseling skills that any caring professional should possess.
4. understand their own limitations and have a knowledge of appropriate referral sources so that referral can be accomplished properly.

RAPE—CONSIDERATION OF PROPER HEALTH CARE OF THE VICTIM

This section deals with rape, the most heinous of sexual assaults. One must be aware that sexual assault or oppression is a spectrum that includes any act that invades the physical and emotional space of a person. Examples include voyeurism, exhibitionism, obscene phone calling or writing, and sexual relationships by subterfuge with compliance but not true consent.

Rape is most commonly defined as unlawful carnal knowledge of a woman forcibly and against her will. Most jurisdictions limit the crime of rape to a person other than the victim's husband and require that genital penetration occur. In 1975, there were 56,000 reported rapes in the United States, a 48% increase over the 1970 statistics from the Federal Bureau of Investigation. The actual number of rapes is estimated to be 5–10 times those reported; thus, rape is the most unreported serious crime in our country. There are many reasons for the victims' unwillingness to report. Many victims desire secrecy, considering themselves personally and socially degraded. Few assailants are identified, and even fewer are convicted. The victim of this serious sexual assault may, from fear, wish to keep it from peers, family, and the entire social milieu, sometimes not seeking care.

Optimal health care for the rape victim encompasses medical care (which includes psychologic care) and attention to the legal needs of the client. The physician who accepts responsibility for the care of a rape victim must have some special skills and knowledge. The immediate needs are an accurate history and physical examination to identify any traumatic injury needing acute care. The physician should have an understanding of necessary prophylaxis against venereal disease and pregnancy and an ability to bring empathy, support, and understanding to the victim. The doctor must also know what is required in evidence gathering—an unusual role for the medical professional. When all of these responsibilities of care are assumed by a limited number of professionals, the victim is better served. When reporting the crime requires an "officer of the court" to be involved in evidence gathering, the woman may be unnecessarily subjected to redundant and traumatic interviews and examinations.

It is imperative that the physician understand that this role is not to make a judgment as to whether or not the woman was raped. His role is to record data (history, physical examination, and laboratory procedures) objectively so as to maximize the protection of the victim if she wishes to report the crime. The care of the patient should be complete regardless of her desire to report.

The issue of whether or not to report is complex. As stated above, the first impulse of many victims with their acute feelings of shock, alienation, degradation, and loss of personal control is a desire to hide the attack from peers, family, and society. With objective care and proper gathering of evidence, a decision about reporting can be delayed until the victim has had time for thought, rest, and proper counseling about reporting. The fact that some jurisdictions will furnish care only to the woman who wishes to report the alleged rape is appalling.

Legal issues are quite complicated, and some recent changes in the laws in some areas have been a two-edged sword. Until recently, most laws required "corroboration," meaning evidence of violence, evidence of penetration, and even of a substantiating witness or witnesses. This is ridiculous since rape most often occurs out of the view of any possible witnesses. Frightened victims frequently have bathed, changed clothes, and delayed reporting for care until some time after the attack, all mitigating against corroboration. Some states have now changed their laws so that there is not an insistence on corroborative evidence. Defense attorneys have often responded by resorting to abusive and intense questioning of the victim as to prior sexual experience, number of partners, and such issues in an attempt to demean her as a credible person before the jury. The sensitive judge will allow such questioning only if it is "relevant," but one must face the reality that some judges may take a sexist view of the victim's contribution to the crime, whether it is a short skirt, being in an "inappropriate" environment, or even acting out a fantasy. This is an issue of human legal rights, but the fact that less than 4% of arrests end in the conviction of the rapist does not speak well for the evenhandedness of our judicial system.

Adequate care for the rape victim requires a specialized environment and personnel. The physical facilities should be separated from other service areas to allow immediate care of the victim in private. Registration formalities and counseling can be done in this area. An adequate examining area with laboratory equipment and supplies should also be in this area.

The professional personnel, including the intake nurse, should create a warm, accepting, and caring interface with the woman. The victim should be allowed to verbalize her feelings, if she so desires, but if not, personnel should not be intrusive or directive in their questioning. The attitude of the personnel should convey a genuine care and concern and should not be judgmental. The physician should have the skill to convey this attitude and may wish to involve a patient advocate at the initial interface. One or more persons involved in the woman's care at this time should be female.

The physician should obtain the pertinent history and perform a physical examination to identify and meet the immediate needs, such as surgical repair of trauma. The doctor should also gather necessary evidence in case the patient should wish to report the rape to authorities.

Several published protocols give useful details of proper data gathering. At the outset, the patient should be told of the procedures that are necessary. Informed consent must be obtained. There should be no implied threat or coercion to report the crime to authorities. After consent and history, exam, and laboratory collection are achieved, the patient may be adequately counseled in a decision to report or not. Reporting, even without pressing charges, is an option that may be helpful to authorities since many rapists are recidivistic and a report may help identify a rapist by reinforcing a previously reported case. The woman should be apprised of what is involved in reporting and pressing charges so that she may make an informed decision as to becoming involved in that process.

The above-cited protocols should be used as references for thorough care. Some essential elements of the basic objective data base are:

1. History—name, address, telephone number, age, contact address, date and time of the exam, an account of the assault in the woman's own words, menstrual history and last menstrual period, last prior voluntary coitus, and present contraceptive practice. A history of whether the patient has bathed, douched, or changed clothes since the attack is important.
2. Physical examination—a complete exam for evidence of trauma; a pelvic exam with a nonlubricated speculum to note any trauma and to assess vaginal fluids.
3. Laboratory collections—photographs of any physical trauma, trimming and combing of the pubic hair and shaving of the specimens, aspirations and swabs of the posterior vaginal fornix, gonorrheal cultures of the cervix and of the pharynx and rectum if appropriate by history. Serology, pregnancy test, and fingernail scrapings may also be of importance.
4. Medical treatment—immediate surgical repair of any laceration or other trauma may be necessary, particularly if violent sexual abuse has been involved; this may represent orthopedic care or medical care other than gynecologic genital care.
5. Prophylaxis—though only 10% of rape victims will contract gonorrhea, it seems prudent to utilize prophylaxis against both gonorrhea and syphilis by administering procaine penicillin 4.8 million units plus 1 gm of probenecid orally. If the woman is allergic to penicillin, alternative prophylaxis may be used. It also seems judicious to utilize postcoital or interceptive drugs, such as diethylstilbesterol 25 mg twice daily by mouth for five days combined with an antinauseant. Should the woman not have a menstrual period following this interception, menstrual extraction or suction abortion is recommended. Obviously, this is unnecessary for a woman who has had a previous sterilization procedure or is using adequate contraception at the time of the assault.

Having taken care of consent, history, exam, laboratory collections, and prophylaxis in an environment of support and empathy, one can focus on the emotional care of the victim. In general, rape victims go through an acute phase characterized by anger, alienation, degradation, damage to self-image, and shock following a life-threatening experience. Support and allowing the patient to verbalize her feelings are important. An important aspect of care is being available to answer questions about injury and the issues of reporting and pressing charges.

Counseling availability is of importance so that the professional team may interface with the family, peers, or partner of the patient. Be this husband, partner, or family, it is essential that they be given the opportunity to verbalize their anger and to be counseled to a stance of support and caring for the victim. All too often, the myth of the victim contributing to the crime must be diffused and counseled against.

The woman's long-term emotional needs must be considered. Help is often necessary for the next stage of the usual emotional response to sexual assault. That stage is incorporation or resolution. The professionals involved must be capable of short-term follow-up to detect pregnancy and venereal disease, and most importantly, to recognize the need for continued support. Such support prevents long-term sexual, social, and interpersonal dysfunctions.

Rape must be dealt with by the health-care system in a circumspect manner that satisfies the medical, psychologic, and legal needs of the woman. The increasing incidence of this assault makes it a priority item for the furnishing of adequate, sensitive health care.

SUGGESTED READING

Barbach LG: *For Yourself*. Garden City, N.Y., Doubleday and Company, Inc., 1976.

Beach FA, et al: *Human Sexuality in Four Perspectives*. Baltimore, The Johns Hopkins University Press, 1976.

Braen GR: *The Rape Examination (1976)*. North Chicago, Ill., Abbott Laboratories, 60064.

Erikson E: *Childhood and Society*, ed. 2. New York, W.W. Norton & Co, Inc., 1963.

Gadpaile WJ: *The Cycles of Sex*. New York, Charles Scribner's Sons, 1975.

Lo Piccolo LJ, Heiman JR: *Becoming Orgasmic: A Sexual Growth Program for Women*, three 16 mm films. New York, Focus International Inc., 1976.

Sadock BJ, Kaplan HI, Freedman AM: *The Sexual Experience*, Baltimore, Williams & Wilkins, 1976.

Suspected Rape. The American College of Obstetricians and Gynecologists Technical Bulletin #14, Revised April 1972. Chicago, Ill. 60601.

QUESTIONS

1. The earliest biologic determinant of sexuality is:
   A. the development of the gonad
   B. internal ductal differentiation
   C. the hormonal milieu affecting the hypothalamus
   D. the genetic composition of the zygote (XX or XY)
   E. the sex identified on the birth certificate
2. The hormonal milieu in the fetus determines:
   (1) the structure of the internal genitalia
   (2) the structure of the external genitalia
   (3) the nongenital body habitus of the prepubertal person
   (4) the pattern of gonadotrophin output
   Choices:
   A. all of the above
   B. (1) and (2) only
   C. (2), (3), (4) only
   D. (1), (2), (4) only
3. Gender identity is firmly imprinted by:
   A. 1 year of age
   B. 3 years of age
   C. 10 years of age
   D. the latency period
4. Characteristics of secondary sexual development include:
   A. change in nongenital body habitus
   B. rapid genital growth
   C. characteristic gonadotrophin release
   D. emotional evolution of individual value sets
   E. all of the above
5. Sexual orientations are characterized by:
   A. staying the same
   B. psychopathology if not conforming to socioculturally defined norms
   C. are determined by the hormonal milieu

D. all of the above
E. none of the above

6. The health-care professional should:
   (1) be aware of his/her feelings on commonly encountered sexual issues
   (2) have a basic fund of knowledge relevant to sexuality
   (3) should have much sexual experience
   (4) should handle almost all sexual problems without referral
   (5) have a realistic grasp of his/her competence in this area
   Choices:
   A. all of the above
   B. (1), (2), (4) only
   C. (2), (4), (5) only
   D. (1), (2), (3) only
   E. (1), (2), (5) only
7. Which sexual dysfunction should be referred to a skilled psychotherapist:
   A. arousal problems—secondary
   B. dyspareunia—primary or secondary
   C. primary nonarousal
   D. bisexuality
8. The prognosis in sexual dysfunction is generally better:
   A. with good patient motivation
   B. with secondary problems
   C. with a functional partner relationship
   D. all of the above
9. The following are factual concerning rape:
   A. its incidence has more than doubled in the first half of this decade
   B. once identified, most rapists are convicted
   C. victimization is usually reported
   D. none of the above
   E. A and C only
10. Optimal medical care for the rape victim should include:
   A. objective history
   B. thorough physical exam
   C. segregation from family and police
   D. no threatening discussion of the episode
   E. A, B, and D
   F. A and C
   G. all of the above

11. Optimal care of the victim should be furnished regardless of reporting. TRUE FALSE
12. The primary-care professional in the care of the rape victim must have some special professional knowledge and skills
    TRUE FALSE
13. Prophylaxis must consider:
    A. venereal disease
    B. pregnancy
    C. psychologic sequelae
    D. A and C
    E. A and B
    F. all of the above

ANSWERS

1. D
2. D
3. B
4. E
5. E
6. E
7. C
8. D
9. A
10. E
11. TRUE
12. TRUE
13. F

chapter four

abnormal reproductive endocrine function and infertility

OBJECTIVES

1. Be familiar with the differential diagnosis of genital ambiguity and the proper steps to evaluate the condition in the newborn.
2. Know the causes of precocious puberty in males and females.
3. Know the causes of delayed puberty in males and females.
4. Know the causes of secondary hypogonadism in males and secondary amenorrhea in females.
5. Know the significance of galactorrhea in males and females.
6. Be familiar with the pathophysiology of hirsuitism and virilization in females.
7. Be familiar with endocrine and nonendocrine causes of infertility in males and females.
8. Know the basic components of a complete infertility evaluation in males and females.

Abnormal function of the reproductive endocrine system may manifest itself in a number of ways, including genital ambiguity, precocious puberty, delayed or absent pubarche, primary or secondary amenorrhea and hirsutism in the female, secondary hypogonadism in the male, galactorrhea, and infertility.

It is convenient to discuss reproductive endocrinopathies in relation to the time of life at which they are most likely to be manifest. From the point of evaluation and differential diagnosis, it may be more meaningful to discuss endocrine malfunction in terms of the hypothalamic-pituitary-gonadal axis and the level at which the basic abnormality of function may lie. Finally, the topic could be discussed in terms of symptomatology. An attempt will be made to amalgamate all three methods of approach.

Some conditions have been discussed in detail elsewhere and will only be mentioned in passing. To the extent that there is overlap, male and female infertility will be considered in relationship to endocrine malfunction. Nonendocrine causes of infertility will be covered at the end of the chapter.

GENITAL AMBIGUITY

Although genital ambiguity is noted at birth, it reflects alterations in the fetal endocrine environment that existed at the time of differentiation of the external genitalia at the end of the first trimester of pregnancy. Maternal causes of genital ambiguity include therapy with synthetic progestational agents that usually cause masculinization of the genitalia of female fetuses but may cause incomplete differentiation of the genitalia of male fetuses. Presence of a steroid secreting tumor in the mother may have similar effects.

Intrinsic fetal causes of genital ambiguity are related to errors of gonadal or adrenal function and fall into two major categories: structural abnormalities of the gonad or deficiency of a critical enzyme in steroid biosynthetic pathways.

Structural Abnormalities of the Gonad

Structural abnormalities of the gonad that result in genital ambiguity almost invariably occur in individuals with an XY karyotype. This predilection is explained by the fact that an endocrinologically competent testis is necessary for normal differentiation of male external genitalia, whereas differentiation along female lines occurs in the absence of testicular function. Syndromes have been identified in which the degree of masculinization ranges from labioscrotal fusion with an apparently normal clitoris to normal penile development with an empty scrotal sac. Exploration has revealed no identifiable testicular tissue. This spectrum of disorders may represent testicular dysgenesis that has occurred at varying stages of fetal life. Occasionally, individuals with XY gonadal dysgenesis and streak gonads will exhibit some degree of genital ambiguity. It must be inferred that functioning testicular tissue was present at some time in early development.

Enzyme Deficiencies in Steroid Biosynthesis

Genital ambiguity with gradations of expression can also be seen in infants with deficiencies of enzymes critical to steroid biosynthesis. Almost all these conditions involve adrenal malfunction, expressed as hypersecretion of those end products of steroid biosynthesis not blocked by the enzyme deficiency. All result in some degree of genital ambiguity. In some cases, the corticosteroid deficiencies are so severe that survival of the infant is threatened if appropriate replacement therapy is not initiated promptly.

Enzyme deficiencies that may affect genital development include 20 hydroxylase, 17α-hydroxylase, 3β-ol-dehydroxylase, 17β-ol-dehydrogenase, 21 hydroxylase, 11β-hydroxylase and 5α-reductase. Only the last of these does not affect adrenal function.

Conversion of cholesterol to pregnenelone is catalyzed by 20 hydroxylase. Deficiencies of all steroids are present. These infants have a female phenotype regardless of chromosomal sex because of the inability to produce androgens.

The conversion from Δ^5 to Δ^4 steroids is controlled by 3β-ol-dehydrogenase. Absence of this enzyme results in deficiencies of cortisol, aldosterone, Δ^4 androgens, and estrogens. Both sexes have ambiguous genitalia because of overproduction of dehydroepiandrosterone, a weak androgen strong enough to induce labioscrotal fusion but not strong enough to induce adequate penile development.

Conversion of progesterone to 17-OH progesterone and pregnenelone to 17αOH pregnenelone, both obligatory precursors of androgen, estrogen, and cortisol biosynthesis, is catalyzed by 17α-hydroxylase. Infants of both sexes present with hyperaldosteronism. Males have ambiguous or female genitalia. Females display no genital ambiguity but persist with infantile genitalia and absent secondary sex characteristics as the age of puberty approaches

17β-ol-dehydrogenase catalyzes conversion of androstenedione to testosterone and estrone to estradiol. Affected males display genital ambiguity because of absent testosterone, whereas females do not.

The step in cortisol and aldosterone biosynthesis that involves 21 hydroxylation of progesterone and 17αOH-progesterone, thus producing desoxycorticosterone and desoxycortisol, respectively, is controlled by 21-hydroxylase; 11β-hydroxylase controls the hydroxylation at the 11 position of the two previous substrates to produce corticosterone, the immediate precursor of aldosterone, and cortisol. Affected female infants present varying degrees of masculinization of the external genitalia. Males have normal genitalia but undergo precocious puberty if undiagnosed in infancy. With severe deficiency, a salt-losing condition exists in infants of both sexes.

The conversion of testosterone to dihydrotestosterone is catalyzed by 5α-reductase. The latter is important for complete masculinization of external genitalia. Affected males have labioscrotal fusion, hypospadias, and undescended testes. Adrenal function is not involved. These individuals masculinize at puberty. This deficiency appears to have no clinical importance in females.

Evaluation of Genital Ambiguity

Initial evaluation of infants with genital ambiguity should include measurement of serum electrolytes since salt-losing enzyme deficiency states create a life-threatening situation that requires prompt replacement therapy. Serum cortisol is low in all varieties of congenital adrenal hyperplasia. Once the immediate metabolic status of the infant has been established, evaluation of sex chromosomal complement should be undertaken. Cytogenetic studies should include peripheral blood karyotype and buccal smears for sex chromatin bodies.

PRECOCIOUS PUBERTY

Precocious puberty refers to the attainment of sexual maturation in girls before age eight and in boys before age ten. In true precocious puberty, full reproductive function of the hypothalamic-pituitary-gonadal axis is attained, whereas in pseudoprecocious puberty, development of secondary sex characteristics occurs without gonadal maturation. True precocious puberty results from neuroendocrine malfunction, whereas pseudoprecocious puberty occurs as the result of hypersecretion of adrenal or gonadal steroids. Precocious puberty occurs twice as often in females as in males, but the occurrence is more ominous in males than in females; 80% of precociously developed females fall into an "idiopathic" category because no abnormality of the brain or hypothalamic-pituitary-ovarian axis can be demonstrated. In males, 60% have some serious underlying disorder. In addition to the idiopathic category, other causes of precocious puberty in females include hypothyroidism, fibrous dysplasia of the bone, and ovarian tumors (about 15% of all cases). Only about 4% of precocious puberty in girls results from intracranial tumors. In males, intracranial tumors account for 20%, adrenal hyperplasia 25%, and testicular tumors 10% of cases. Hypothalamic hamartomas, gliomas, astrocytomas, and pinealomas are the most common intracranial lesions involved.

Evaluation of precocious puberty in children of both sexes should include skull films and thyroid function studies. Beyond this point, evaluation depends on the sex of the affected child and associated physical findings. For instance, the presence of patchy hyperpigmentation suggests fibrous dysplasia of the bone. In such a case, skeletal roentgenograms should be obtained. In males, cortisol and 17-ketosteroid values are important to delineate congenital adrenal hyperplasia. The presence of neurologic signs in children of either sex mandates a thorough search for intracranial lesions.

DELAYED PUBERTY AND DELAYED MENARCHE

Puberty integrates elements of both adrenal and gonadal maturation with resultant maturation of the genitalia and nongenital secondary sex characteristics. Puberty is considered delayed in males if not initiated by age 15. Serious problems are considered to exist if full sexual maturation is not reached by age 20. In females, less attention is paid to the onset of puberty and more to the onset of menses in determining delay of maturation. First puberal changes generally occur about two and a half to three years before the onset of menses so that an abnormality could be suspected well before the age of 18, the age beyond which a diagnosis of delayed menarche is made.

Causes of delayed puberty in both sexes may be either hypothalamic, pituitary, or gonadal in origin. In females, delay of menarche with normal puberal maturation suggests an aplastic or an obstructive disorder of the Müllerian system.

Since adrenal androgen secretion is responsible for the earliest growth of pubic hair, the constellation of sexual infantilism with short stature and absence of pubic hair suggests some degree of panhypopituitarism. If early pubic hair development is normal, but secondary sex characteristics fail to develop, lack of gonadal function or hypogonadism is suggested. It is then left to determine whether the gonad is not secreting the proper amount of steroid because it is incapable of responding to hypothalamic-pituitary stimulation (hypergonadotrophic hypogonadism) or because it is not being properly stimulated by the hypothalamic-pituitary portion of the axis (hypogonadotrophic hypogonadism). The simplest way to identify the level of the axis at which the abnormality exists is to measure gonadotrophin levels. If follicle stimulating hormone (FSH) and luteinizing hormone (LH) are markedly elevated (over 50 mIU/ml) gonadal failure is suggested. The differentiation is of important prognostic significance. Individuals with gonadal failure are hopelessly infertile, whereas females with hypothalamic-pituitary failure are potentially fertile if appropriate gonadotrophin replacement therapy is given. Thus far, fertility has not been achieved in males with full-blown hypogonadotrophic hypogonadism. This failure presumably reflects a lack of complete understanding of the relationships between FSH and LH secretion and normal spermatogenesis.

Syndromes of hypergonadotrophic hypogonadism are almost invariably related to a dysgenetic or absent gonad. If a Y chromosome is included in the karyotype of the affected individual, phenotypic expression may range from infantile female to genital ambiguity to immature male. If a Y chromosome is not present, phenotypic expression is almost invariably infantile female.

The most common cause of hypergonadotrophic hypergonadism in phenotypic males is Klinefelter's syndrome. The chromosomal complement in these individuals is 47, XXY. Affected individuals generally have an eunuchoid body habitus, small penis, small firm testes, and, in some cases, gynecomastia. They are azoospermic, and testicular biopsies exhibit seminiferous tubule failure and Leydig cell dysfunction. Less common forms of hypergonadotrophic hypogonadism in genotypic males include: XY gonadal dysgenesis, characterized by persistent Müllerian derivatives, ambiguous or female external genitalia, and streak gonads; congenital bilateral anorchia, characterized by normal prepubertal penile development, accompanied by an empty scrotal sac and absent gonads on abdominal inguinal exploration; Reifenstein's syndrome, which is characterized by XY karyotype, hypospadias, testicular atrophy, eunuchoidism, gynecomastia, azoospermia, seminiferous tubule hyalinization, and Leydig cell hyperplasia with decreased testosterone synthesis; and agonadia, which has been discussed previously under the topic of genital ambiguity.

In phenotypic females, dysgenetic gonads are most commonly encountered in individuals with a 45,X karyotype—Turner's syndrome. Dysgenetic gonads are also found in individuals with an XX or XY karyotype, several varieties of mosaicism, and various abnormalities of the X chromosome, including ring formations, deletions, and isochromosomes. Conditions similar to agonadia or bilateral anorchia have not been reported in the female.

Hypogonadotrophic hypogonadism may occur in both sexes in conjunction with panhypopituitarism, combined gonadotrophin and growth hormone deficiency, isolated gonadotrophin deficiency, and isolated deficiency of either FSH or LH. Isolated gonadotrophin deficiency is one of the best characterized of these disorders. It may occur in conjunction with anosmia or hyposmia in both sexes. In males, mental retardation is also a common concomitant, but in females, intelligence is generally normal. Affected females are potentially fertile, whereas males are not for reasons mentioned previously. The constellation of hypogonadotrophic hypogonadism with anosmia has been called Kallman's syndrome. In males, it appears to be transmitted as an X-linked disorder, whereas in females, transmission is most consistent with an autosomal recessive mode of inheritance. Isolated deficiencies of either FSH or LH have been rarely reported in individuals of both sexes.

Although most enzyme deficiency states in steroid biosynthesis appear at birth as either genital ambiguity or serious metabolic derangement, occasional genotypic females are detected because of delayed puberty and delayed menarche. The most likely deficiency to appear in such fashion is 17α-hydroxylase deficiency. Patients with this disorder have markedly elevated serum progesterone and deoxycorticosterone concentrations. They exhibit sexual infantilism because they are incapable of synthesizing either androgens or estrogens.

Anatomic defects associated with primary amenorrhea in phenotypic females include imperforate hymen, transverse vaginal septum, Müllerian agenesis, cervical aplasia, and testicular feminization. All of these disorders are characterized by normal development of secondary sex characteristics with absence of menarche at the appropriate time. Patients with imperforate hymen, transverse vaginal septum, and cervical atresia generally experience cryptomenorrhea and give a history of cyclic abdominal pain. If diagnosis is delayed, a midline pelvic mass (the uterus engorged with menstrual blood) may be palpable. Once the obstruction to outflow of menstrual blood is relieved, cyclic menses are noted. All such patients are potentially fertile, although those with cervical aplasia generally have a poor prognosis. Patients with Müllerian agenesis cannot be made fertile, but a vagina may be developed either by surgery or the use of dilators. These patients have a fairly high incidence of adnexal torsion that may require surgical treatment. Patients with testicular feminization fall into a slightly different category. These individuals have an XY karyotype, and circulating testosterone levels are in the normal range for males, but their appropriate target organs are insensitive to androgen. They have excellent breast development and female external genitalia and scanty axillary and pubic hair, presumably secondary to their androgen insensitivity. A normal or shortened vagina ends in a blind pouch. The testes may be intraabdominal or in the inguinal canal and should be removed after puberty because of a tendency to undergo malignant degeneration.

SECONDARY HYPOGONADISM AND SECONDARY AMENORRHEA

Secondary Hypogonadism—Male

Most males with hypergonadotrophic hypogonadism will not have delayed puberty but will instead go through relatively normal puberal development. However, they may fail to obtain full sexual development or may show signs of regressive change after full sexual development is achieved. Individuals with these findings are invariably infertile, and if regressive changes are not present, most are first diagnosed in the course of an infertility evaluation. Symptoms may include impotence and decreased libido, but these symptoms may also occur in males with normal testicular function and severe depression. Physical examination may reveal some regression of secondary sex charactristics with decrease in muscle mass, increase in estrogen dependent fat deposition, and gynecomastia, but these findings may take a long time to develop. The testes are invariably small and soft. Semen examination reveals absent or markedly defective spermatogenesis. Gonadotrophin levels are generally elevated or become so as the process progresses. Testicular biopsy is similar to that seen in Klinefelter's syndrome, but the karyotype is XY. A large percentage of patients with Klinefelter's syndrome have secondary hypogonadism.

Several types of male hypogonadism have isolated destruction of the seminiferous tubules or selective absence of the germinal epithelium.

Leydig cell function is generally normal so that secondary sex characteristics are not altered. LH and testosterone values are normal, but FSH values are high. Spermatogenesis is absent or markedly abnormal. Causes include the rare Sertoli cell only syndrome, mumps orchitis occurring in the pubertal or adult male, acute gonococcal orchitis, and irradiation.

Adult Leydig cell failure may also occur in some of the conditions described in the preceding paragraphs. In such cases, LH values are elevated, testosterone values are low, and testicular biopsy shows interstitial hyalinization. Symptoms include impotence, decreased libido, hot flashes, and depression. Very rarely, idiopathic Leydig cell failure occurs in the absence of any predisposing insult.

Hypogonadotrophic hypogonadism developing after puberty in males may be idiopathic or secondary to pituitary neoplasms. Physical findings are similar to those seen with hypergonadotrophic hypogonadism, but laboratory findings differ in that gonadotrophins are depressed rather than elevated. Chorionic gonadotropin therapy may be of some benefit if the diagnosis is made early, but once hyalinization of the seminiferous tubules and degeneration of the Leydig cells take place, spermatogenesis cannot be restored.

Occasional cases of isolated LH deficiency of various degree have been described. Testosterone values are low. Because FSH levels are normal, early spermatogenesis is observed. Such patients are amenable to therapy with hCG. Individuals having the most severe form of this disorder show findings of the "fertile eunuch" syndrome. Affected individuals have eunuchoid body habitus, gynecomastia, and underdeveloped male secondary sex characteristics.

Secondary Amenorrhea

Secondary amenorrhea in females is a symptom of a much wider variety of disorders than secondary hypogonadism is in the male. By definition, secondary amenorrhea is the cessation of menses for at least one year after spontaneous menarche has occurred. The causes of secondary amenorrhea do not fall as neatly into hypo- or hypergonadotrophic categories, as do those for male hypogonadism, but an attempt at rough etiologic classification has been made through the use of the progestin challenge.

The progestin challenge test is based on the observation that estrogen exposure is necessary before target tissues will respond to progestins. Such a state is referred to as "estrogen priming." The progestin challenge test is valuable in the evaluation of secondary amenorrhea because once secondary sex characteristics have developed, it is often difficult to determine when the patient was last exposed to adequate levels of estrogen. The test consists of administration of a standard dose of synthetic or natural progestin (oral medroxyprogesterone acetate, 10 mg daily for five days, or 150 mg progesterone in oil by intramuscular injection). If the individual is estrogen primed, and the endometrium is capable of responding, vaginal bleeding should ensue within three to five days of the last dose of medication. Bleeding occurs because the progestin mimics the effect of corpus luteum function. The bleeding is referred to as "progestin withdrawal bleeding."

While the progestin challenge test is of great value in the evaluation of secondary amenorrhea, it has limited usefulness in the evaluation of primary amenorrhea. Many patients with primary amenorrhea are not estrogen primed. Many of those who are estrogen primed have no Müllerian structures or an obstruction to the outflow of menstrual products. Thus, with careful history and physical examination, progestin challenge can be eliminated in cases of primary amenorrhea. One of the most widely used schemata for classification of secondary amenorrhea based on the progestin challenge is that of Kletsky et al. This classification also utilizes gonadotrophin values measured on a blood sample obtained prior to the administration of the progestin. An abbreviated form of this scheme is found in Table 4–1.

Patients are initially divided into two major groups, depending on their response to the progestin, estrogen-primed and estrogen unprimed. Those patients who demonstrate priming as evidenced by vaginal bleeding are further subdivided into two groups according to LH levels. Those with elevated LH are considered to have polycystic

Table 4–1. Interpretation of the Progestin Challenge Test for Secondary Amenorrhea

I. Patient bleeds (estrogen primed endometrium)
   A. Elevated LH = polycystic ovary disease
   B. Normal or low LH = hypothalamic amenorrhea
II. Patient does not bleed (not estrogen primed or pregnant[a])
   A. Elevated FSH = gonadal failure (hypergonadotrophic hypogonadism)
   B. Low FSH = pituitary failure (hypogonadotrophic hypogonadism)

[a]The progestin challenge test should not be used as a pregnancy test.

ovary disease, while those with normal or low LH are considered to have "hypothalamic amenorrhea," although an early pituitary neoplasm must be ruled out. Those who do not bleed in response to progestin administration, and hence are not estrogen primed, are further classified according to FSH level. A low level of FSH indicates pituitary failure (hypogonadotrophic hypogonadism), and a high level indicates gonadal failure (hypergonadotrophic hypogonadism).

The use of such a scheme for classification is intended to facilitate rapid and inexpensive initial evaluation of patients who have secondary amenorrhea. It is not a substitute for thinking or for the performance of an adequate history and physical examination. It should be used to confirm one's initial diagnostic impression not to formulate a diagnosis. This simplified scheme is most useful in patients who have the single complaint of secondary amenorrhea. When galactorrhea and/or hirsutism are among the complaints, the approach must be modified.

Conditions that may have secondary amenorrhea as the primary symptom include polycystic ovary disease, emotional stress (so-called hypothalamic amenorrhea), poststeroid hypothalamic-pituitary suppression (following either use of oral contraceptives or pregnancy), pituitary neoplasms, pituitary necrosis, anorexia nervosa, the empty sella syndrome, premature ovarian failure, and the unresponsive ovary syndrome. Other conditions that have secondary amenorrhea or oligomenorrhea among the symptoms also will be discussed below.

Although the above-mentioned conditions all have secondary amenorrhea, all do not have consistent findings in relationship to estrogen priming from case to case. Thus, repeated emphasis must be placed on the importance of history and physical examination, with lesser emphasis placed on the results of the progestin challenge test.

Polycystic Ovary Disease

The syndrome of obesity, hirsutism, oligomenorrhea or amenorrhea, and enlarged polycystic ovaries was originally described by Stein and Leventhal in 1928. Since that time, a spectrum of disorders known as polycystic and sclerocystic ovary syndromes has been described. These disorders have in common the findings of anovulation and increased ovarian androgen production, which may or may not be expressed as hirsutism. The classic polycystic ovary is at least twice normal size, with a thick, smooth, white capsule, absence of the usual rugae on the ovarian surface, and presence of translucent areas, suggesting multiple cysts buried beneath the capsule. The term "oyster ovaries" has also been used to describe their appearance.

Polycystic ovary disease appears to result from multiple etiologies. In some instances, several members of a family are affected, suggesting an autosomal dominant mode of transmission. In the past, there has been great controversy over the malfunction in the hypothalamic-pituitary-ovarian axis that results in polycystic ovary disease. Originally, it was thought that abnormal ovaries were the cause of hypothalamic-pituitary malfunction. Then it was thought that hypothalamic-pituitary malfunction produced the abnormal ovaries. Current opinion admits to either possibility, with additional modulation of the function of the axis from extra-axis sources such as the adrenal, central nervous system, skin and fat.

A recent model for the genesis of polycystic ovary disease takes into account these multiple possibilities. The theory rests on the following premises and experimental data: A nadir in circulating estrogen is required to stimulate FSH release at the beginning of the cycle, and the source of estrogen that inhibits FSH release need not be of ovarian origin because peripheral tissues such as skin and fat are capable of converting androstenedione secreted by the ovary or adrenal to estrone and testosterone. In the normal individual, approximately 1.5% of secreted androstenedione is coverted to estrone. In obese individuals, this amount is increased. With rapid weight gain and increase in body fat, peripheral conversion of androstenedione to estrone may add enough estrogen to the circulating pool to bind hypothalamic receptor sites and thus block the stimulus to FSH secretion. Continued LH stimulation without FSH to stimulate follicular growth promotes elaboration of small amounts of estradiol by appropriately developed tertiary follicles and elaboration of more androstenedione by the ovarian stroma, thus feeding the established peripheral conversion system. Continued stimulation of immature follicles without growth may result in production of multiple small follicle cysts. The increased output of ovarian androstenedione increases the pool of precursor available for peripheral conversion to androgens as well as estrogens, thus explaining the finding of hirsutism in patients with polycystic ovaries.

Presumably, any steroidogenic anovulatory state could lead to the development of a polycystic ovary. Although the percentage of patients with classic phenotype of obesity and hirsutism described by Stein and Leventhal is small, the availability of endoscopic techniques has demonstrated the presence of polycystic ovaries in many patients with primary and secondary amenorrhea and oligomenorrhea who do not fit the classic description. Occasionally, patients with regular menses and mild hirsutism will also have polycystic ovaries.

Laboratory findings in polycystic ovary disease vary somewhat. LH is generally high normal or elevated, while FSH is normal or low. The LH/FSH ratio is generally 2:1 or greater regardless of the absolute level of LH. An attempt has been made by some investigators to correlate the level of LH with ovarian size, but this correlation has not always been consistent. What remains consistent is an inappropriate tonic LH secretion with no discernible midcycle peak. Urinary 17-ketosteroid and serum testosterone levels may be normal or elevated. However, the percentage of unbound testosterone may be significantly greater than that seen in normal women, thus explaining the tendency toward hirsutism.

Hypothalamic Amenorrhea

A history of recent emotional stress, perhaps accompanied by weight loss or weight gain, suggests a diagnosis of "hypothalamic amenorrhea." It is assumed that stimuli from higher centers interfere with normal function of the hypothalamic releasing hormones with resultant anovulation. The stress required to produce such profound alteration in hypothalamic function ranges from mild (going away to college) to severe (internment in a concentration camp).

In the absence of a marked change in weight, stress-related amenorrhea may be generated by mild hyperfunction of the adrenal. The effects of adrenal medullary hyperfunction may be mediated via the pineal and the effects of adrenal cortical hyperfunction mediated via the peripheral conversion of androstenedione to estrone. Both of these pathways are somewhat hypothetical, but animal and some human experimental data suggest intriguing possibilities.

Amenorrhea Related to Marked Shift in Weight

There is a difference of opinion regarding the etiology of amenorrhea related to marked shift in body weight. The same emotional stress that results in compulsive eating or compulsive dieting (anorexia nervosa) may interfere with cyclic hypothalamic function. Women in concentration camps became amenorrheic long before this could be accounted for on the basis of malnutrition. However, as essential amino acids are necessary for the synthesis of gonadotrophins, anovulation may persist for nutritional reasons once the starvation state has been reached. Patients with obesity-related amenorrhea may perpetuate the anovulatory state by chronic peripheral conversion of androstenedione to estrone in the newly accumulated fat.

Reproduction appears to be one of the earliest endocrine functions affected in states of emotional or physical stress because it is nonessential to the continued survival of the individual. Patients with hypothalamic amenorrhea generally have normal to low-normal levels for both FSH and LH. Those patients with established obesity may progress to a polycystic ovarylike picture with an elevated LH.

Post-Oral Contraceptive Amenorrhea

A special category of hypothalamic amenorrhea patients are those with post-pill amenorrhea—failure to resume spontaneous menses after discontinuing oral contraceptives. In the original reports of this condition, the diagnosis was made after a year of amenorrhea. More recent reports have accepted six months as an adequate period of amenorrhea to justify the diagnosis. Many of these women demonstrate estrogen priming confirmed by progesterone withdrawal bleeding, but many do not. Serum levels of FSH and LH are low to low normal. Presumably, these women are unable to recover from the medication-induced hypothalamic suppression.

The incidence of post-pill amenorrhea is difficult to ascertain because of the large number of women taking oral contraceptives and the difficulty of obtaining complete follow-up on a population cohort. The maximum incidence currently quoted is 2.5% of women who have ever taken oral contraceptives. Previous menstrual abnormalities may be a predisposing factor. It has been estimated that as many as 17% of the female population demonstrates anovulation intermittently if not chronically. Most reports of post-pill amenorrhea indicate that at least one-third of the women reported grossly irregular menses prior to taking oral contraceptives. There appears to be no relationship between type of preparation or the duration of ingestion and the occurrence of post-pill amenorrhea. The condition has been reported after as little as three months of medication and with every type of pill available. Therefore, there appears to be no justification for arbitrarily discontinuing oral contraceptives every two years to allow resumption of normal hypothalamic function as advocated by some physicians.

A certain percentage of patients with hypothalamic amenorrhea may be manifesting the earliest signs of hypothalamic-pituitary failure. If they do not respond to treatment or show a progression of symptoms, a more extensive evaluation should be embarked upon even if estrogen priming was originally demonstrated.

Hypothalamic-Pituitary Failure

Patients who fail to have withdrawal bleeding after administration of progestins and who have a low serum FSH and LH fall into the category of hypothalamic-pituitary failure (hypogonadotrophic hypogonadism). Profound suppression of gonadotrophin secretion may result from both organic and functional disorders. The organic lesions that occur most frequently are pituitary neoplasms. Pituitary necrosis, which generally occurs as a sequel to severe postpartum hemorrhage, is encountered with decreasing frequency because of improvement in obstetric care. Rare cases of neoplastic or inflammatory lesions of hypothalamic tissues have been reported. The major conditions that fall into the category of functional hypothalamic-pituitary failure are anorexia nervosa and severe postoral contraceptive or postpartum hypothalamic suppression.

Galactorrhea, inappropriate secretion of breast milk, is a common symptom of hypothalamic-pituitary failure if an ablative lesion of the pituitary is not involved. Galactorrhea occurs either because of neoplastic proliferation of prolactin secreting cells or because of suppression of gonadotrophin-induced inhibition of prolactin release. This latter phenomenon can occur in conjunction with nonfunctioning neoplastic lesions or poststeroid hypothalamic suppression following either pregnancy or oral contraceptives.

In addition to cessation of menses, findings that suggest hypogonadotrophic hypogonadism include visual field abnormalities, especially involving peripheral vision, headache, recent delivery with profound hemorrhage, continued lactation in the absence of breastfeeding, recent use of oral contraceptives, decrease in libido, and entrance dyspareunia. If pituitary necrosis is involved, the patient may have symptoms consistent with mild panhypopituitarism, including lethargy, mental dullness, and generalized weakness.

Physical findings indicating a lack of estrogen include atrophy of the breasts, external genitalia, vaginal mucosa, and uterus. Vaginal cytology is similar to that seen in the postmenopausal woman with a predominance of intermediate and parabasal cells. The vaginal mucosa is dry, erythematous, and lacking in rugations. Cervical mucus is scanty.

Pituitary Neoplasms

Pituitary neoplasms result in secondary amenorrhea either because of destruction of the gonadotrophs or because of suppression of the gonadotrophs as a result of elevated prolactin secretion. Functioning pituitary tumors are generally in the acidophilic granuloma category and primarily secrete growth hormone and prolactin. Basophilic granulomas may secrete elevated levels of ACTH. Compensatory pituitary hypertrophy with elevated TSH secretion has been reported in some cases of profound thyroprivic hypothyroidism, but as yet there have been no reports of either gonadotrophin secreting pituitary tumors or pituitary hypertrophy secondary to ovarian failure. Chromophobe adenomas are among the more common pituitary neoplasms seen in conjunction with secondary hypogonadotrophic hypogonadism. Craniopharyngiomas may also appear in this fashion.

In general, additional presenting symptoms depend on the location and type of lesion. Functioning tumors have symptoms appropriate to hypersecretion of the particular trophic hormone involved (acromegaly in the case of hypersomatrophism, Cushingoid changes accompanying hypersecretion of ACTH, and galactorrhea accompanying prolactin secreting adenomas). Nonfunctioning tumors may have only symptoms of hypogonadism and galactorrhea. Both types of tumors may have symptoms suggestive of encroachment on other systems if there is extension beyond the sella. Headaches and visual field disturbances are both suggestive of extrasellar extension.

Initial laboratory studies that are appropriate for the symptomatology should be selected. The temptation to obtain an immediate broad screen of all pituitary functions should be resisted because of the expense involved. The vast majority of these neoplasms have minimal symptomatology and are very slow growing. In the absence of acute visual changes or severe headache, delay of a few weeks to several months in making a definitive diagnosis has little effect on prognosis. On the other hand, some tests, now regarded as part of a routine evaluation, are neither inexpensive nor innocuous. Cases with florid acromegalic or Cushingoid changes obviously deserve expeditious evaluation.

Basic initial evaluation of patients who have only secondary amenorrhea and hypogonadotrophic hypogonadism should consist of serum FSH, LH, and prolactin. Visual field examination and roentgenograms of the skull for evaluation of the sella turcica may be obtained to rule out either gross intrasellar lesions or upward extension. If visual fields, skull films, and prolactin values are all normal, the possibility of a pituitary neoplasm becomes remote even if gonadotrophin values are quite low. Prolactin values that are above normal but below 100 ng/ml of serum are compatible with prolactin elevation secondary to gonadotrophin suppression. The suspicion of tumor is decreased in such patients if an appropriate cause for gonadotrophin suppression, such as oral contraceptive use, recent pregnancy, or profound weight loss, can be uncovered in the patient's history. Nonetheless, prolactin values should be repeated to confirm that they are consistently in this mildly elevated range. If prolactin values remain consistently elevated above 100 ng/ml in plasma, and gonadotrophins are suppressed, polytomography of the sella should be undertaken even if screening skull films and visual fields are normal. Abnormal polytomography suggests the need for further evaluation of the sella, such as angiography or pneumoencephalography. Both of these may be necessary to diagnose the empty sella syndrome. Computerized axial tomography does not appear warranted unless suprasellar extension is suspected on the basis of visual field defects. This technique does not delineate the sellar contents with any degree of accuracy. Positive evidence of a space-occupying lesion in the sella on any radiographic studies suggests the need for surgical exploration of the sella for excision of the neoplastic tissue. Perfection of microsurgical techniques and use of the transphenoidal approach has made this mode of treatment preferable to irradiation because eventual return of function can be expected from the normal residual pituitary tissue in most cases once the neoplasm has been resected.

Pituitary Necrosis (Sheehan's Syndrome)

Pituitary necrosis occurs almost exclusively in the setting of severe postpartum hemorrhage with profound hypovolemia and impending cardiovascular collapse (Sheehan's syndrome). The condition is now rare because of widely practiced hospital-based delivery, more aggressive use of fluid replacement in labor, and maintenance of intravenous therapy throughout the initial postpartum hours when hemorrhage is most likely to occur. The pituitary is particularly susceptible to infarction because it has undergone progressive hypertrophy during pregnancy. The increase in size results primarily from increase in the size and number of lactotrophs, an estrogen-related phenomenon. As estrogen levels decrease beginning with separation and expulsion of the placenta, involution of the pituitary begins and continues for several days. It is of interest that the first symptoms of pituitary insufficiency generally occur at a time somewhat remote from the hypovolemic episode and appear to be related to the process of involution rather than to the acute insult.

Symptoms may be so mild as to go unrecognized until the patient fails to resume spontaneous menses postpartum. At the other extreme are cases so severe that a picture of profound adrenocortical insufficiency ensues (Simmond's syndrome). In general, patients with Sheehan's syndrome demonstrate progressive genital atrophy and signs of mild panhypopituitarism. An important feature that distinguishes these cases from cases of pituitary neoplasms is a very low level of prolactin. However, history is the single most important factor in establishing the diagnosis of pituitary necrosis.

Anorexia Nervosa

Anorexia nervosa combines features of stress-induced amenorrhea and severe malnutrition. In general, affected individuals are overly concerned with body image. Initially, they may be slightly to moderately above average in weight. There is a difference of opinion over the degree of abnormal behavior that must be demonstrated to substantiate the diagnosis, but these patients appear to have in common an approach to weight loss characterized by near starvation and phobic response to food, especially carbohydrates or anything considered "fattening." In spite of a markedly reduced caloric intake, they may demonstrate inappropriate propensity for physical activity, such as ritual exercising and insomnia. Gonadotrophin and prolactin values are low; growth hormone responses are brisk. Thyroid function is normal.

Premature Ovarian Failure

Premature ovarian failure is defined as onset of menopause prior to age 40. The condition tends to be familial and is strongly correlated with late or delayed menarche. Because of the tendency for prolonged use of oral contraceptives by women who have completed desired childbearing by the late twenties, certain of these patients are thought initially to have post-pill amenorrhea. The hallmark that distinguishes these patients from other patients with secondary hypogonadism is the history of hot flashes. Otherwise, they are clinically indistinguishable from patients with hypogonadotrophic hypogonadism who fail to bleed when given a progestin challenge. Many will have demonstrated normal fertility. The confirmatory laboratory finding is an FSH value consistently elevated above 50 mIU/ml. McDonough has reported a high incidence of X chromosomal mosaicism in patients with premature ovarian failure in their late teens or early twenties. Ovarian biopsy reveals total exhaustion of follicular elements with a marked proliferation of ovarian stroma.

Unresponsive Ovary Syndrome

A small percentage of patients who seem to have premature ovarian failure may have the "unresponsive ovary" syndrome. They have a full-blown picture of secondary hypergonadotrophic hypogonadism, including hypoestrogenism and elevated gonadotrophins. However, unlike patients with premature ovarian failure, ovarian biopsy reveals the presence of unstimulated primary follicles. Primary infertility is usually an accompanying finding. The suggested etiology is either structurally aberrant gonadotrophin(s) or aberrant gonadotrophin receptors. Occasional spontaneous pregnancies have been reported. The dynamics of the condition are poorly characterized. However, prognosis for fertility is not hopeless as it is in patients with bona fide ovarian failure, as demonstrated by virtual absence of follicular units.

Evaluation of Secondary Amenorrhea

The extent to which one pursues the evaluation of secondary amenorrhea depends on the combination of presenting symptoms and the woman's desires. In estrogen-primed patients, thyroid function studies are of value since some of these women have varying degrees of hypothyroidism. Endometrial biopsy may also be valuable because endometrial hyperplasia is a frequent finding in women with chronic estrogen stimulation unopposed by progesterone. Serum gonadotrophin values should be obtained but may be eliminated or limited to LH alone if patients have progestin withdrawal bleeding. If the patient desires pregnancy, clomiphene therapy can be undertaken to induce ovulation without knowing the LH level, but the LH level is a helpful adjunct for predicting response to treatment. If the woman desires regulation of menses only, knowledge of the LH value is helpful in deciding whether oral contraceptives or periodic progestin withdrawal is the treatment of choice. Oral contraceptives may be therapeutic as well as offering symptomatic improvement in women with chronically elevated LH. However, these agents may actually exacerbate the problem of an anovulatory patient with low LH. In the latter case, progestin withdrawal with mechanical contraception, if needed, would be the treatment of choice. In patients with secondary amenorrhea without estrogen priming, a more aggressive approach toward establishing the diagnosis must be undertaken before instituting therapy. FSH levels will be more informative than LH levels because there is essentially no overlap between postmenopausal levels and normal midcycle levels for the former, while there is for the latter. Once the diagnosis of pituitary or ovarian failure has been established, the evaluation should proceed in an orderly fashion as outlined in the sections devoted to these disorders. In general, endometrial biopsy is of little or no value in patients who are not estrogen primed. The endometrium will be either atrophic or insufficient for diagnosis, findings that could be inferred from the patient's failure to demonstrate withdrawal bleeding.

GALACTORRHEA

Galactorrhea, inappropriate secretion of breast milk, is generally a symptom of endocrine malfunction but may occur with stimulation of the suckling reflex. Galactorrhea occurs with greater frequency in women than in men because the breast requires previous priming with estrogen for milk to be synthesized and secreted. Galactorrhea is occasionally seen in newborns because fetal breast tissue has been exposed to high levels of estradiol and progesterone in utero.

Causes of galactorrhea include pituitary neoplasm, oral contraceptive therapy, postpartum hypothalamic suppression, use of certain drugs, estrogen therapy, primary (thyroprivic) hypothyroidism, chest wall injury, breast manipulation, nonpituitary prolactin-secreting tumors, and idiopathic.

In women, galactorrhea is commonly accompanied by amenorrhea, but presence of regular menses should not discourage a preliminary evaluation. Conversely, if an adequate explanation can be found after history, physical examination, and preliminary laboratory study, a full-scale evaluation of the pituitary is not necessary.

Galactorrhea occurs as the result of one of four phenomena: hypersecretion of prolactin by neoplastic tissue that is not responsive to normal inhibition of prolactin release, interference with normal mechanism of prolactin inhibition without hypersecretion of prolactin, direct stimulation of prolactin synthesis and secretion, and stimulation of the suckling reflex.

Secretion of Prolactin by Neoplastic Tissue

Pituitary tumors are the most common neoplasms found to secrete prolactin. Although it was originally theorized that prolactin secretion from these tumors resulted from interference with release of prolactin inhibiting factor by compression of normal tissue or interference with hypothalamic-pituitary tracts, that idea has been dispelled. Histochemical studies indicate that neoplastic tissue secretes prolactin in increased amounts. Dynamic studies of the hypothalamus and pituitary indicate that prolactin secretion from these tumors is poorly suppressed or not suppressed by normal mechanisms that control prolactin inhibition. Galactorrhea has been found in a variable percentage of patients with acromegaly resulting from hypersomatotrophism. Some of these patients have elevated levels of prolactin, whereas in others growth hormone may have a lactotrophic effect. Several recent reports of galactorrhea indicate a high incidence of elevated prolactin levels in patients with pituitary tumors. Many of these tumors fall into the microadenoma category (less than 1 cm in diameter), making compression of normal tissue an unlikely cause of the hyperprolactinemia. Occasionally, patients have been reported who had galactorrhea associated with a pituitary tumor but who have low levels of prolactin. Conversely, occasional female patients have been identified who have markedly elevated levels of prolactin but no galactorrhea. Kleinberg et al reported a 50% incidence of pituitary neoplasms in males with galactorrhea and an 81% incidence of pituitary tumors in females who had amenorrhea and galactorrhea. In general, prolactin levels of 100 ng/ml or greater are strongly associated with neoplastic secretion and warrant a careful evaluation of the pituitary.

Occasionally, neoplasms of nonpituitary origin secrete prolactin. Bronchogenic carcinomas and renal carcinomas are the two tumor types with which this finding had been reported.

Interference with Prolactin Inhibition

Interference with prolactin inhibition may result from failure of the hypothalamus to recover from ovarian steroid induced suppression or from therapy with drugs that block synthesis and/or release of prolactin inhibiting factor.

Although the exact dynamics have not been worked out, gonadotrophin suppression and prolactin release appear to be reciprocally related. When gonadotrophin levels are normal, release of prolactin is inhibited, whereas suppression of gonadotrophins promotes release of prolactin. Physiologic suppression of gonadotrophins occurs during pregnancy, and pharmacologic suppression occurs during oral contraceptive therapy. In spite of the increased secretion of prolactin during pregnancy, lactation does not occur. Experimental evidence indicates that estrogen in high concentration blocks secretion of milk by a local effect on the breast tissue. Once estrogen levels fall, milk is secreted.

Under normal circumstances, gonadotrophin levels begin to rise as steroid levels fall. Spontaneous ovulation ordinarily resumes within 6–20 weeks postpartum in nonnursing mothers and within 3–12 weeks in women discontinuing oral contraceptives. Postpartum amenorrhea-galactorrhea has been designated the Chiari-Frommel syndrome. Severe cases are characterized by persistent lactation with absence of estrogen priming. Genital atrophy may develop with paradoxical breast regression and atrophy of the vaginal mucosa and resultant dyspareunia. An identical picture may be seen after discontinuation of oral contraceptives. Although it has been assumed that this condition has resulted from abnormal hypothalamic function, pituitary neoplasm should be ruled out. Prolactin levels may be normal to moderately elevated. In some patients, this process may represent the earliest sign of a pituitary adenoma.

Drugs that interfere with prolactin inhibition fall into two general categories, psychoactive and antihypertensive. The psychoactive drugs include tranquilizers of the phenothiazine and benzodiazepine groups and tricyclic antidepressants. Antihypertensives with a central effect, primarily rauwolfia alkaloids (reserpine), have also been noted to cause galactorrhea. These drugs appear to be dopamine antagonists. Although dopamine is not thought to be the prolactin inhibiting factor (PIF), it is the monoamine neurotransmitter that stimulates synthesis and release of PIF. In general, prolactin levels are normal to modestly elevated in patients on these medications.

Stimulation of Prolactin Synthesis

Agents specifically noted to stimulate prolactin synthesis and release in addition to the yet uncharacterized prolactin releasing factor are thyrotrophin releasing hormone (TRH) and estrogens. The observation that galactorrhea and varying degrees of menstrual dysfunction are occasionally associated with primary hypothyroidism led to the study of prolactin response to stimulation by TRH. In males, TSH and prolactin respond about equally to TRH stimulation, whereas in females the response of prolactin secretion to TRH stimulation may be even more brisk than the response of TSH. In primary hypothyroidism, TSH levels are elevated and thus presumably so are TRH levels. In spite of galactorrhea, prolactin levels are not generally elevated beyond the upper limits of normal.

Estrogen presumably stimulates prolactin secretion through direct action on the lactotrophs. Hyperplastic growth of lactotrophs during pregnancy appears to be an estrogen effect. Estrogen-related galactorrhea in females occurs most commonly in women taking oral contraceptives. It must be assumed that the level of estrogen is not high enough to inhibit the breasts directly, as is noted during pregnancy. Prolactin levels are normal to slightly elevated but not in the tumor range. Because pill-related galactorrhea may be the first sign of an adenoma, the medication should be discontinued to see if the galactorrhea ceases. If it does not, more extensive evaluation should be undertaken.

Galactorrhea is rarely reported with estrogen replacement therapy in postmenopausal females; however, it is reported in hypogonadal males receiving estrogen therapy for conditions such as prostatic neoplasia. These men generally exhibit gynecomastia that predates the galactorrhea, emphasizing the need for priming of the breast tissue before galactorrhea can occur.

Stimulation of the Suckling Reflex

Galactorrhea secondary to stimulation of the suckling reflex occurs in two settings: chronic breast manipulation and chest wall injury. The suckling reflex is normally initiated by nipple stimulation. The impulse generated is transmitted via the intercostal nerves. The ultimate result of stimulation is dual, release of prolactin from the anterior pituitary with resultant synthesis of milk and release of oxytocin from the posterior pituitary with resultant milk letdown. Breast manipulation generally occurs during precoital stimulation or in an attempt to squeeze all of the milk out of the breast if a discharge is noted. Chest wall injury presumably involves injury to the intercostal nerves with chronic stimulation of the reflex.

Evaluation of Galactorrhea

Initial evaluation of patients with galactorrhea should include history and physical examination with emphasis on areas that suggest any of the above-mentioned mechanisms. Minimal evaluation should include serum prolactin determination. If menses are normal, prolactin is normal, and there is a clear-cut history of manipulation with no other symptoms, evaluation can stop at that point.

If drug therapy is implicated, it should be discontinued to see if galactorrhea ceases before undertaking further evaluation. Similarly, if hypothyroidism is present, thyroid replacement therapy may be undertaken before continuing with a more extensive evaluation.

If prolactin values are markedly elevated, or if galactorrhea does not respond to change in habit or medication, evaluation for pituitary neoplasm as outlined above should be undertaken.

HIRSUTISM AND VIRILIZATION

Of all gynecologic endocrine abnormalities, hirsutism is perhaps subject to the greatest controversy. The term hirsutism refers to excessive growth of hair. As hair growth is an androgen dependent phenomenon, hirsutism presumably represents a state of androgen excess or perhaps excessive end organ sensitivity to normal amounts of androgen. The degree of expression of such androgen imbalance may range from a mild growth of dark hair on the upper lip to frank virilization.

The term hirsutism refers solely to excessive hair growth; virilization encompasses a constellation of changes associated with a masculine body habitus. Signs of virilization include marked hirsutism, clitoral enlargement, temporal balding, enlargement of the larynx, deepening of the voice, increase in muscle mass, and depletion of estrogen-dependent fat. Hirsutism may result from mild to moderate hypersecretion of androgens from normal tissue, but virilization almost always reflects a neoplastic process. Hirsutism alone does not constitute virilization regardless of the heaviness of the hair growth.

Confusion surrounding the classification of hirsutism reflects the complexity of androgen metabolism in the female. Normal sites of androgen synthesis are the adrenal cortex and the ovarian stroma. In addition, conversion of adrenal and ovarian precursors of active androgens takes place in peripheral sites, including skin, muscle, and liver. In the normal female, 99% of serum testosterone is bound and thus inactive. Decreased levels of binding globulin result in excessive levels of free testosterone even though daily production rates are within normal range. Finally, expression of androgenicity may be determined by the amounts of critical enzymes and androgen receptors in target tissues.

The classification most frequently used categorizes hirsutism as idiopathic or constitutional, ovarian hyperandrogenism, adrenal hyperandrogenism, and combined adrenal and ovarian hyperandrogenism. These distinctions may be artificial. From a pathophysiologic point of view, a classification based on metabolic alteration rather than source of androgen may be more useful.

The major androgens secreted in the female include dehydroepiandrosterone (DHEA), androstenedione (A), and testosterone (T). In normal females, DHEA is almost totally of adrenal origin. Androstenedione is about half ovarian and half adrenal in origin, and testosterone is up to 20% ovarian in origin, 30% adrenal, and 50% from peripheral conversion of androstenedione.

Excessive Androgen Secretion

Excessive androgen secretion may result from hyperstimulation of secretory tissues or hyperplastic or neoplastic change in these tissues with blunting or loss of feedback control. In cases of hyperstimulation and hyperplastic change, a self-perpetuating cycle may be established such that the initial point of alteration may never be identified. Moreover, there may be such a significant interrelationship between steroid synthesis in the adrenal and ovary that each organ may show secondary effects from primary changes in the other.

Clear-cut sources of hyperstimulation of androgen-secreting tissues include excessive ACTH secretion seen in congenital adrenal hyperplasia and Cushing's disease and elevated LH secretion seen in polycystic ovary disease. Some cases of ovarian stromal hyperplasia seen in postmenopausal women may also be a response to elevated LH levels in the menopause.

Primary hyperplastic and neoplastic change in androgen-secreting tissues is seen in adrenal adenomas and carcinomas and in polycystic ovary disease, hyperthecosis, and a variety of ovarian neoplasms. In adrenal hyperplasia and neoplasia, DHEA is the androgenic steroid most commonly elevated. Elevated levels of A and T may be seen. In ovarian hyperandrogen secretion, A and T, especially the latter, are elevated. These distinctions may be important in determining the primary site of origin of the disorder, especially if a malignancy is involved. Both DHEA and A are metabolized directly as 17-ketosteroids, while T is not. Thus, 17-ketosteroids will tend to be elevated in adrenal disorders, whereas plasma T tends to be elevated in ovarian disorders. These distinctions are not absolute but may be a helpful preliminary guideline.

When malignancy is suspected, accompanying symptoms may be taken into consideration when determining the site or origin. Patients with adrenal malignancies generally exhibit varying degrees of hypercortisolism and hyperaldosteronism, including obesity, hypertension, glucose intolerance and disturbances of sensorium, while those with ovarian malignancies do not. Ovarian tumors that may secrete androgens include hilus cell tumors, adrenal rest cell tumors, gynandroblastomas, dysgerminomas, fibrothecomas, and some granulosa cell tumors. Ovarian stromal hyperplasia may also be a source of excess androgen in postmenopausal patients. Patients are generally virilized without signs of adrenocortical dysfunction. As some of these tumors are quite small, adnexal enlargement may not be present. Serum testosterone levels are generally elevated above 300 ng/dl of plasma, but 17-ketosteroid levels may be normal. Abdominal exploration with ovarian biopsy is generally necessary to confirm the diagnosis. In difficult cases, differential ovarian and adrenal vein catheterization to determine the source of excess androgens may be necessary. Angiography alone may reveal an adrenal neoplasm.

The mechanism by which excess androgens are secreted in polycystic ovary disease (PCO) is not perfectly elucidated. Polycystic ovaries generally exhibit hyperplasia of theca externa, the possible site of androstenedione synthesis. Elevated LH levels have been observed in many individuals with polycystic ovaries, and FSH levels have been low. The following phenomena have been observed in in vitro situations: FSH may be necessary for the aromatization of testosterone to estradiol; LH may favor the activation of 5α-reductase with preferential formation of dihydrotestosterone (DHT), which, in turn, is a local antagonist to aromatization. Thus, conditions that exist in polycystic ovary patients may favor the production of androgens rather than estrogens as the secretory product of the ovaries. Androstenedione may still be converted peripherally to estrone, thus keeping FSH suppressed and LH chronically elevated. Peripheral conversion of T to DHT may favor the development of hirsutism in some PCO patients. Androgens of adrenal origin, if excessive in amount, may affect the ability of the ovary to carry out aromatization, thus contributing to the development of PCO. Conversely, androgens block 11-β-hydroxylase activity in vitro. Thus, elevated ovarian androgens could conceivably produce a secondary cortisol-deficient adrenal hyperplasia.

Disorders of Androgen Binding

The problem of excess androgen activity becomes more complex when binding properties are studied. It is entirely conceivable that some women with normal androgen secretion rates and elevated free testosterone have a congenital deficiency of testosterone-estrogen binding globulin (TEBG). Such a phenomenon has been noted in the case of thyroid binding globulin. The problem with identifying such a group lies in the effect of androgens and estrogens themselves on binding globulin levels. Estrogens tend to elevate levels of TEBG, while androgens tend to depress them. Once a state of decreased TEBG with concomitant elevated free T has been established, it may be perpetuated without the necessity for increased androgen production. This condition naturally occurs as a secondary phenomenon in cases of identified androgen hypersecretion. It is possible that in some cases idiopathic or constitutional hirsutism may be due to congenital TEBG deficiency.

End Organ Sensitivity

End organ sensitivity to androgens can be expressed in terms of relative availability of enzymes or receptors. Although no clear examples of receptor or enzyme excess have been recorded, deficiency states have been reported. The most dramatic of these deficiency states involves 5α-reductase, which is responsible for conversion of T to DHT. These individuals have ambiguous external genitalia if male with some virilization at puberty. Body hair may be relatively scant. Testicular feminization is an androgen insensitivity syndrome that may involve defective or absent androgen receptors. The amount of body hair is noted to be an ethnic characteristic (heavy in Mediterranean peoples and scanty in Orientals); it is likely that hirsutism related to end organ hypersensitivity is genetically determined.

Evaluation

The enthusiasm with which the evaluation of hirsutism is pursued tends to depend on the degree of accompanying endocrine malfunction. Hirsute women with normal menses are generally considered to have constitutional hirsutism, and extensive evaluation is embarked on only if marked menstrual aberration or virilization is present. Such an approach does disservice to the woman with normal menses and a cosmetic problem. At a minimum, a serum testosterone level, preferably with free and total values, should be obtained. If free testosterone is elevated, this indicates that deficient binding globulin is the problem. If total and free testosterone are normal, an end organ response problem is more likely.

When menstrual dysfunction accompanies hirsutism, serum testosterone levels are more likely to be elevated. Although there have been numerous reports on differential ovarian and adrenal suppression tests attempting to identify one organ or the other as the major source of androgen, current evidence indicates that the ovary is almost always the source of excess androgen secretion except in undiagnosed cases of congenital adrenal hyperplasia. There is still some debate over whether or not such an entity as adult-onset congenital adrenal hyperplasia exists. Results of differential suppression tests are often confusing because cross-reactivity exists between the adrenal and ovary. Differential catheterization studies do not appear warranted unless a tumor is suspected and its site of origin is not clear.

Patients who are truly virilized almost certainly have a tumor. In such cases, suppression tests may be of great value, as tumors are generally autonomous and not subject to feedback control. Differential suppression tests require initial suppression of the adrenal with dexamethasone, a synthetic corticosteroid. This is followed by a maintenance dose of dexamethasone and one cycle of a combined oral contraceptive to suppress the ovary. Serum testosterone and urinary 17-ketosteroid levels are measured before suppression and at the end of each suppression period. Failure to suppress by at least half the pretreatment values suggests autonomy of the steroid-secreting tissue. Differential stimulation tests utilizing metyrapone to stimulate the adrenal and hCG to stimulate the ovary have been used but are gradually being abandoned because of inconsistent results.

INFERTILITY

Infertility is defined as the inability of a couple to achieve pregnancy after one year of unprotected coitus. The condition is referred to as primary if there has been no history of a previous pregnancy and secondary if there has been a previous pregnancy. For practical purposes, infertility could also be defined as the inability to carry a pregnancy to the stage of fetal viability or repeated loss of viable infants in the perinatal or neonatal period.

Infertility occurs in approximately 15% of couples. About 65% of couples achieve conception within three months of regular unprotected coitus, another 10% within six months, and 5% within one year. A final 5% achieve conception within the second year of exposure but are still infertile by definition.

In evaluating infertile couples, it is important that the condition be viewed as a joint rather than an individual condition. Although texts state that the female partner is "at fault" in 60% of cases and the male in 40%, there are many instances in which several abnormal findings in each partner add up to an infertile state that might not exist if each were married to another, more fertile individual.

At the outset, one must ascertain that the couple is practicing proper coital technique with reasonable frequency. Failure to penetrate the vagina or impotence may be the cause of infertility in some couples. The couple having intercourse less than four times per cycle may experience difficulty in conceiving. One study reports optimal fertility with more than two to less than seven episodes of coitus per week. However, high frequency of coitus should never be considered the cause of infertility until other factors are excluded.

Infertility in Patients with Endocrine Malfunction

It has been estimated that endocrine factors are responsible for infertility in approximately 25% of women who present for evaluation. In the individual patient, endocrine malfunction manifests itself by anovulation or oligo-ovulation.

The prognosis for fertility in patients with malfunction of the reproductive endocrine system is directly related to the state of the gonad and to the extent of malfunction in nonreproductive endocrine systems.

Pituitary, thyroid, and adrenal malfunction may have primary or secondary effects on the reproductive endocrine system that will revert to normal once the basic condition has been treated.

In both males and females, absolute destruction of germ cells renders the individual hopelessly infertile. Females with ovarian failure are also rendered hypogonadal, whereas males may have selective destruction of the germ cells without effect on sex steroid synthesis (Sertoli cell-only syndrome). However, the female with intact germ cell units and aberrant gonadotrophin synthesis is much more responsive to therapy than the male with comparable problems.

In general, the woman who is anovulatory but estrogen primed responds to hypothalamic stimulation, whereas the woman with hypogonadotrophic hypogonadism requires substitutional gonadotrophin for induction of ovulation. Conditions such as hypothyroidism, hyperprolactinemia, and hyperandrogenism must be adequately treated prior to institution of gonadotrophin therapy. In some cases, the patients will be refractory to treatment, and in others, the untreated underlying cause may be exacerbated in pregnancy with serious consequences.

In males, correction of adrenal, thyroid, and pituitary disorders is important for the restoration of fertility. However, disorders of germ cell maturation are much less amenable to treatment than in the female.

A substantial number of infertility problems have been related to disorders of follicle maturation, timing of ovulation, and corpus luteum function. As more sophisticated methods have been developed for the detection of pituitary gonadotrophin levels, information has begun to accumulate regarding aberrations in cyclic function of the hypothalamic-pituitary-ovarian axis. The average normal menstrual cycle is approximately 28 days in length, with ovulation occurring on or about the 14th day. Occurrence of ovulation significantly earlier (day 7–10) or significantly later (day 17–21) may result in release of "unripe" or "overripe" ova. It is postulated that malfunction of the corpus luteum may result from inappropriate or unsatisfactory follicular maturation caused by improper FSH secretion and that the disorder may be further potentiated by abnormal patterns of LH secretion. The presence of such a disorder may be detected in several ways. Follicular maturation can be estimated by progressive changes in vaginal cytology and in the quantity and quality of the cervical mucus. Both of these reflect increasing estrogen production. The most precise measure of follicular function is obtained through daily serum estrogen determinations.

Timing of ovulation is generally determined by the occurrence of the thermal nadir in basal body temperature (BBT) recordings. It can also be established by serial LH determinations, change in the quality of the cervical mucus, and change in vaginal cytology.

Luteal function can be assessed by observation of the BBT curve, sequential vaginal cytology, timed endometrial biopsy, and daily serum progesterone determinations.

Abnormal luteal function may manifest itself as inadequate progesterone production throughout a normal length luteal phase or as adequate daily progesterone production with a shortened luteal phase. Hormonal findings may be correlated with daily BBT, which is affected by the thermogenic properties of progesterone. In the normal individual, there will be a rise of one-half to one degree in BBT in the 18–24 hours following ovulation. The elevation will be sustained until a day or two prior to the onset of menstruation when progesterone production declines. Inadequate progesterone production may be indicated by a flattening of the BBT curve or a peak with a rapid drop-off rather than a sustained rise.

Histologic changes in the endometrium can be correlated with progesterone levels and BBT. In the case of inadequate progesterone levels, the endometrium may appear normal, but menstruation will consistently occur earlier than predicted by biopsy findings.

The end result in either type of defective luteal function may be failed implantation, early disruption of implantation, or repeated early abortion. The presumption is that adequate progesterone support of the endometrium is a necessary prerequisite for satisfactory placentation. Since many ectopic pregnancies seem to function quite satisfactorily without any endometrium at all in which to implant, it seems probable that the greatest importance of the observed histologic changes is as an indicator of available progesterone.

Mechanical Factors

In addition to the hormonal status, one must evaluate possible mechanical factors involved in infertility. The most obvious mechanical cause is obstruction of the Fallopian tubes. The most common process causing total or partial tubal occlusion is gonococcal endosalpingitis with or without associated superinfection by organisms of the coliform or enterococcal type. The resultant scarring may affect the tube at any level. Constriction of the tubal lumen permitting passage of sperm but not of the fertilized ovum may occur, leading to ectopic pregnancy. Tuberculosis salpingitis may lead to the same end result. The most common noninfectious process leading to tubal problems is endometriosis. In this condition, endometrial implants may be widespread enough to cause extensive adhesions of the pelvic organs to one another as well as to the bowel and omentum, interfering with tubal patency as well as mobility and motility. In other instances, scattered endometrial implants, apparently too small to cause gross mechanical problems, may be the only findings. The role of endometriosis in causing infertility remains to be elucidated, although altered tubal function has been suggested as a possible cause.

Mechanical factors involving the uterine cavity are more often a cause of secondary than of primary infertility and should be suspected in cases of repeated abortion when normal luteal function has been demonstrated. Uterine septa or submucous or intramural myomas may cause repeated abortion or premature delivery either because they interfere with satisfactory placentation or because they prevent expansion of the uterine cavity to accommodate the growing fetus. A totally duplicated cervix and uterus is probably not a cause of repeated abortion but may result in fetal malpresentation or premature labor.

A rare mechanical cause of infertility involving the uterine cavity is endometrial sclerosis or Asherman's syndrome. The condition may result from tuberculosis, septic or therapeutic abortion, or curettage in the postpartum or postabortal state. The predisposing factor leads to the destruction of the endometrium with the formation of synechiae between the anterior and posterior uterine walls with total or partial obliteration of the uterine cavity. Patients so affected have amenorrhea or hypomenorrhea and a history of one of the predisposing conditions mentioned. Absolute diagnosis is made by hysterosalpinography or hysteroscopy. A biphasic BBT curve in the face of amenorrhea is suggestive of this condition.

Cervical factors may be of great importance in the genesis of infertility. A quasi-mechanical-chemical factor is that of hostile cervical mucus, which appears to possess spermicidal or sperm immobilizing properties. Hostile mucus is thick, opaque, and gummy in consistency. This type of mucus may result from chronic infection of the endocervical glands or from insufficient amounts of estrogen. In some individuals, sperm antibodies are found in the cervical mucus. These antibodies may be detected in serum. In such instances, mucus may be normal in appearance but immunologically lethal to the partner's sperm.

Vaginal anomalies such as longitudinal or transverse septa may act as mechanical barriers to conception or satisfactory coitus, but patients will generally complain of either amenorrhea or dyspareunia rather than infertility as a chief complaint.

Evaluation of infertility in the male is less complex than in the female because the male reproductive tract is responsible only for the production and transport of sperm as far as the cervical os. Assuming that consistent intravaginal ejaculation occurs, performance of one laboratory examination, semen analysis, will implicate or exculpate the male in cases of infertility. Causes of male infertility include: faulty spermatogenesis—failure of the testis to produce an adequate number of normal spermatozoa; faulty sperm transmission—partial or complete obstruction and congenital malformation of the excretory ducts, including the epididymis, vas deferens, ejaculatory duct, and urethra; and alteration of the biochemical composition of the prostatovesicular secretions, resulting in impaired spermatozoal motility.

Evaluation

Evaluation of the infertile couple should take place as rapidly as possible. In the past, it was the practice of some gynecologists to have a patient record BBT for several cycles to demonstrate her degree of motivation. Following this, one test might be carried out per menstrual cycle until the evaluation was complete, a total time of approximately six months. Most patients tend to be discouraged rather than encouraged by such an approach.

At the termination of the initial visit, laboratory determinations, such as blood count, thyroid function, glucose tests, FSH and LH levels, may be obtained or scheduled as indicated by the patient's history and physical examination. A blood count is generally obtained on all patients. FSH and LH levels should be obtained only if anovulation is suspected. T_3 and T_4 should be obtained if overt or subtle thyroid malfunction is suspected and evaluation of glucose metabolism if diabetes is suggested by family history or previous unexplained fetal loss.

The remaining examinations are performed on every patient regardless of the suspected etiology. Multiple etiologies must always be considered. For maximum information in the ovulatory patient, BBT should be recorded for one cycle prior to proceeding. All examinations can be performed in the subsequent cycle. BBT should also be recorded during the evaluation cycle. In the anovulatory patient, all tests can be performed at the same visit as soon as the patient is appropriately prepared.

The timing of the following studies is given both for a 28-day cycle and in relation to specific events that occur during an ovulatory cycle regardless of the length.

Tubal patency studies should be performed between days 8 and 11 or after the cessation of menses and well before ovulation. The tests commonly employed are the Rubin test—CO_2 insufflation of the Fallopian tubes and/or a hysterosalpingogram, a dye contrast radiographic study of the uterus and tubes. The Rubin test may be performed in the physician's office. If the Rubin test shows no tubal patency after a CO_2 pressure of 200 mm Hg is achieved, or if there is a history suggestive of tubal disease or a uterine anomaly, a hysterosalpingogram should be performed.

A postcoital mucus test should be performed on day 12 or 13 or during the two- to three-day period immediately prior to ovulation to coincide with the preovulatory estrogen peak. This is known as the Huhner-Sims test and is ideally performed within two hours of coitus. The cervical mucus is examined for color, clarity, viscidity, pH, and ability to form a thread (Spinnbarkeit). The mucus is also examined for actively motile sperm and the presence of a fern pattern when the mucus is permitted to dry on a glass slide.

Serum progesterone should be performed between days 19 and 23 or from five to nine days after the presumed time of ovulation. Progesterone levels are at a peak during this period, and even single samples give an accurate qualitative reflection of luteal function.

Endometrial biopsy should be performed between days 23 and 25 of the cycle or from 9 to 11 days from the presumed time of ovulation in order to yield maximal information regarding discrepancies between expected and observed secretory changes. Biopsies performed between days 16 and 22 or day 26 and the onset of menses will indicate that ovulation has occurred but may be less helpful in evaluating defective luteal function. In the anovulatory patient, endometrial biopsy is of value in indicating the patient's response to unopposed estrogen stimulation.

Semen analysis should be obtained as early as possible in the course of evaluation. Unsatisfactory or abnormal analysis is repeated twice before initiating further investigation. By convention, most semen samples are requested after a two-day period of sexual abstinence. Repeat samples should be obtained at intervals of four to six weeks to allow for recovery from intercurrent infections that may cause transient depression of spermatogenesis. Postcoital tests showing immobile sperm indicate the need for sperm antibody determinations even if the semen analysis is normal.

If all tests are reported as normal, or if tubal malfunction is ascertained, an endoscopic procedure should be performed. The timing of this procedure may depend on ancillary information and should be tailored to each individual case.

SUGGESTED READING

Baker HWG, Bremner WJ, Burger HG, et al: Testicular control of follicle-stimulating hormone secretion. *Recent Prog Horm Res* 32:429, 1976.

Hansson V, Calandra R, Purvis K, et al: Hormonal regulation of spermatogenesis. *Vitam Horm* 34:187, 1976.

Kleinberg DL, Gordon LN, Frantz AG: Galactorrhea: a study of 235 cases, including 48 with pituitary tumors. *N Engl J Med* 296:582, 1977.

Means AR, Fakunding JL, Huckins C, et al: Follicle-stimulating hormone, the Sertoli cell, and spermatogenesis. *Recent Prog Horm Res* 32:477, 1976.

Speroff L, Glass RH, Kase NG: *Clinical Gynecologic Endocrinology and Infertility* ed. 2. Baltimore, Williams & Wilkins, 1978.

Williams RH (ed): *Textbook of Endocrinology,* ed. 5. Philadelphia, WB Saunders Co, 1974.

Yuen BH, Keye WR, Jr., Jaffe RB: Human prolactin: secretion, regulation and pathophysiology. *Obstet Gynecol Surv* 28:527, 1973.

QUESTIONS

1. Each of the following enzyme deficiency diseases results in genital ambiguity in infants of one or both sexes. Indicate in which sex ambiguity is generally expressed by using the symbols M, F, or M & F.
   A. 20-hydroxylase
   B. 3-β-ol-dehydrogenase
   C. 17α-hydroxylase
   D. 17β-ol-dehydrogenase
   E. 21 hydroxylase
   F. 11β-hydroxylase
   G. 5α-reductase
2. Which of the following studies is necessary in the evaluation of genital ambiguity to determine if a life-threatening emergency exists?
   A. buccal smear
   B. karyotype
   C. urinary 17-ketosteroids
   D. urinary 17-hydroxysteroids
   E. serum electrolytes
3. Precocious puberty in females more commonly reflects a severe underlying disorder than in males. TRUE FALSE
4. Delayed puberty in males is defined as failure to initiate puberty by age ________.
5. Delayed menarche in females is defined as failure to initiate menses by age ________.

6. Which of the following causes of delayed menarche in females is accompanied by elevated levels of gonadotrophins?
   A. testicular feminization
   B. gonadal dysgenesis
   C. Müllerian agenesis
   D. hypogonadotrophic hypogonadism
   E. imperforate hymen
7. Which of the following is the most common cause of hypergonadotrophic hypogonadism in males?
   A. Turner's syndrome
   B. Swyer's syndrome
   C. Klinefelter's syndrome
   D. Reifenstein's syndrome
   E. Kallman's syndrome
8. Name three environmental agents that cause destruction of the germinal epithelium.
9. Adult Leydig cell failure may be difficult to distinguish from emotional stress. Which one of the following symptoms would be characteristic of the former but not of the latter?
   A. impotence
   B. decreased libido
   C. hot flashes
   D. depression
10. Which class of patients with secondary amenorrhea cannot be rendered fertile with appropriate therapy?
    A. polycystic ovary disease
    B. hypothalamic amenorrhea
    C. pituitary adenoma
    D. premature ovarian failure
    E. Sheehan's syndrome
11. What is the significance of a positive progestin challenge test?
12. What are the possible conditions if the patient has a negative progestin challenge?
13. Which gonadotrophin patterns are characteristic of the following disorders?

A. polycystic ovaries	1. High FSH, high LH
B. hypothalamic amenorrhea	2. Low FSH, high LH
C. Sheehan's syndrome	3. Normal FSH + LH
D. premature ovarian failure	4. Low FSH + LH

14. List nine causes of galactorrhea.

15. Explain the difference between hirsutism and virilization.
16. Name four sources of excessive androgen secretion in women.
17. Of 100 couples practicing unprotected coitus, which percentage will be infertile at the end of one year?
    A. 5%
    B. 65%
    C. 15%
    D. 20%
    E. 10%
18. Match the tests below with the appropriate time during the menstrual cycle for performing the test.

A. tubal patency	1. during menses
B. postcoital test	2. after menses but prior to ovulation
C. evaluation of luteal function	3. just prior to ovulation
	4. just postovulation
	5. nine days postovulation

19. Name three broad generic categories into which male infertility may be divided.

ANSWERS

1. A. M
   B. M & F
   C. M
   D. M
   E. F
   F. F
   G. M
2. E.
3. FALSE
4. 15
5. 18
6. B
7. C
8. 1. mumps
   2. gonorrhea
   3. irradiation

9. C
10. D
11. The patient has had sufficient estrogen priming of the endometrium for proliferation to have occurred.
12. 1. pregnancy
    2. insufficient estrogen priming
    3. inability of the uterus to respond to estrogen
    4. obstruction to the outflow of uterine contents
13. A. 2
    B. 3
    C. 4
    D. 1
14. 1. prolactin-secreting pituitary tumor
    2. profound primary hypothyroidism
    3. breast manipulation
    4. chest wall injury
    5. psychotrophic agents
    6. oral contraceptives
    7. estrogen
    8. nonpituitary prolactin-secreting tumors
    9. reserpine
15. Hirsutism refers to excessive hair growth only. Virilization includes temporal balding, increase in muscle mass, clitoral enlargement, and deepening of the voice.
16. 1. adrenal tumors
    2. ovarian tumors
    3. polycystic ovaries
    4. adrenal hyperplasia
17. C
18. A. 2
    B. 3
    C. 5
19. 1. faulty spermatogenesis
    2. faulty sperm transmission
    3. alterations of the chemical composition of the seminal plasma

chapter five

contraception, sterilization, and pregnancy termination

OBJECTIVES

1. Describe the ideal contraceptive method.
2. List four event related types of contraceptives and know the advantages and disadvantages of each.
3. Describe the mechanism of action of oral contraceptives and their effectiveness rate.
4. Know major and minor complications of oral contraceptives.
5. Know complications and effectiveness of intrauterine devices (IUD).
6. List contraindications to the use of oral contraceptives and IUD's.
7. List several contraceptive methods currently under investigation.
8. List methods of male and female sterilization.
9. Know the failure rate and major complications of each method of sterilization.
10. Name appropriate abortion techniques when the uterus is less than 13-week size and greater than 13-week size.
11. Discuss advantages, disadvantages, and complications of each method of abortion.

CONTRACEPTION

The ideal family planning method should be effective, safe, simple, easy to dispense, inexpensive, reversible, and acceptable. Several methods in use approach this ideal, yet none achieves it.

Twenty years ago, the most widely used methods of contraception in the United States were the condom, diaphragm, rhythm, and various types of vaginal spermicidal agents. Since the 1950s, contraceptive research and development have provided highly effective hormonal contraceptives and intrauterine devices (IUD). Older, traditional methods of contraception have become less popular and have been downgraded by family planning personnel as being inconvenient to use, event related, and less effective than the newer products. Trends in contraceptive use are demonstrated by data collected in the National Fertility study of 1970. Methods used by contraceptors surveyed were: pills, 34.2%; sterilizations (male and female), 16.3%; condoms, 14.2%; IUD, 7.4%; foam, 6.1%; diaphragm, 5.7%; and other methods, 16.3%. Individuals surveyed in the study were married women living with their husbands. Recently, data from Planned Parenthood World Federation show the diaphragm use among American women has increased. In 1976, 10.4% of patients chose the diaphragm, compared with 5.7% in 1975. Preference for the IUD remained constant at 7.7% for 1975 and 1976, and oral contraceptives declined in popularity from 79.2% in 1975 to 72.4% in 1976.

Condom

The condom is the oldest contraceptive device still widely used. Condoms are available without prescription. Standards set and enforced by the U.S. Food and Drug Administration (FDA) assure that American condoms are well constructed and free of defects. They are packaged dry or prelubricated in various thicknesses and colors. The thinner the condom, the more sensitive and more expensive. Pregnancy rates vary from 4 to 36 pregnancies per hundred women per year (HWY) of exposure.

Failure of the condom method of contraception is usually because of rupture of condom (1 in 150–300), vaginal penetration, and ejaculation before condom is applied, failure to unroll the condom the full length of the penis with subsequent regurgitation of semen into the vagina, or loss of the condom in the vagina.

Instructions should include application of the condom the full length of the erect penis prior to vaginal penetration, leaving a pocket in the condom sheath to catch the ejaculate, using adequate lubrication to prevent tearing of condom, and withdrawal of the condom and penis from the vagina before the penis becomes flaccid.

Advantages of condoms include prevention of venereal disease, especially gonorrhea. Condoms do not prevent transmission of syphilis from the external female genitalia to unprotected areas of the male. Condoms may be helpful in treating premature ejaculation by dulling sensation during intercourse.

Couples may use contraceptive foam, cream, or jelly in conjunction with the condom to provide lubrication as well as a defense against spilled semen. To improve acceptability of the method, the female may apply the condom during foreplay.

Diaphragm

Development of vulcanized rubber enabled Mensigna in 1882 to produce a dome-shaped rubber device with the rim incorporated in a round flat watch spring—the diaphragm. Diaphragms range in size from 50 to 105 mm in diameter. When properly fitted and positioned, the diaphragm lies behind the symphysis anteriorly and against the posterior fornix posteriorly, completely covering the cervix. It is a mechanical form of contraception but is usually used with spermicidal jelly or cream. Pregnancy rates of 6–13/HWY have been reported.

Success with this method depends on the motivation of the sexual partners. The diaphragm must be inserted prior to coitus, but it can be placed well before intercourse occurs. It is not popular with all women because some prefer not to manipulate their genitalia.

The proper-size diaphragm is selected by estimating the diagonal length of the vagina and inserting the diaphragm estimated to fit best. If this is too large or too small, the next size is tried until the proper one is found. The contraceptor choosing the diaphragm must be taught self-examination to check proper positioning of the device.

It is recommended that the diaphragm be left in place for six to eight hours following coitus in order not to uncover the cervix until sperm in the vagina have become immobile.

The diaphragm should be checked for holes or deterioration before use by holding it to light and stretching the rubber along the rim. It should be washed after removal with warm water and mild soap and stored dry with cornstarch or a nonmedicated powder. With proper care, a diaphragm should last two to three years, but the size should be rechecked annually and after pregnancy or pelvic surgery.

Vaginal Chemical Contraceptions

These preparations have appeared in many forms, including jellies, creams, pastes, solutions, powders, suppositories, and foaming powder or tablets. In 1961, aerosol contraceptive foam became available and has become the most popular form of vaginal spermicide. Vaginal foam has been favored because patients consider it to be less messy and easier to handle than the other preparations. It leaves little residue and does not tend to drain from the vagina.

Pregnancy rates with jellies, cream, and suppositories range from 1.3 to 38 per HWY. In most of the studies, the pregnancy rates were less than 10, and higher rates were usually obtained with less reliable groups of patients.

Rhythm

The "safe period" or rhythm method of birth control restricts intercourse to the period of physiologic infertility in each menstrual cycle. It is based on the premise that the ovary expels only one ovum per cycle and that this ovum is available for fertilization for only 24–48 hours. Spermatozoa, once they have entered the uterus, can remain viable for at least 48–72 hours. Thus, there is a minimum period of about five days in each cycle during which conception is possible. If intercourse is omitted during this time, pregnancy will be avoided.

In practice, few women ovulate with great regularity. Based on what is known about the menstrual cycle, four major methods for practicing rhythm to avoid conception have been established:

1. The calendar method
2. The temperature method
3. The combined calendar-temperature method
4. The ovulation method

The calendar method assumes ovulation occurs 14 days *before* the next menstrual period. This is subject to a variation of two days in either direction, so days 12–16 are regarded as possible ovulatory days and are the basis of the fertile period. To these are added days 17 and

18 because an ovum ovulated on day 16 might be fertilized until day 18. Days 11–14 are also unsafe because sperm deposited then could fertilize an ovum ovulated on day 14. These eight days constitute the fertile period. Women with regular cycles, whether of 20, 25, or 30 days' duration, all have a potential fertile period of eight days per cycle. It must be remembered that these days bear a relationship to onset of menses that cannot be predicted. For contraceptive purposes, a woman must know the earliest and latest possible dates a period could occur. She does this by keeping and reviewing a record of her periods for at least six months and preferably for one year and identifying the cycle that was the shortest in length and the cycle that was the longest in length. The rule for determining the fertile days is: Subtract 18 from the shortest recorded cycle and 11 from the longest recorded cycle. The fertile days are represented by the range of the differences between these calculations.

For example, if the shortest cycle during the recorded menstrual history is 24 days, the first fertile day after the start of the period is day 6 (24 − 18 = 6); if the longest cycle is 32 days, the last fertile day after the start of any period is day 21 (32 − 11 = 21). Therefore, the fertile or unsafe period is day 6 through day 21 (16 days) of any menstrual cycle for this patient. The calendar method used without baseline dates is ineffective because deviations in the menstrual pattern that reflect the time of ovulation would not be considered.

An alternative to calendar technique is the temperature method. The woman takes her BBT by mouth on awakening and before she begins any activity. An upward temperature shift indicates that ovulation has occurred. Estrogen tends to lower the temperature, while progesterone raises it. Temperatures are lower just before ovulation. One cannot predict when ovulation will or did occur from the BBT. The temperature chart indicates the end of the possible fertile time when coitus becomes safe. A simple rule is "three consecutive (daily) temperatures must be at least 0.5° to 0.7°F higher than the highest reading of the low level of the previous five days." After that, conception is unlikely. Illness and activity may cause the temperature to rise and lead to misinterpretation of the curve. Temperature and calendar methods may be combined. The first fertile day is calculated on the basis of the calendar method (the shortest cycle minus 18 days). The last fertile day is determined by the temperature method. This provides as effective control of conception as the other rhythm methods and minimizes the number of days in each cycle when coitus is prohibited.

Another natural method of fertility control is based on changes in cervical mucus. Estrogens cause the mucus to become thin, clear, slippery, and abundant. These mucus characteristics peak at the time of ovulation and enhance the sperm's movement to the uterus (sperm transport). After ovulation, progesterone causes the mucous to become thick, cloudy, and tenacious. This mucus is hostile and impenetrable to spermatozoa. A woman who develops the ability to recognize changes in cervical mucus can predict fairly accurately when ovulation has occurred; hence, she knows when intercourse becomes safe.

The theoretical effectiveness of rhythm is 14 pregnancies per HWY. Use effectiveness is 35–40/HWY. Rhythm is the only method that has the unqualified approval of the Roman Catholic Church, and Catholics may wish to rely on it. The method is inexpensive and offers satisfactory protection against unwanted pregnancies to the woman who has regular cycles, who is carefully instructed, and who has the cooperation and motivation of her male partner. They must rigorously follow the prescribed instructions and abstain from sexual intercourse during the fertile period. Rhythm requires no mechanical aids associated with the sexual act and is suitable in couples who prefer not to use them.

Coitus Interruptus

Even though this is a very unreliable method of contraception, it is the oldest method known, and the provider of family planning services should be aware of it. This method is also known as "withdrawal," "coitus incompletus," and "being careful." If coitus interruptus is to be successful, the man must withdraw his penis when he feels ejaculation is imminent, and he must ejaculate well outside the vagina. About one-half of the couples who practice coitus interruptus find it an effective method. The psychological impact on both partners may have tremendous effects.

Oral Hormonal Contraceptives

Regulation of reproductive function by hormones is not new. The concept that estrogens could inhibit ovulation was suggested in 1940. The value of hormones as antifertility agents was ignored for a number of years until the Pincus group undertook their first antifertility studies in Puerto Rico with progesterone and later with synthetic progestogens. The results of the first clinical studies with progesterone in large daily doses suggested it had antifertility properties, but it failed to give adequate control of the menstrual cycle. Work demonstrating ovulation suppression in women with a combination of two synthetic progestogens, norethynodrel and norethindrone, and the estrogen mestranol was published by Pincus and Rock, 1958.

There are several conclusions that may be derived from studies with conventional hormonal contraceptives.

1. Combination progestogen-estrogen has a high order of efficacy.
2. Ovulation inhibition is the principal reason for the antifertility effect of the agents.
3. The hormonal contraceptives suppress the midcycle surge of LH and FSH, although the effect on LH is more consistent.
4. The ovary responds to exogenously administered gonadotropic hormones, which suggests anovulation occurs secondary to suppression of the pituitary gland.
5. Because of the pituitary suppressive effects, gonadotropic and ovarian hormone control of the menstrual cycle has been replaced by the contraceptive hormones.
6. Besides the above effects on reproductive functions, these agents influence other organs and systems of the body. These are in part due to estrogens, in part to progestogens, and, in some cases, the causative agent has not been defined.

There are more than 20 different oral contraceptives available in the United States. Two estrogens and six progestins are used in different combinations and dosages in these products and represent different but comparable medications. There are three basic types of preparations. These types are:

1. The "combination pill": These pills contain estrogen and progestin administered together throughout the cycle. One tablet is taken daily for 20–21 days, beginning on the fifth day of the cycle.
2. The "sequential pill": These pills provide only estrogen for the first 15 or 16 days and then estrogen plus progestin for five or six days. These products, which represented about 5–10% of all oral contraceptives used in 1975, were withdrawn from the market voluntarily after the FDA pointed out several known and suspected serious disadvantages of the sequential pills and asked the manufacturers to show that there was a population in whom well-defined benefits outweighed these disadvantages.
3. The "mini-dose pill": Three "mini"-dose pills have been released for sale in the United States. These pills contain progestin alone in a

dose one-tenth to one-half that employed in combination pills. A pill must be taken at exactly the same time each night throughout the cycle.

Progestogens are used in oral conceptives instead of progesterone because progesterone is inactive when orally ingested. The progestogens used in oral contraceptive drugs are modified mainly at the C-17 and C-3 positions so that they are effective when administered orally.

There are two major chemical prototypes of progestogens. One is chemically related to progesterone, and the other has 19 nortestosterone as its basic component. Various modifications of this latter type of progestogen are used in the only oral steroidal contraceptives now available in the United States. Various progestogens differ in potency, depending on their chemical structure. One cannot compare pharmacologic activity of the various progestogens solely on the weight of the compound. Biologic activity of the various progestogens has to be considered. There are biologic assays in animals as well as in humans to determine the potency of these compounds.

The estrogenic component of oral contraceptive pills consists of orally active synthetic estrogens instead of natural estrogen. Two types of estrogenic compounds present in oral contraceptives are ethinyl estradiol and mestranol. These two estrogens have different biologic activity in the human. Ethinyl estradiol is considered slightly more effective in preventing ovulation, while mestranol is more potent in preventing breakthrough bleeding and in causing estrogenic side effects.

The mechanism of action of oral contraceptives is inhibition of ovulation. In addition, cervical mucus is altered so that it remains thick and viscid and interferes with sperm transport. The endometrium is altered so that the glands do not produce sufficient glycogen to support optimally the blastocyst prior to implantation. The ovary is changed so that it is less responsive to normal amounts of gonadotropins. Also, there are effects on the oviduct and the uterine muscle that may interfere with sperm and ovum transport.

Oral contraceptives must be taken every day to be effective because they are metabolized in 24 hours. If a woman misses one or more tablets, an increase in production of gonadotropins may occur, causing ovulation. Secondary contraceptive mechanisms mentioned above may protect against pregnancy even if ovulation occurs. For example, the watery cervical mucus normally present at ovulation is transformed into a scanty, thick, cellular mucus—a physical barrier to sperm penetration. At the same time, glandular changes in the endometrium occur earlier and with greater intensity than in the spontaneous cycle, so the endometrium is out of phase with ovulation and is less able to support and nurture the ovum if it is released and fertilized.

Progestogen-only preparations have some of these secondary effects, and they usually do not suppress ovulation. Predictably, they have a higher failure rate than the combination pills.

Side effects and complications of oral contraceptives can be divided into those that are minor and those that are major. The minor side effects have mostly a nuisance value but may be symptoms of a hormone excess that could lead to a more serious complication and may cause some patients to discontinue using oral contraceptives. The minor effects are symptoms of pseudopregnancy and symptoms related to the menstrual cycle.

The symptoms of pseudopregnancy include nausea, vomiting, dizziness, headaches, enlarged and tender breasts, edema or fluid retention, and weight gain. These are usually attributed to estrogen. Reliable measurement of these side effects is difficult. In general, the symptoms that resemble those of early pregnancy tend to disappear after first few months of use, when the woman's endocrine system has had an opportunity to adjust to the new regimen.

Most women taking oral contraceptives experience some changes in the menstrual cycle. For the majority of normal women, oral contraception produces more regular cycles of shorter duration with less bleeding, less discomfort, and partial or total relief from premenstrual tensions. Women with more serious disorders, such as dysfunctional bleeding, menorrhagia, or endometriosis, often obtain some relief with oral contraceptives. This is not surprising since oral contraceptives eliminate the hormonal peaks and troughs that occur during menstrual cycles and provide more even hormonal levels. The Royal College of General Practitioners in Britain found that "scanty menstruation" occurred nearly three times as often among pill users as nonusers. This decrease in menstrual blood loss is more likely to occur with the lower dosage pills. Reduced menstrual bleeding is an important health benefit. There is a lower incidence of anemia among pill users. Studies from England, Sweden, Egypt, and the United States, using different techniques, show blood loss reduced by 50–70% in women taking oral contraceptives. Other minor side effects are spotting, breakthrough bleeding (BTB), loss of libido, acne, chloasma, and amenorrhea.

Many patients experience a brief delay in resuming their normal menstrual pattern after stopping the pill, and a few develop true secondary amenorrhea. In most instances, these are patients who had irregular or anovulatory cycles before medication was started and who might have developed secondary amenorrhea in any event. The suppressive effect of oral contraceptives on gonadotropin secretion creates an additional problem. In some patients where the hypothalamic-pituitary-ovarian axis is compromised, the pills must be considered partially responsible for amenorrhea when it occurs. One of the most confusing aspects of oral contraceptives' side effects is that their occurrence is so variable. Side effects occur in some women more often than others and with some pills more often than others. The same pills may cause symptoms of hormone excess in one patient and symptoms of hormone deficiency in another patient.

Serious complications have been observed in women taking oral contraceptives, the occurrence of thromboembolic disorders being the most significant. Retrospective studies in Great Britain reported in 1968 and 1969 suggested that the risk of death from thromboembolism in oral contraceptive users was 1.3 per 100,000 women as compared with 0.2 per 100,000 women in the same age group not taking oral contraceptives. Comparable data released in 1969 for the United States indicate that the incidence of thromboembolic disease, including thrombophlebitis, pulmonary embolism, and cerebral thrombosis, is four times greater in pill users than in nonusers, though in actual numbers this represents only 3–6 per 100,000 women per year. Furthermore, in a prospective study involving 80,000 women using oral contraceptives for more than 2 million cycles, no increased incidence of thromboembolic phenomena could be demonstrated.

Evidence of a small, often reversible rise in blood pressure in some women taking oral contraceptives has also emerged in several careful studies. A rise in plasma renin and angiotensin levels, as well as an increased aldosterone level with fluid retention, has been observed and may be the mechanism involved. In most instances, these patients had exhibited slightly elevated blood pressures before taking oral contraceptives. The development of hypertension in previously normotensive women has only infrequently been observed, and in nearly all such cases, systolic and diastolic pressures returned to normal once the medication was stopped. Nevertheless, women who are even mildly hypertensive should use some other form of contraception.

Another serious side effect was reported in 1975. A study in Great Britain indicated an apparent increased risk of both nonfatal and fatal myocardial infarctions in oral contraceptive users. In women in the "30–35" age bracket, the combined incidence of fatal and nonfatal coronary occlusions among those using oral contraceptives was 11 per

100,000 compared with an incidence of 4 per 100,000 in age-matched controls. For women in the age bracket "40–44" who were using oral contraceptives, the incidence was 112 per 100,000 as compared with 22 per 100,000 for nonusers. Thus, the average overall annual additional risk of a fatal or nonfatal myocardial infarction for women using oral contraceptives is about 1 per 1,000 in users over the age of 40. Women with other coronary risk factors such as cigarette smoking, diabetes, hypertension, a history of preeclampsia, hypercholesterolemia, and obesity are at greater risk. Long-term prospective studies will be necessary to resolve the controversial issue that a cause-and-effect relationship exists between the use of oral contraceptives and heart attacks. The risk of mortality in pregnancy (22.8 deaths per 100,000 pregnant women from all causes associated with pregnancy) is also increased with age. One must bear in mind that deaths from oral contraceptives are approximately 0.3–3.0 per 100,000 women per year, while deaths from elective abortions are 3.2 per 100,000 procedures and from auto accidents are 27 per 100,000 persons per year. Oral contraceptives should not be used by patients who have a history of phlebitis or thromboembolic disorders or in women over 40 with additional coronary risk factors. Some authorities even deny the increased risk associated with age to be significant when considering the magnitude of the risk of pregnancy in the women over 40 years of age.

Neuro-ocular complications, including retinal thrombosis and optic neuritis, have also been encountered in patients using oral contraceptives. Although no statistically significant cause-and-effect relationship has been established, immediate discontinuation of oral contraceptives is indicated if these complications occur.

Before starting any woman on oral contraceptive pills, a thorough history and physical examination, including but not limited to blood pressure, weight, and examination of the breasts, neck, heart, chest, abdomen, and pelvis, including cervical cytology and culture for gonorrhea, should be performed. Routine laboratory evaluation consists of complete blood count, urinalysis, and serology. Information, education, counseling, and follow-up evaluation of women adopting this method are of utmost importance.

Absolute contraindications to oral contraceptives include a history of thrombophlebitis, thromboembolic disorders, stroke, impaired liver function, known or suspected carcinoma of breast, known or suspected estrogen-dependent neoplasia, undiagnosed abnormal genital bleeding, sickle cell anemia, and known or suspected pregnancy. Relative contraindications to oral contraceptives include epilepsy, migraine, asthma, cardiac or renal dysfunction, diabetes mellitus, hypertension, uterine leiomyomata, psychic depression, varicose veins, sickle cell trait, obesity, and age 40 years or older.

Changes in liver function tests as well as the occasional occurrence of cholestatic jaundice have been observed in some oral contraceptive users. The progestogen component appears to interfere with bile secretion in these women, leading to increased bromosulfophthalein (BSP) retention and, less often, jaundice. Women with preexisting liver disease should not use oral contraceptives. There also appears to be an increased tendency toward the development of gallstones by oral contraceptive users.

Some changes in blood chemistry in women using oral contraceptives are noted. An increase in T_4 and a decrease in T_3 resin uptake are secondary to an increase of thyroxine-binding proteins. These changes are also seen in pregnancy and with estrogen administration. Tests return to normal two to four months after stopping the pills. A decrease in glucose tolerance with an increase in fasting blood sugar has been noted in 10–25% of oral contraceptive users. An increase in plasma triglyceride and plasma phospholipid levels has also been found in oral contraceptive users. This is a dose-dependent estrogen effect and not related to the progestin. Patients who already have hypertriglyceridemia become worse, and this change may be associated with acute vascular accidents.

There have now been reports from several centers suggesting a potential teratogenicity of progestogen-estrogen agents because of a small but definite increased incidence of multiple congenital anomalies of the type covered by the acronym VACTERL (vertebral-and-cardiac-tracheal-esophageal-renal-limb) in the offspring of women taking combined progestogen-estrogen compounds during early pregnancy. Such exposures usually occurred by accident or while under hormone therapy. Also, it has been reported that women who stop the pills in order to become pregnant may experience a higher incidence of abortion when conception takes place immediately after cessation of the contraceptives. Fetal defects have been reported (limb reduction and cardiac defects) in case of immediate pregnancy after pill discontinuation.

Small numbers of oral contraceptive users have experienced prolonged amenorrhea or oligomenorrhea. In most instances, normal ovulatory menses return. In case the difficulty persists, it is anticipated that the basic ovulatory disturbance may have antedated the contraceptive drug therapy. Amenorrhea or oligomenorrhea related to use of the pill is due to hypothalamic-pituitary suppression by the estrogenic and progestational agents in the contraceptive. Treatment by cyclic therapy with estrogen and progesterone to produce artificial periods is pointless since such treatment only enhances hypothalamic inhibition. Expectant treatment, reassurance, and watchful waiting are the best management in most such cases. Treatment with clomiphene to induce ovulation might be considered in some cases, provided a complete endocrine evaluation has excluded other possible causes for the amenorrhea. Persistent lactation sometimes accompanies amenorrhea and is presumed due to the release of inhibition of prolactin when hypothalamic suppression occurs.

The incidence of benign breast disorders and functional ovarian cysts has been observed to be lower than expected in pill users. However, there have been reports of an increased incidence of the previously uncommon benign hepatic adenomas, the majority of cases now occurring in women taking oral contraceptives. Spontaneous liver rupture and often massive intraabdominal hemorrhage may occur and require emergency partial liver resection.

The incidence of cervical dysplasia and carcinoma in situ has been claimed by a few reports to be increased among women taking oral contraceptives. It is generally agreed that the significance of these observations cannot be properly evaluated because of the lack of an adequate control group.

The woman begins taking the contraceptive pills on the fifth day of the menstrual cycle, taking one tablet daily for 20 or 21 days. Usually within 48 hours of cessation of the drug, she will experience an "artificial period." On the fifth day after the onset of this withdrawal flow, she begins another 20- or 21-day treatment cycle. In the case of some products, the woman simply stops the medication for 7 days after completing a 21-day pill cycle and then automatically begins the next pill cycle, regardless of when, or if, her artificial withdrawal period occurs. Most manufacturers also supply 28-tablet packages that include 7 placebo tablets at the end of the 21-day active drug supply in each package, thus allowing the woman to take a tablet every day during each 28-day cycle; hopefully, this avoids mistakes in timing and develops the habit of daily pill taking.

There has been a steady trend to lower doses of both the progestational agent and especially the estrogenic component used in oral contraceptives, with a resulting decrease in side effects. The low-dose preparations may not be satisfactory for many women because of an increased incidence of breakthrough bleeding during each cycle. Pills may be started immediately postpartum or in two to three weeks. In a double-blind study of postpartum patients, the low-dose pills did not prevent lactation, but there was a significant decrease in milk as evidenced by the need for more supplemental feedings and by a lower weight gain for the infants. In addition, protein, fat, and the inorganic constituents (sodium, potassium, calcium, and magnesium) were all decreased. In rare cases, the development of jaundice in the newborn may occur due to the steriods excreted in breast milk. It is better to avoid use of oral contraceptives by breast-feeding women.

Postcoital interception or the "morning after" pill utilizes the properly timed administration of an adequate dose of estrogen to women to prevent implantation of the fertilized ovum. This method of contraception is not suitable for use in a routine, long-term contraceptive program. It is supposed to be an emergency type of contraceptive and is of great value in cases of rape, ruptured condom or diaphragm, or unplanned, unprotected intercourse under a variety of circumstances. The estrogen compounds utilized are: diethylstilbestrol (DES), 25–50 mg daily; diethylstilbestrol diphosphate, 50 μg daily; ethinyl estradiol, 1–5 mg daily; or conjugated estrogens, 25–30 mg daily.

To be effective, they must be given immediately after the potential fertilization but before the stage of implantation. If intercourse occurred at or after midcycle, treatment should be started preferably within 24 hours and not later than 72 hours. The patient should be warned not to have subsequent unprotected coitus in case ovulation should occur later in the cycle when the course of estrogen has been completed.

Possible mechanisms of action include altered tubal transport of the ovum, abnormal endometrium, or direct effect on the blastocyst. The end result is failure of the implantation process.

The pill regimen consists of a five-day course of estrogen compound. Transient side effects due to excessive amounts of estrogen include fluid retention, breast soreness and swelling, and subsequent, irregular, prolonged menstrual flow. When properly administered and timed, the postcoital estrogen therapy is nearly 100 percent successful in preventing pregnancy. A slightly increased incidence of ectopic gestation is noticed in the rare instances in which this method fails to prevent implantation. The pregnancy rate is about 0.3%, including method and patient failure; 10% of these pregnancies were ectopic. Because of the teratogenic potential of these agents, if interception fails, the pregnancy should be terminated. Another approach to postcoital interceptions immediately after intercourse has recently been tried: the insertion of a copper-bearing IUD. Early results suggest a high degree of protection against subsequent pregnancy. The copper presumably acts by preventing nidation. This technique offers a potential and highly effective alternative for some patients, many of whom might elect to retain the IUD as a permanent contraceptive.

Injectable steroid contraceptive formulations have undergone extensive clinical trials but are not yet marketed in the United States. Depomedroxyprogesterone acetate (DMPA) is administered in a dose of 150 mg every three months. Other preparations have also been studied, but not as extensively. In contrast to the metabolic effects associated with the combination estrogen-progestin oral contraceptives, no apparent changes in liver function, lipid metabolism, or blood pressure are noted during DMPA treatment. There is some evidence, however, that DMPA in a dose of 150 mg every three months causes some deterioration of glucose tolerance and an increase in plasma insulin levels. Nearly all clinical studies demonstrate a significant gain in body weight during therapy.

Patients receiving this drug suffer complete disruption of the normal menstrual cycle with a totally irregular bleeding pattern. During the three months after the first injection, the majority of patients bleed between 8 and 30 days of each 30-day period. Thereafter, the incidence of amenorrhea gradually increases in direct proportion to the time the patient has received the drug, while the incidence of increased bleeding steadily diminishes. At the end of one year, approximately 10–15% of women using this preparation discontinue its use because of irregular bleeding. An additional 5% discontinue its use because of other reasons, such as headache, nervousness, and weight gain. Resumption of regular menses is delayed for more than a year in 20–25% of women, although ovulation and fertility resume in the majority of women within one year after stopping treatment. This unpredictable delay in the time of onset of spontaneous ovulation after discontinuing DMPA limits its use to women who are not planning a future pregnancy.

Intrauterine Contraceptive Devices (IUD)

The first physician to use an IUD for pregnancy prevention appears to have been Richard Richter, a German village practitioner. The IUD devised by Richter in 1909 was a flexible ring fashioned of silkworm gut. In 1928, Graefenberg used a silkworm gut ring but with a silver wire twisted about it to make the device radiopaque. In 1961, the Population Council of New York began to fund IUD studies around the world. Between 1965 and 1970, the percentage of married women using an IUD for contraception increased from 1.2 to 7.4%.

The mechanism by which the IUD prevents pregnancy is unknown.

It is generally accepted that the contraceptive action is associated with the production of a local sterile inflammatory reaction with polymorphonuclear leucocyte infiltration caused by the presence of a foreign body in the uterus. It has been shown in rabbits that the inflammatory reaction changes the receptivity of the endometrium for nidation of the blastocyst. The same effect is believed to occur in humans. Cytotoxic substances and phagocytosis may damage the zygote and engulf spermatozoa, respectively. In 1972, it was reported that the efficacy of the IUD is determined by the surface area of the device. Devices engineered to provide optimal endometrial surface contact have lower pregnancy rates.

The IUD prevents tubal pregnancies at a rate of only about 90%, as compared with a 98% prevention rate for intrauterine pregnancy. The incidence of congenital defects in babies born to mothers who had any type of IUD, including copper-containing devices, is no greater than that of the general population. The incidence of fetal death in such pregnancies is not increased, but the incidence of spontaneous abortion is significantly higher, ranging from 30 to 50%. The pregnant patient should have the IUD removed since the abortion rate is lower after removal of the device and the possibility of a septic abortion will be abrogated. Several long-term studies indicate that the IUD is not associated with an increased incidence of carcinoma of the cervix or endometrium.

IUDs rank with oral contraceptives as highly effective at preventing pregnancy. The pregnancy rate for IUDs is about 1.5–3.0 per HWY during the first year of use and is lower in subsequent years. About three-fourths of pregnancies in IUD users occur with the device in place. Once inserted, no further motivation or effort is required for continuing contraception. IUDs are not event related contraceptives such as diaphragms, foams, or condoms. They are valuable for patients who have trouble following directions or remembering to use a method. Their lack of systemic effects appeals to some, although major complications, including death, have been reported.

Contraindications to IUDs include:

1. Known or suspected uterine pregnancy
2. Acute or subacute pelvic inflammatory disease
3. Incapacitating dysmenorrhea or polymenorrhea
4. Uterine hypoplasia, cervical stenosis, or distortion of uterine cavity
5. Developmental anomaly such as bicornuate uterus

6. Unexplained genital bleeding
7. Known or suspected uterine or cervical malignancy
8. Valvular heart disease
9. Endometriosis
10. Endometrial polyps or endometrial hyperplasia
11. In case of copper-containing devices, known or suspected allergy to copper or defective copper metabolism—Wilson's disease.

The most common complaints related to IUD use are bleeding or spotting and pain, including cramps and backache. Medical reasons are usually responsible for removal of IUDs. Recent studies show the IUD does not affect gonadotropins or steroids hormone patterns during the menstrual cycle. Nevertheless, the device does exert a local effect on the endometrium, causing menses to begin when steroid levels are higher than in the control cycles, about two days earlier than the onset would normally occur. This early onset of menses may be produced by a premature and increased release of prostaglandin brought about by the presence of an intrauterine foreign body. The stimulation of uterine contractions by excessive prostaglandin levels may prolong the duration of the menstrual period. The amount of blood lost in each menstrual cycle is significantly increased in women wearing an IUD. Histologic studies of endometrial specimens obtained by biopsy or hysterectomy have shown two types of lesions in association with IUDs. Vascular erosions have been seen in areas of direct contact with the IUD, and increased vascular permeability has been seen in areas not in direct contact with the IUD. Both types of lesions could cause increased menstrual blood loss.

Transient bleeding will occur in most women the first two months after IUD insertion and will persist in some. Reassurance and supplemental oral iron may be provided, and women are counseled to return if excessive bleeding continues so the IUD may be removed. If women still desire to use IUD contraception, another device may be inserted, but a switch to a copper or progesterone-releasing IUD should be made, as these types are associated with less blood loss than the larger plastic devices.

Pelvic inflammatory disease associated with the IUD has been interpreted as a reactivation of preexisting chronic disease brought on by the insertion procedure. Mishell and Moyer reported that in the first 24 hours following insertion of an IUD, the normally sterile endometrial cavity is infected with bacteria. Natural host defenses destroy these bacteria in 80% of patients within 24 hours. Cultures obtained after one month were sterile. The incidence of acute salpingitis in IUD users is greatest in the first two weeks following insertion but diminishes rapidly. It has not been resolved whether insertion of an IUD in women with healthy pelvic organs can produce pelvic inflammation or not. The majority of infections are mild and can be successfully treated with antibiotics without removing the device. If endometritis or salpingitis is of a significant degree, the condition is probably best treated with antibiotics and removal of the IUD. The incidence of gonorrheal PID is probably the same in women with IUDs in place as in the population at large and varies with the group studied. Maternal sepsis in the midtrimester of pregnancy was reported in IUD wearers, especially those with Dalkon shields. Characteristic signs and symptoms of pregnant IUD wearers who develop midtrimester infections are of generalized sepsis with few if any localized signs of pelvic or uterine infection and with intact membranes.

Uterine perforation is a potentially serious complication, and its incidence depends on the skill of the person doing the insertion and the type, shape, and size of the device. It is reported that IUDs inserted during the early postpartum period as well as immediately postabortal are associated with the highest incidence of perforations. Perforation occurs during insertion and is usually asymptomatic except for pregnancy, which often results from the absence of contraceptive protection. This complication can be minimized by proper patient selection and careful insertion, considering uterine position and length. Sometimes partial perforation occurs when the distal portion of the device migrates into the peritoneal cavity. Perforation should be suspected when the IUD string is not visible in the cervix and the user denies expulsion. The physician should rule out pregnancy and then probe the uterine cavity for the device. If the device cannot be felt, x-ray or sonar may be used to locate the IUD. If the device is visualized outside the uterus, it should be removed. Copper IUDs and Dalkon shields should be removed as soon as possible, as they produce severe peritoneal reaction leading to adhesions and subsequent bowel obstruction.

Laparoscopy is the method of choice for removing extrauterine devices unless severe adhesions have developed. Hysteroscopy has been used for removal of tailless IUDs from the uterine cavity when other methods have failed.

The mortality rate for IUDs is three to five deaths per million users per year, mainly due to infection. The IUD is believed to be as safe or safer than other methods of contraception.

When an IUD is chosen for contraception, a pelvic examination is performed. A decision is made to insert the device based on the absence of contraindications. The patient should be scheduled for insertion during a menstrual period. This is a good general policy because the cervix is slightly softened and dilated and therefore insertion is easier. Bleeding or spotting that usually follows insertion will not be evident during menses, and menstruation verifies that the woman is not pregnant. Bimanual pelvic examination is performed to ascertain the size and position of the uterus. A vaginal speculum is inserted, exposing the cervix, which is cleansed with an antiseptic solution and grasped with a tenaculum. Traction on the cervix is applied to straighten the cervicouterine angle, thereby reducing both the force necessary for insertion and the risk of perforation. Sterile technique involving all instruments that go into the uterus is essential. The uterus is sounded, and the direction, the depth, and the pressure required for successful insertion are determined. If the uterus sounds to less than 5.5 cm, another method must be considered.

All IUDs currently marketed in this country come in presterilized packages. Most must be drawn into introducers just before insertion so that the plastic will not lose its "memory." The exact method of insertion will vary with the type of device used. The string attached to the device is left long enough to make it easy for the patient to check and to facilitate future removal. Pain and syncope immediately after insertion are not uncommon. Paracervical blocks may be used to prevent this.

Methods Under Investigation

Intravaginal rings (IVRs) containing norethindrone, medroxyprogesterone acetate, or norgestrel have been reported as a vaginal contraceptive method. The intravaginal rings are made of polysiloxane in 9-mm diameter cylinders, molded into a ring shape with an outside diameter of 60 mm. Prior to molding, the designated amount of steriod is mixed with the polysiloxane to form an elastomer in which the steroid is evenly dispersed throughout the polymer. Each IVR is initially inserted on the fifth day after the onset of menses and left in place for 21 days. At that time, the woman removes the device herself and reinserts it or a new ring one week later for another three weeks. The method has a high acceptance rate and few side effects, but further investigation is needed.

Administration of norgestrel to prevent conception following coitus may have special appeal to many women. It was reported that the higher the dose, the more efficacious is contraception. Doses of 150,

250, 300, 350, and 400 μg of d-norgestrel, taken by women three hours after sexual intercourse, have been studied. A failure rate of 2.2 pregnancies per HWY with postcoital administration of 350 μg of d-norgestrel was reported. The side effects were similar in nature to those reported for mini-pills.

Subdermal implants of progestin have been proven effective for periods of nine months to one year and are in an advanced stage of product development. The method requires the insertion and removal of silastic tubes or rods containing the hormone. Clinical trials are proceeding, but the optimal compound has not yet been selected.

Other methods that are still under investigation include: monthly pill or suppository to be used to induce menstruation, monthly pill to suppress ovulation, monthly pill, injection, or suppository to regularize ovulation as a basis for improved "rhythm," weekly pill to disrupt endometrial development, and immunization of women with an antigenic compound of chorionic gonadotropin. Methods for use by men include a combination-steriod method to arrest sperm production, based on oral, injectable, or subdermal implant routes of administration, and an antiandrogen to prevent epididymal sperm maturation, based on oral or subdermal implant routes of administration.

Sterilization: A Method of Permanent Contraception

Sterilization, for the purposes of our discussion, is any surgical procedure in males or females that results in permanent infertility. Sterilization has had increasingly wide acceptance as a means of fertility control. In the United States, in 1970, a National Fertility Survey indicated that in 16.5% of married couples aged 20–39, at least one partner had been sterilized. The survey also noted that sterilization was the most popular method of preventing pregnancy for couples in which the wife was between 33 and 44 years of age. There are no restraining laws on voluntary sterilization of normal individuals. In most hospitals, sterilization has become a matter of decision between the physician and the individual couple. In federally funded programs, the Department of Health, Education and Welfare (HEW) introduced certain requirements before sterilization is performed and reimbursed.

Sterilization, for any reason, should have the written consent of the patient. Probably the only exception to this rule is the case in which an emergency surgical procedure that will result in sterility must be performed under conditions where the patient's consent cannot be obtained. In most states, written consent is sufficient if the patient is legally competent. There are, however, legal dangers in failing to inform the patient of the nature of the sterilization, its intended effect, all possible consequences, and the possible failure to prevent pregnancy. It is desirable that both husband and wife sign an informed consent for sterilization. Even though pregnancies have occurred after either the male or female has had reversal of a sterilization operation, no sterilization should be undertaken as a temporary or reversible contraceptive method. The couple who views sterilization as a "cure-all" for their problems may need psychiatric or marriage counseling rather than sterilization. Young people under age 30, especially when single, divorced, or with unstable marriages, should carefully consider other alternatives to sterilization.

Vasectomy

Vasectomy is a procedure done on males for the purpose of becoming sterile. Cutting the vas deferens on both sides in a male disrupts the normal passage of spermatozoa from the testicles to the male urethra. An estimated 700,000 vasectomies were performed in the United States in 1972, usually as office or outpatient procedures.

There are several techniques for this surgery, and the patient should be informed what to expect. In one technique, local anesthetic is used, and an incision is made to the sheath of the vas deferens, making an opening in the sheath. The vas deferens is then clamped or clipped, and a quarter-inch to half-inch segment is removed. The ends of the vas deferens are then tied. Some surgeons fold the proximal end of the vas deferens back on itself with double ligation, and some ligate the stump.

A man is not sterile immediately after vasectomy. Semen analysis should be performed periodically after the operation to be sure no sperm are present before the couple is advised that they can discontinue other contraceptive methods. Following azospermia, it is advisable that a semen specimen be examined in six months and then yearly since there is a 0.5–1.0% chance of recanalization of the vas deferens with return of fertility. Physical effects following vasectomy include postoperative pain, which is usually minimal, and hematoma or infections, which are rare. Spermatic granulomas or epididymitis sometimes occur. Long-range, postoperative complications have not been well evaluated but appear to be minimal. Development of circulating antibodies to sperm is under evaluation. Psychological upsets have greatly decreased since vasectomy has become more commonly used and accepted. Proper counseling is important for couples who tend to have dependent personalities and who look to the operation as a "cure-all" for problems ranging from sexual inadequacy to financial insecurity. These couples still have many adjustment problems following the operation. Men with healthy egos and self-confidence accept sterility without sexual dysfunction.

Vasectomy is considered irreversible, and patients must understand that having the procedure means permanent sterility. However, attempts to reestablish the continuity of the vas deferens have resulted in about 50% successes. Recently, there have been attempts to make the procedure reversible by inserting small plastic valves in the vas deferens at the time of operation. There is hope that a return of fertility can be accomplished later by opening the valve and allowing sperm to reenter the semen.

Vasectomy provides some advantages that no other birth control method can offer:

1. Effective as a one-time procedure that eliminates the need to buy and use contraceptives
2. Safe, with slight morbidity and almost no mortality
3. Simple, requiring minimal extra training for most physicians
4. Short, taking only 10–15 minutes
5. Convenient, since only local anesthesia is required
6. Inexpensive, compared with female sterilization which requires more expensive surgery and equipment
7. Culturally acceptable in many countries, especially where the man makes the crucial decisions on sexual activity or reproduction
8. Culturally preferable in countries where women hesitate to go to a male doctor and female doctors and paramedics are scarce

Vasectomy, in comparison with other methods of birth control, has a low failure rate—about 0.15 per 100 person years. Failures are usually caused by recanalization of the ends of the vas, sexual intercourse before the reproductive tract is cleared of sperm, division of a structure other than the vas, or the rare but not impossible presence of more than one vas on one or both sides.

Female Sterilization

One and one-half million American women have been sterilized, and about 100,000 women per year choose to be sterilized. Most female sterilizations are accomplished by operations on the oviduct. Tubal sterilization, which involves disruption of both Fallopian tubes, is accomplished by over 100 different surgical techniques. Tubal ligation is the most commonly employed method of producing sterility by permanently destroying an essential part of the reproductive tract. With this method, childbearing ability is terminated without altering ovulatory or menstrual function. Types of tubal sterilization include simple ligation with sutures, clips, or bands, ligation with partial resection of the tubes, cornual resection, bilateral salpingectomy, burial of the tubal stumps in the broad ligament, transuterine cauterization of the uterine cornua by hysteroscopy, injection of sclerosing solution into the tubes, fimbriectomy, and electrocoagulation. Other methods of female sterilization have been abandoned except for hysterectomy. As an elective sterilization procedure, hysterectomy has the advantage of removal of a potentially troublesome organ. The major advantage is that hysterectomy assures sterility with possible correction of concomitant pelvic relaxation problems. Hysterectomy may be of prophylactic value for subsequent gynecologic disorders. Following tubal ligation, further pelvic surgery may be necessary in up to 25% of patients.

Tubal sterilization may be performed at the time of cesarean section, at the time of laparotomy for some other reason, or as the primary reason for surgery. Tubal ligation may be performed easily in the early postpartum period or may be done soon after a menstrual period. These choices of time ensure that the woman is not pregnant at the time of surgery.

Although there are many techniques for tubal sterilization, only two will be discussed in detail. The Pomeroy technique is simple, safe, and commonly used. It is most frequently performed at the time of cesarean section or in the early puerperium when the uterus and adnexa are abdominal organs easily accessible by a small incision. With the Pomeroy technique, the midportion of each tube is tented and a loop of tube ligated with absorbable suture material. The tented, ligated portion is then excised and submitted for pathologic confirmation. As the suture is absorbed, the excised ends of the tube separate and close by scar formation. The failure rate of this procedure is low (0.4%), and morbidity is minimal.

Laparoscopic tubal sterilization procedures have come into vogue recently, and a number of techniques are currently employed. Local or general anesthesia may be used. Carbon dioxide is insufflated into the peritoneal cavity. A laparoscope is then introduced through a small infraumbilical incision and the pelvic organs visualized. The actual sterilization may be accomplished by electrocoagulating the oviducts, electrocoagulation and division, or, more recently, by clipping or banding procedures. Failure rates similar to those of the Pomeroy procedure have been reported. Morbidity is usually low, with the patient able to return to her daily routine the same day. Several complications specific to laparoscopy have emerged, however, and include bowel burns from the coagulation current, hemorrhage from the adnexa or from introduction of the laparoscope, and problems relative to the introduction of several liters of carbon dioxide into peritoneal cavity. The overall mortality rate from laparoscopic sterilizations is 2.5 per 100,000 procedures.

PREGNANCY TERMINATION—ABORTION

There is justification for the statement that nature is the greatest abortionist since about 15% of recognized pregnancies terminate early by spontaneous abortion. At least 35% of such fetuses have been found to have chromosomal abnormalities that reduce their viability. This chapter will only deal with induced abortion. Induced abortions are those initiated voluntarily with intent to terminate the pregnancy; all other abortions are called spontaneous. The medical definition of abortion is termination of a pregnancy before the fetus has attained viability. This usually means a pregnancy of less than 20 weeks or of a fetus weighing less than 500 gm. In recent years, laws restricting abortion were challenged in the courts, resulting in the 1973 U.S. Supreme Court decision that during the first trimester, abortion is a matter between a woman and her physician; that during the next three months abortion may be regulated in the interest of maternal health; and that after viability has been reached (24–26 weeks of gestation), abortion may be prohibited except to preserve the life or health of the mother. Liberalization of the abortion laws has had many effects. There has been an increase in the number of legal abortions, presumably with a corresponding decrease in extralegal abortion and a drop in the number of abortion-related maternal deaths. There has been a decline in the birth rate and a decline in the number of illegitimate births. Both New York City and California have reported near disappearance of septic abortion, with a marked drop in maternal abortion deaths. Immature and premature deliveries have decreased as well as the neonatal mortality rate. The number of "spontaneous" abortions was reduced, confirming the idea that a large number of spontaneous abortions were actually induced. It was recently reported that 85% of women who are denied legal abortions find ways to terminate an unwanted pregnancy, a finding that indicates the futility of attempts to prevent abortion by legislation. The impact of abortion availability on population growth is emphatically documented in Eastern Europe as well as in Japan. It is also observed in the United States, where the most rapid decreases in birth rate have occurred in geographic areas where abortion is most accepted and where the largest number of pregnancy terminations are performed.

Abortion is still associated with a moral stigma in the American culture. Until this stigma is erased, together with the moralistic and religious pressure that creates it, it will be necessary to provide counseling service for abortion applicants. This work is done to alleviate, as much as possible, postabortal psychological sequelae. The two most important factors in determining the latter are the indication for the abortion and the preabortal psychological status of the patient. The main purposes of counseling are to explore alternatives if the patient is at all ambivalent about abortion; to allay apprehension about the procedure and its consequences; to combat guilt imposed by outside social forces; and to prevent recidivism by stressing the preferability of contraception. Precounseling must be conducted by skilled, experienced, sympathetic personnel. As a form of crisis intervention, abortion does serve as a method of locating women in dire need of contraception. It was reported that about 80% of women seeking abortion were using no birth control method at the time of conception, and 7% reported a previous abortion. The repeaters report a higher incidence of contraceptive use and failure than women being aborted for the first time. This leads to the conclusion that women who have had pregnancy termination once will seek this option again when contraception fails.

Methods currently used for the termination of pregnancy can be discussed under three headings: instrumental uterine evacuation by vaginal route, stimulation of uterine contraction, and major surgery (hysterotomy or hysterectomy). Vaginal evacuation is the method of choice for termination of pregnancy in the first trimester, although in some countries, such as the United Kingdom, it is often used in the second trimester as well. The surgical procedure involves dilating the cervical canal by insertion of a series of dilators of sequentially larger sizes. Cervical dilatation may be difficult before cervical softening at about eight weeks of pregnancy. The use of laminaria tents for preabortion cervical dilatation may reduce the incidence of injury to cervix and thereby the risk of premature births due to cervical incompetence in subsequent pregnancies. They are about two inches long and range in thickness from 3 to 5 mm. Laminaria expands to five times its dry size over 24 hours by absorbing water from cervical secretions. After cervical dilatation, the contents of the uterus are removed with an ovum forceps, followed by scraping with a sharp curette. In recent years, vacuum methods have superseded sharp curettage because of ease, speed, and decreased blood loss. Suction abortions are usually done on an outpatient basis, using local anesthesia.

Perforation of the lower uterine segment or uterine fundus is a serious complication. Its incidence ranges between 1.4 and 2.5 per 1,000 procedures. Approximately 30–50% of such patients require exploratory laparotomy. Surgical exploration is mandatory when intestinal injury has occurred or when the perforation has extended into the broad ligament, making uterine artery or vein laceration likely. Hemorrhage and infection add to the morbidity of dilatation and evacuation. Both of these complications commonly indicate incomplete evacuation of the uterus. Retained products of conception frequently require another surgical evacuation to control either bleeding or associated infection. The incidence of repetition of surgical evacuation is around three per 1,000. An infection rate of 15 per 1,000 and a hemorrhage rate of 2.1 per 1,000 in association with curettage are reported.

Menstrual regulation, also known as endometrial aspiration or minisuction, is a variant of suction procedure that is limited to the first two weeks after a missed period. As a rule, neither anesthesia nor cervical dilatation is required. A flexible plastic canula with a diameter of 4–5 mm is used to remove the uterine contents. The two most common problems are failure to terminate the pregnancy and a pain-synscope syndrome lasting 15–30 minutes, which occurs in 15–20% of patients. A certain number of women on whom this procedure is performed are not pregnant.

Second-trimester pregnancy termination may be initiated by intraamniotic instillation of hypertonic solutions or prostaglandins. The most commonly used abortifacient is 20% sodium chloride. Hypertonic solutions have been used to induce elective abortion as early as 12 weeks and as late as 28 weeks. In Texas, midtrimester therapeutic abortion is only done before 20 weeks of gestation. Transabdominal introduction of a needle into the amniotic sac is difficult prior to the 14th week of gestation, and the chance of failure remains high until the 16th week, after which the chances of success exceed 95%. Under local anesthesia, the amniotic sac is entered by means of a needle through the abdominal wall, and amniotic fluid is withdrawn and replaced by an equal or larger amount of solution. All agents used for midtrimester pregnancy termination have advantages and disadvantages. Hypertonic glucose has been associated with maternal infection and death in some instances as well as with a long instillation to abortion interval. Hypertonic saline is a rapid, reliable abortifacient but causes problems concerned with electrolyte imbalance, coagulopathy, and if intravascular injection occurs, cardiovascular collapse. Hypertonic urea instillation reliably produces abortion, but some oxytoxic agent must also be administered, or the injection to abortion interval may be prolonged. Intravascular injection of urea solutions does not have dire consequences. Prostaglandin injection into the amniotic sac is effective in producing uterine contractions and emptying but may do so with such force that cervical lacerations are produced. Side effects of prostaglandins may prove troublesome to patients. These include nausea, vomiting, abdominal cramps, diarrhea, fever, and asthma. Prostaglandins are contraindicated in patients with a history of asthma. In addition to the amniotic route, prostaglandins successfully produce abortions when administered intravenously, by vaginal suppository, of by injection into the uterus between the endometrium and the amniotic sac.

Complications of midtrimester abortions are more frequent than those after first trimester abortions. In addition to complications already mentioned, retention of the placenta or placental fragments may be associated with hemorrhage or infection. Curettage is often necessary after delivery of the fetus to be certain the uterus is empty. Cervical lacerations can occur with any method and can cause morbidity during the procedure or in future pregnancies.

Major surgical procedures such as hysterectomy or hysterotomy have the highest maternal morbidity and mortality. Anesthesia must be utilized, and the patient may require prolonged hospitalization. Hysterotomy, a miniature cesarean section, is seldom indicated. The uterine scar produced increases the woman's chance for morbidity and death in future pregnancies. Abortion by hysterectomy is sometimes chosen when the woman has a reason for hysterectomy, such as myomas or carcinoma in situ of the cervix.

As determined by the Center for Disease Control, the death-to-case rate for legal abortion in the United States for the three years 1972–1974 was 3.9 per 100,000 procedures. Duration of pregnancy proved to be the most important risk factor. Abortion performed at eight weeks' gestation or less carried minimal risks with death-to-case ratio of 0.4 per 100,000 procedures. Mortality risk increased approximately twofold with each gestational interval at 11–12 weeks, 13–15 weeks, and 16–20 weeks. Between 1972 and 1974, the death-to-case ratio for pregnancy and childbirth (exclusive of abortion deaths) was 14.8 per

100,000 live births. The death-to-case rate for abortions induced in the first trimester was 1.7 per 100,000 legal abortions. Therefore, legal abortion in the first trimester was nearly nine times safer than carrying the pregnancy to term. Mortality for different methods of abortion has been analyzed for 402,000 abortions performed in New York in 1970–1972. Uterine aspiration had the lowest rate, 1.1 deaths per 100,000, which might be due partly to factors such as health status and age. Gestational age, abortion type, parity, and health status are important factors influencing morbidity. Some late sequelae attributed to induced abortion are reduced future fertility, more premature births, higher perinatal mortality, more spontaneous abortions, and increased risk of Rh isoimmunization.

SUGGESTED READING

Bernstein J: Conventional methods of contraception: condom, diaphragm and vaginal foam. *Clinical Obstet Gynecol* 17:21, 1974.

Cates W, Grimes DA, Smith JC, et al: Legal abortion mortality in the U.S.A. Epidemiologic Surveillance, 1972–1974. *JAMA* 237:452, 1977.

Dickey RP: The pill—physiology, pharmacology, and clinical use, in Isenman AW, Knox EG, Tyrer LB (eds): *Seminar in Family Planning.* Chicago, Ill., The American College of Obstetricians and Gynecologists, 1972.

Edstrom KG: Early complications and late sequelae of induced abortion: a review of the literature. *Bull WHO* 52:123, 1975.

Mackay EV, Khoo SK, Clarke, PBB: A review of recent developments in human fertility control. *Med J Aust* 1 (special suppl.):1, 1976.

Population Report: Vasectomy—old and new techniques. Series D, Number 1, December 1973.

Shepard MK: Female contraceptive sterilization. *Obstet Gynecol Survey* 29:739, 1974.

Tatum HG: Clinical aspects of intrauterine contraception: Circumspection 1976. *Fertil Steril* 23:3, 1977.

Tietze C, Murstein MC: Induced abortion: 1975 fact book. *Rep Popul Fam Plann* (14):1, 1975.

QUESTIONS

1. Patients who are recovering from childbirth (one to five days after delivery) are too uncomfortable and busy to be counseled and advised regarding contraception. TRUE FALSE

2. Condoms, foams and jellies, and diaphragms are called "event-related" contraception because they are used in reference to certain dates of the menstrual cycle. TRUE FALSE
3. Condoms, foams, and jellies may be purchased only by persons 21 and over. TRUE FALSE
4. Patients need to be examined and fitted for a contraceptive diaphragm. TRUE FALSE
5. The chief advantage of "event related" methods is their medical safety. TRUE FALSE
6. "Event related" methods are generally not as effective as the pill and the IUD. TRUE FALSE
7. Oral contraceptives are the most effective temporary method of contraception available today. TRUE FALSE
8. Patients taking the pill have an increased likelihood of developing hypertension. TRUE FALSE
9. IUDs require anesthesia for insertion. TRUE FALSE
10. Abortion is legal in the United States. TRUE FALSE
11. "Suction abortion" refers to removal of the products of conception via the vagina. TRUE FALSE
12. The most frequent kinds of permanent contraception (sterilization) are tubal ligation (female) and vasectomy (male). TRUE FALSE
13. Most surgeons require the consent of the spouse prior to sterilizing a married man or woman. TRUE FALSE
14. Sterilization procedures are reversible, with better than 50% success. TRUE FALSE
15. Examination of family planning patients should include but not be limited to:
    A. breast examination
    B. cervical cytology and gonorrhea culture
    C. bimanual pelvic exam
    D. all of the above
16. Failures of diaphragms may be due to:
    A. ballooning or dilatation of vaginal vault during orgasm
    B. improper size
    C. multiple mounting or intromission
    D. lack of motivation, concurrence, and empathy between partners
    E. prolapse of uterus
    F. any of the above

17. Rhythm method of contraception depends on:
    A. time of ovulation
    B. safe period
    C. regularity of cycle
    D. motivation
    E. all of the above
18. Rank the following methods of birth control on a five-point scale, with one being the most effective and five being the least effective:
    A. diaphragm
    B. coitus interruptus
    C. intrauterine device
    D. oral contraceptives
    E. postcoital douching
19. Effectiveness of a contraceptive method is based on:
    A. number of pregnancies per 1000 woman years of exposure
    B. number of pregnancies per 100 woman years of exposure
    C. number of pregnancies per 10 woman years of exposure
    D. none of the above
20. Changes in thyroid function that occur in women receiving oral contraceptives are probably due to:
    A. increased thyroid binding globulin
    B. increased renal clearance of T_4
    C. activation of liver enzymes with increased degradation of thyroid hormones
    D. all of the above
21. The mortality rate of oral contraceptives secondary to thromboembolic disorders is probably about:
    A. 20 per 100,000 per year
    B. 10 per 100,000 per year
    C. 3 per 100,000 per year
    D. There is no mortality rate
22. Nursing mothers should not use oral contraceptives because:
    A. the drugs are secreted in the breast milk and may cause some end organ effect in the infant
    B. there is an increased risk of hypertension during lactation
    C. drugs may inhibit lactation
    D. A and B above
    E. A and C above
23. A 30-year-old female, para 4, was known to be normotensive prior to starting oral contraceptives one year ago. Her blood pressure is now 160/110. She should be:
    A. reassured and left on the pills

B. taken off the pills and observed
C. hospitalized immediately for hypertensive workup

24. Side effects or complications of IUDs are:
A. perforation of uterus
B. irregular bleeding, cramps, and expulsion
C. two to three pregnancies per HWY
D. ascending infection
E. all of the above

25. Contraindications to IUD insertion are:
A. salpingitis, endometritis, or both
B. developmental abnormality of the uterus
C. pregnancy
D. submucous fibromyoma of the uterus
E. all of the above

26. Side effects and/or complications of oral contraceptives include all the following except:
A. nausea, vomiting, abdominal cramps, and bloating
B. breast changes, weight changes, and rise in blood pressure
C. thrombophlebitis, pulmonary embolism, and cerebral thrombosis
D. bacteremia
E. mental depression, migraine, and amenorrhea during and after treatment

27. The failure rate of tubal ligation is:
A. 20%
B. 10%
C. 5%
D. less than 1%

28. Complications of tubal ligation include:
A. failure of procedure with recanalization
B. infection and hemorrhage
C. psychological sequelae
D. all of the above

29. Complications of the vasectomy include:
A. hematoma
B. infection and spermatic granuloma
C. accessory vas missed
D. epididymitis
E. all of the above

30. Which is the safest method of performing a therapeutic abortion?
A. dilation and suction curettage
B. hypertonic saline

C. hypertonic glucose
D. hysterotomy

31. Match the following:

A. Hypertonic saline	1. Association with high incidence of nausea and vomiting
B. Hysterectomy	2. Safest method at 12 weeks' gestation
C. Prostaglandins	3. Hypofibrinogemia
D. Dilatation and vacuum curettage	4. Highest morbidity rate

32. Ten weeks after the last menstrual period, abortion is usually accomplished by:
    A. suction curettage
    B. hypertonic saline
    C. hysterotomy
    D. all of the above are good methods

ANSWERS

1. FALSE
2. FALSE
3. FALSE
4. TRUE
5. TRUE
6. TRUE
7. TRUE
8. TRUE
9. FALSE
10. TRUE
11. TRUE
12. TRUE
13. TRUE
14. FALSE
15. D
16. F
17. E

18. A-3, B-4, C-2, D-1, E-5
19. B
20. D
21. C
22. E
23. B
24. E
25. E
26. D
27. D
28. D
29. E
30. A
31. A-3, B-4, C-1, D-2
32. A

chapter six

changes in menstruation

OBJECTIVES

1. Know the vocabulary used to describe normal and abnormal uterine bleeding.
2. Know three important questions to be answered in the evaluation of abnormal bleeding.
3. Know physical findings to be looked for during the examination of women with abnormal uterine bleeding.
4. Characterize four groups of patients with abnormal bleeding that can be identified after the initial clinical assessment.
5. Describe the further evaluation and management of patients in each of the four groups.
6. Define menopause and climacteric.
7. Discuss the response of the pituitary, uterus, vagina, cervix, bladder, and breasts to estrogen loss after the menopause.
8. Discuss four psychologic stresses that are associated with menopause and climacteric.
9. Discuss symptoms and signs of menopause.
10. List four errors commonly made in treating menopausal women.
11. Discuss the role of estrogens in treating menopausal symptoms.
12. List the three entities that comprise trophoblastic disease.
13. Discuss symptoms and signs of molar pregnancy.
14. List three ways to diagnose molar pregnancy.
15. State the incidence of molar pregnancy in the United States.

16. List three factors that are associated with the highest risk for progression of hydatidiform mole to malignant disease.
17. Discuss the management of a patient who has just aborted a hydatidiform mole.
18. Discuss the management of trophoblastic malignancy.

ABNORMAL UTERINE BLEEDING

Except when clearly related to trauma, genital bleeding is presumed to be of uterine origin. Occasionally, patients are uncertain whether the bleeding is coming from the genitalia or the urinary or anal orifice. Examination at the time of bleeding may be necessary to make this determination. Similarly, examination of urine and feces and of the interior of the bladder and rectum may be helpful. Atrophic vaginitis in postmenopausal women may produce bleeding that can be identified on examination. With these exceptions, however, genital bleeding must be considered uterine in origin and evaluated accordingly.

The vocabulary used to describe normal and abnormal uterine bleeding is frequently confusing and often imprecise. Table 6–1 contains a glossary of terms that will be used to describe menstruation and its disorders in this book.

The pathophysiology of abnormal uterine bleeding relates to the tissues involved: endometrium, endocervix, and exocervix. The endometrium may bleed as a result of hormonal withdrawal. If this occurs in a regular fashion, it is termed cyclic bleeding; if irregularly, it is termed acyclic bleeding. The endometrium may also bleed when there is a sustained estrogenic stimulus at a constant level. Here the endometrium becomes sufficiently overgrown that the amount of hormone becomes insufficient to sustain it and bleeding ensues. Inflammatory, gestational, and neoplastic changes may also produce acyclic bleeding from the endometrium. Abnormal uterine bleeding in association with chronic pelvic congestion is poorly understood. In the cervix, inflammation and neoplasia are the usual processes resulting in abnormal uterine bleeding.

In the assessment of the complaint of abnormal bleeding, the duration of the symptom should be clearly noted. Three important questions to be answered are:

1. What are the characteristics of the bleeding?
   a. The quantity of blood lost can be estimated by the number of

Table 6–1. Vocabulary and Definitions—Uterine Bleeding

Definition	*Term*	*Normal Range*
1. Onset of menses	Menarche	9–17 years
a. before age 9	Precocious puberty	
b. no menses by age 18	Primary amenorrhea	
2. Cessation of menses	Menopause	40–55 years
a. temporary (over 90 days)	Secondary amenorrhea	
b. prior to age 40	Premature menopause	
c. after age 55	Delayed menopause	
d. bleeding one year after menopause	Postmenopausal bleeding	
e. bleeding within a year after presumed menopause	Perimenopausal bleeding	
3. Interval, first day of menses to first day of menses	Menstrual cycle	21–35 days
a. less than 21 days	Polymenorrhea	
b. over 35 days	Oligomenorrhea	
4. Duration of menstrual bleeding	Menstrual period	2–7 days
a. less than 2 days	Shortened menses	
b. more than 7 days	Prolonged menses	
5. Blood loss—normal	Normal period	30–150 ml[a]
a. less than normal	Hypomenorrhea	
b. more than normal	Hypermenorrhea or menorrhagia	
6. Bleeding between menses	Intermenstrual bleeding or metrorrhagia	None
a. light	Spotting	
b. just prior to menses	Premenstrual bleeding	
c. following intercourse	Postcoital bleeding	

[a]Thirty ml of menstrual blood contains 14 mg of iron.

days of bleeding, the number of pads or tampons used, and whether the blood soaks through clothing or runs down the legs.

b. Relationship to the menstrual cycle should be determined. Is this bleeding at the time of the menses, does it prolong the menses or occur between menstrual episodes?
c. Is the bleeding associated with other symptoms such as cramps, backache, abdominal pain, or fainting?
d. Is the bleeding related to bodily functions such as coitus, urination, or defecation?

2. How does abnormal bleeding bother the woman?
   a. Is the patient anxious about cancer or inability to conceive?
   b. Is she weak and easily fatigued?
   c. Is the bleeding a nuisance problem because of the necessity to use a pad or tampon? Does it interfere with coitus?
   d. Is there a volume problem with the patient fearful of soaking her clothing?
3. Is the bleeding characteristic of a disease entity such as cancer, pregnancy, anovulation, or coagulation defect?

In addition to answering these three basic questions, the patient should also be asked about medications, past history of abnormal bleeding, and problems during previous pregnancies. A review of systems and family history is always important.

The presence of clots in the menstrual blood signifies brisk bleeding. The color of the blood is often mentioned by patients but has no particular significance.

Physical examination of the woman with abnormal bleeding should be carried out in order to obtain as many objective data about the complaint as possible. The physician should look for signs of chronic illness such as pallor, wasting, jaundice, and bruisability. Abdominal and pelvic viscera should be carefully evaluated by examination of the abdomen, pelvis, and rectum of all patients of any age who complain of abnormal bleeding. Finally, the historical assessment is often inexact in describing the amount of blood lost, so it may be helpful to examine the woman while she is bleeding in order to estimate blood loss.

Initial assessment of patients with abnormal bleeding can be made as a result of the data base obtained through interview and examination. Patients can be divided into four groups (Table 6–2).

Group I. These patients have demonstrable disease that is most often related to pregnancy. Threatened or incomplete abortion as well as ectopic pregnancy may be detected on the initial assessment. Gross lesions of the cervix such as polyps, ectropion, inflammation, or cancer may cause abnormal bleeding. Uterine leiomyoma are readily detected on physical examination and managed as discussed in Chapter 10. Abnormal uterine bleeding may be the initial complaint of women with diseases such as leukemia, immunologic thrombocytopenic purpura, thyrotoxicosis, von Williebrand's disease, or cirrhosis. Ovarian hormone production may be altered by neoplasms, endometriosis, or chronic salpingitis producing abnormal uterine bleeding.

In group I patients, abnormal uterine bleeding is a by-product of a

Table 6–2. Initial Assessment of Patients with Abnormal Uterine Bleeding Based on Clinical Data

GROUP I *Demonstrable Disease*	*GROUP II* *Hyperpolymenorrhea of Adolescence*
1. Pregnancy complication 2. Gross cervical lesions 3. Leiomyomas 4. Systemic illness 5. Other pelvic pathology Evaluate and treat according to diagnosis	Evaluation—blood count thyroid evaluation Treatment—as for anovulation
GROUP III *No Demonstrable Disease* *Confusing and/or Uncertain History*	*GROUP IV* *No Demonstrable Disease* *History of Persistent Nature or Suspicious for Cancer*
1. Short history 2. Minor variations 3. Relationship to contraception Evaluation—Blood count Cervical cytology Menstrual calendar Reevaluate in 1–2 Months Resolution If no resolution, manage as group IV	1. Patient over age 40 2. Abnormality persistent over 90 days 3. Postcoital bleeding 4. Abnormal bleeding within four months of pregnancy Evaluation—Blood count and other lab Cervical cytology Endometrial Biopsy D&C (biopsy or D&C should be done premenstrually or on first day of bleeding) Treatment—according to Diagnosis

detectable abnormality. Evaluation and management vary with the diagnosis.

Group II. Adolescents often exhibit hyperpolymenorrhea due to anovulatory bleeding. This is usually a self-limiting condition, and the menses become normal when cyclic ovulation is established. These patients are placed in a separate group because the clinical setting is so characteristic. Evaluation consists of blood tests for anemia and blood dyscrasias and possibly thyroid evaluation. Patients in this group alone may be treated for anovulatory bleeding without an endometrial tissue sample obtained by biopsy or cervical dilatation and uterine curettage (D and C).

Group III. This group consists of women with no demonstrable disease and a confusing or uncertain history. Patients in this group include those with a short history (a single prolonged menstrual period or one or two episodes of postcoital or intermenstrual bleeding), those with minor variations from normal (a gradual increase in the duration of flow from three to five days), and those in whom the abnormality is related to contraception (hypermenorrhea after IUD insertion or intermenstrual spotting shortly after starting to take oral contraceptives).

Evaluation of group III patients consists of cervical cytology, blood count, and a menstrual calendar. By recording the days in which bleeding occurred and the amount of bleeding, the patient and physician can make a more objective evaluation of the problems. Women in this group should be seen again in one or two months. By this time the problem has frequently resolved; if it has not, the patient should be evaluated as in group IV.

Group IV. Patients in this group have no demonstrable disease, but the symptoms are persistent in nature or suspicious for cancer. Patients should be assigned to this category if they are over age 40, if the abnormality has persisted for three months, if there is postcoital bleeding, or if they have terminated a pregnancy in the last four months.

Unless there is a specific contraindication, these patients should be evaluated by cervical cytology, blood count, and other indicated laboratory tests and endocervical and endometrial biopsies. They should also be scheduled for a D and C. Either the endometrial biopsy or the D and C should be done premenstrually or on the first day of bleeding to assess the endometrium from which the patient bleeds. Results of this evaluation will reveal neoplasm, polyps, infection, pregnancy, anovulatory endometrium (proliferative or hyperplastic), secretory endometrium, or systemic illness.

An accurate diagnosis is necessary prior to institution of treatment. The most serious violation of this principle is the use of hormonal therapy when it is unsuited or harmful. In the evaluation of group IV patients, curettage may be therapeutic, and no further treatment may be necessary for patients with proliferative or secretory endometrium. If indicated, hormonal preparations should be used to obtain a specific response—lengthening the cycle, shortening the cycle, converting proliferative endometrium to secretory, or reducing the amount of flow. Hormone therapy is of no use in treating postcoital bleeding unless related to atrophic vaginal changes in a postmenopausal woman.

Climacteric and Menopause

The climacteric (critical period) is the time in the later years of reproductive life when changes occur in association with diminished gonadal function. The climacteric has neither a precise beginning nor end and may extend over many years. Menopause is the cessation of menses due to failing ovarian function.

In spontaneous menopause, ovarian failure is heralded by a decrease in the frequency of ovulatory cycles with resultant menstrual irregularities, abnormal bleeding, and infertility. Examination of the endometrium shows the bleeding to be coming from a nonsecretory endometrium. After some time, insufficient estrogen is produced to stimulate endometrial proliferation. This reduction is usually gradual and manifested by decreasing flow (hypomenorrhea), increasing interval between menses (oligomenorrhea), and, ultimately, cessation of menses (amenorrhea). The ovary appears to contain an inherent "life expectancy" that varies from woman to woman and is unrelated to age of menarche or the use of oral contraceptives.

The pituitary responds to ovarian failure by basophil hyperplasia and increased activity. This results in elevated levels of follicle stimulating hormone (FSH) and increased production of growth hormone. The pituitary may also stimulate increased androgen production via the adrenal cortex.

The uterus responds to ovarian failure in several ways. The endometrium remains proliferative since progesterone is not available, and if the amounts of estrogen are normal, it may become hyperplastic. When estrogen decreases in amount, the endometrium becomes atrophic. The myometrium evidences no significant change for some time after estrogen is no longer abundant. Gradual involution then occurs until the uterus becomes small due to decreased muscle mass. Leiomyomas, if present, frequently diminish in size postmenopausally. Cervical glands become atrophic and produce little secretion. The cervical canal narrows and may become stenotic. The exocervix also decreases in size and often becomes flush with the vagina.

The vagina exhibits little change for several years after menopause. Coital activity continues, unaffected by estrogen reduction. Eventually, vaginal cytology reveals few superficial cells, and the vagina becomes more rigid. It loses folds, fornices, and adaptability. The epithelium appears pale and has an increased susceptibility to infection. Supporting tissue of the vagina, particularly fascia between the vagina and bladder or rectum, loses strength. Small asymptomatic cystourethroceles and rectoceles may enlarge and become symptomatic. Some women develop atrophic urethritis and trigonitis, and the external urethral meatus becomes reddened.

Following estrogen decrease, the glandular tissue in the breasts is gradually replaced by fibrous tissue, and elastic tissue becomes rigid and inelastic. Breasts gradually lose fat and become atrophic and smaller after estrogen loss.

Osteoporosis develops in approximately 25% of women prematurely deprived of estrogen. This is a result of a decrease in matrix formation with continued bone resorption due to stimulation by adrenal corticoids. This process usually takes 10–20 years following estrogen loss. The same process is noted in women who have menopause at the usual age, but then it is usually ascribed to the general aging processes. Other factors related to osteoporosis include activity and calcium metabolism.

Estrogen loss produces a diminution of subcutaneous fat, particularly in areas where fat deposits occurred at puberty—breasts and hips. The relative increase in androgen production causes a gradual redistribution of hair to a more masculine pattern—the mustache and chin hairs of an older lady, for example.

Vasomotor instability (hot flashes) occurs in many women who are deprived of estrogen after long exposure to it. This instability does not occur in very young women deprived of their ovaries or in women with ovarian agenesis. Hot flashes occur in more than half of women having spontaneous menopause and in nearly all women having premature surgical menopause. The reason for vasomotor instability is not understood. Some believe elevated FSH levels are responsible, but most evidence points toward estrogen deficiency as the cause. Hot flashes may be promptly relieved by exogenous estrogen without a change in FSH levels.

The effect of estrogen deprivation on the development of arteriosclerosis and its sequellae is poorly understood. It is known that prior to menopause, women have less coronary artery disease than men of the same age and that there is a higher incidence of coronary artery disease in women who had premature menopause. One study revealed four times the expected incidence of coronary artery disease in women 15 years after early castration. There is no good evidence that exogenous estrogen can prevent these changes.

The climacteric and menopause occur coincidentally with many environmental changes in the middle years. It is difficult to identify cause and effect, but four specific psychologic stresses can be identified that are associated with menopause and climacteric:

1. Changes in the family content with resultant loss of purpose. This problem becomes apparent in women whose lives have been centered around their children for many years. With this responsibility gone, an individual may feel purposeless and useless.
2. Goal realization or failure. For those women fortunate enough to have achieved all or most of their life goals by their midforties, life may be thought to hold little future. On the other hand, when few or none of their hopes and desires have been realized, the sense of failure and frustration can become overwhelming. The realization that "what has been achieved already is probably all that will be" may result in bitterness and resentment.
3. Diminished self-esteem. This often comes about as a result of unavoidable changes in appearance as well as decreased ability to perform. The tendency to increased weight, decreased physical strength and staying power, the beginning of facial lines, and hair changes all may lead to negative feelings about self.
4. Loss of sexual confidence. For those women who identify menstruation and reproduction with femininity, the loss of these two functions may imply a lack of sexual attractiveness and result in insecurity in their relationships with men. The physical changes previously mentioned also play a role in a woman's concept of her sexual attractiveness. If, at the same time, her husband is undergoing climacteric changes, such as a reduction in sexual interest and capability, the woman may consider herself sexually deficient. Finally, some individuals may believe coital activity ceases after menopause.

Perhaps 85% of American women exhibit some symptoms during the climacteric. In somewhat less than half of these, treatment is sought. Fewer than 10% of women have severe disability. In some patients, the symptoms last over long periods of time, but in most patients, the significant symptoms exist for less than five years.

Hot flashes are a specific sign of menopause. Typical hot flashes or flushes occur in the upper part of the body and are accompanied by sweating, a red blush, and a feeling of suffocation. They customarily last one to three minutes and occur more frequently when the woman is quiet, such as in bed or sitting at a table writing, playing cards, or eating. At times, the perspiration can be so profuse as to require a change of clothes, and the episode may be noticed by those about her. Often associated with this is a significant change in the woman's "thermostat." Women in the climacteric may be consistently warmer than people around them. Later in the climacteric, as old age approaches, this "thermostat" change may be reversed. Significant hot flashes occur in perhaps 50% of menopausal women.

Amenorrhea is the primary menstrual symptom of the menopause and, with the hot flash, constitutes the basic menopausal syndrome. Anovulatory cycles preceding the menopause may produce oligomenorrhea or hypomenorrhea. Hypermenorrhea and/or polymenorrhea cannot be considered normal phenomena of the menopause, although they do occur frequently in the climacteric.

The previously described changes in the vagina and introitus may produce vaginal dryness, painful coitus (dyspareunia), and/or postcoital bleeding. Burning, itching, and discomfort may be caused by atrophic vaginitis. Relaxation of the pelvic fascia may result in pressure symptoms as well as urinary incontinence. Dysuria and urgency occur with atrophic urethritis and/or cystitis.

Emotional disturbances run the entire spectrum of common emotional complaints. Apathy, insomnia, eating irregularities, anxiety, self-abasement, unreasoned hostility, musculoskeletal discomforts, unrealistic suspicions, morning fatigue, and irritability are some of the more common problems. It is frequently impossible to understand cause-and-effect relationships. It is clear, however, that the impacts of the stresses of the middle years and estrogen deprivation are intimately related to the quality of mental health existing in the years preceding. Severe emotional disturbances with severe melancholia may be expected for those women who have a long history of difficulty in making life adjustments. Contrariwise, menopause may be almost insignificant to a woman who has been serene and who has adjusted easily to the vicissitudes of living.

Diagnosis of menopause is often made erroneously. To make the clinical diagnosis with certainty, signs of estrogen deficiency should be observed. There should be marked diminution of menstrual activity, oligohypomenorrhea with episodes of amenorrhea. The endometrium should be resting or atrophic. The vaginal epithelium may look normal, but vaginal cytology that reveals estrogen deficiency tends to confirm a diagnosis of menopause. In difficult cases where it is uncertain whether or not amenorrhea is associated with estrogen deficiency, FSH levels may be helpful. An elevated FSH confirms the diagnosis of ovarian failure and is most useful for women under the age of 40.

Four errors commonly made in treating menopausal women are:

1. Far too often, abnormal bleeding by women in their forties is labeled "menopausal," whereas a more thorough investigation reveals a neoplasm. It is a common experience in a referral center to receive patients who have been treated with estrogens for menopausal bleeding only to find on examination cervical, endometrial, or ovarian carcinoma. The possibility of cancer must be evaluated whenever abnormal uterine bleeding is associated with the menopause.
2. Hot flashes, despite normal menstrual activity, cannot be presumed due to estrogen deficiency. Hot flashes associated with stress or tension can occur at any age.
3. Amenorrhea may be considered to be due to the menopause when it is a result of pregnancy or other causes.
4. Women with estrogen deficiency are often treated for vaginitis without an attempt being made to restore the vaginal epithelium by exogenous estrogen. The vaginitis of the menopausal and postmenopausal lady is usually related to atrophy of the vaginal epithelium.

Treatment of the menopause is directed to making the woman comfortable by reducing hot flashes, restoring the vaginal epithelium, or assisting her through emotional difficulties. In addition, patients prematurely deprived of estrogen (prior to the 40th year) require estrogen replacement until age 50 or so for the prevention of bone and perhaps vascular changes. The question of treatment past the 50th year for all women is unanswered as yet. There is no good evidence that it is helpful in any way.

Hot flashes and vaginal atrophy respond quickly and dramatically to exogenous estrogen. In both cases, moderate doses are employed in order to reduce the discomfort quickly and control symptoms. Thereafter, dosage is reduced until a level is reached that achieves the desired goals. Both synthetic and natural estrogens are effective; thus, the choice of agent is based on cost and incidence of side effects. Once the symptom is under control, therapy is continued until such time as the patient or her physician decides that it should be withdrawn. In the absence of side effects, this generally would not occur prior to the 50th year. The original symptom may or may not reappear. When treating prophylactically to prevent bone and vascular changes, the lowest possible dose of the drug is employed in order to avoid unnecessary side effects. Prophylactic treatment is continued until the 50th year and perhaps, in some high-risk patients, longer.

Estrogen will not effect a good response for emotional symptoms of menopause. It is often necessary to use mild psychotherapeutic drugs. Short-term goals such as temporary correction of insomnia, the lifting of an episode of depression, or the calming of anxiety are most appropriate. Sedatives and tranquilizers may make a depressed patient more depressed.

Patients who have had their uterus removed but retain functioning ovaries may or may not require treatment later in life when ovarian failure occurs. Therapy is directed toward relief of any specific symptoms, and the absence of the uterus reduces the likelihood of bothersome side effects of estrogen. The relationship of postmenopausal estrogen to endometrial adenocarcinoma is discussed in Chapter 10.

Trophoblastic Disease

Some changes in menstruation are due to trophoblastic disease. Three entities comprise trophoblastic disease: hydatidiform mole, invasive mole (chorioadenoma destruens or villous choriocarcinoma), and choriocarcinoma (avillous choriocarcinoma).

In hydatidiform moles, chorionic villi are transformed into masses of clear vesicles (cluster of grapes) that vary in size up to a centimeter in diameter. Histologically, the vesicles demonstrate swelling of the villous stroma with absence of blood vessels and proliferation of the trophoblast covering the villus. A fetus rarely coexists with a molar pregnancy. Hydatidiform moles are generally benign, however, 7–20% may progress to a more serious lesion. The pathologic appearance of a hydatidiform mole gives no clue to the likelihood that it will progress to malignancy.

In the early stages, a molar pregnancy simulates normal pregnancy, but with time numerous differences appear. Trophoblast in hydatidiform moles secretes chorionic gonadotropin (hCG) like normal trophoblast, but the levels of hCG are much higher than the range for normal pregnancy. Continued high-level hCG stimulation is associated with the development of theca-lutein cysts in the ovaries. These cysts do not require treatment unless some accident befalls them, for they regress after molar evacuation.

Abnormal bleeding is frequently noted in molar pregnancies. It may vary from spotting to hemorrhage, and women with molar pregnancies are often anemic. Absence of fetal heart beat and fetal movements at a time they should be present may be the first clue that a woman has a molar pregnancy. The uterus is often larger than expected for the duration of pregnancy. Hydatidiform mole is also associated with the development of toxemia of pregnancy prior to 20 weeks' gestation and may be associated with hyperthyroidism because of the production of a substance similar to TSH by the trophoblast. Hyperemesis gravidarum may cause the woman to seek medical care.

The diagnosis of hydatidiform mole is easy if the patient is seen to be aborting a "bunch of grapes." Prior to this, however, there may be some confusion, and the condition may be misdiagnosed as twins, intrauterine fetal death, or threatened abortion. Sonotraphy is currently the most helpful modality in diagnosing molar pregnancy. The uterus is seen to be filled by a "snow storm" of echoes on sonography. If sonar is unavailable, hCG titers may be of value, but hysterography is probably quicker and more accurate. With this technique, water-soluble radiopaque substances are injected into the uterus transabdominally or transcervically and roentgenograms obtained. A characteristic "honeycomb" appearance is diagnostic of molar pregnancy.

Treatment of hydatidiform moles consists of evacuation as soon as the diagnosis is made. Many techniques have been used, but the preferred method is suction aspiration of the uterine cavity with the same instruments used for suction abortion.

Villous and avillous choriocarcinoma are forms of trophoblastic malignancy. Villous choriocarcinoma or invasive mole is usually confined to the uterus and pelvis, whereas the avillous form metastasizes widely, especially to lungs and brain.

The incidence of all trophoblastic disease is 1 per 2,000 deliveries in the United States and Europe. The Far East, Philippines, Mexico, and Africa have much higher incidences—up to 1 per 100 deliveries. This increased incidence seems to relate more to socioeconomic conditions than to geographic location. The indigent population in San Antonio, Texas, has an incidence of trophoblastic disease of approximately 1 per 400 deliveries.

Fifty percent of patients with trophoblastic malignancy have a history of passage of a benign hydatidiform mole. Others have a history of recent abortion with no mention of molar changes or of a recent normal pregnancy and delivery. Rarely, the patient has no past history of pregnancy. Patients aborting a benign mole are at risk of progression to malignant disease, and they cannot be identified by histologic examination of the tissue. Women at highest risk for malignant progression are those above age 40 or with a parity of four or more. Women with systemic disease, especially tuberculosis, are at great risk for malignant progression.

The management of women after passage of a benign mole has been revolutionized within the last few years. Trophoblastic neoplasia is the first carcinoma that can be accurately diagnosed and monitored biochemically. Radioimmunoassay for hCG relates directly to the presence of trophoblast remaining in the body. With this technique, residual tumor or metastatic disease can be detected prior to its clinical appearance and far earlier than it could be detected with other diagnostic modalities.

Trophoblastic malignancy is the first malignancy where chemotherapy is the treatment of choice and is curative. In a disease that was once uniformly fatal, cure rates of over 90% have been obtained when treating patients with no clinically evident metastasis (diagnosed by elevated hCG alone). Cure rates of over 70% are obtained when metastatic disease is clinically apparent. The most frequently used chemotherapeutic agents to treat trophoblastic malignancy are methotrexate and actinomycin D.

Patients who have aborted a mole are at risk for development of malignant trophoblastic disease and must be followed after evacuation of the mole; 80% of patients have normal levels of hCG within 60 days. Initially, hCG titers are obtained every two weeks until they are in normal range. Thereafter, they are obtained monthly for a year. Contraception, preferably with oral contraceptives, should be practiced so that an early pregnancy will not be confused with trophoblastic disease because of elevated hCG titers. If a woman has an elevated hCG titer 60 days after evacuation, or if a woman with a previously normal titer shows a rise, she is identified as having trophoblastic malignancy, and metastatic evaluation is performed. If no metastatic disease can be found, she can expect a 90% cure rate with chemotherapy. Should metastatic disease be detected, her cure rate falls to 70%. These patients are treated with chemotherapy until the hCG titer is normal, and then they receive two more courses of treatment. They are followed with monthly hCG titers for a year to be sure a cure was achieved.

SUGGESTED READING

Novak ER, Woodruff JD: *Novak's Gynecologic and Obstetric Pathology*, ed. 7. Philadephia, W.B. Saunders Co., 1974, pp 599–631.

QUESTIONS

For the first four questions, circle the correct term.

1. The onset of menstruation is termed (menarche, puberty).

2. Menstrual cycles occurring every 45–75 days are best described as (oligomenorrhea, hypomenorrhea).
3. Menstrual cycles occurring every 16–18 days are best described as (hypermenorrhea, polymenorrhea).
4. Excessive menstrual flow is (hypermenorrhea, polymenorrhea, menorrhagia).
5. List four etiologies for abnormal bleeding.
6. All of the following are important features in the medical history of a patient with abnormal bleeding *EXCEPT*:
   A. quantity of blood lost
   B. color of the blood
   C. relationship of bleeding to the menstrual cycle
   D. relationship of bleeding to body functions
7. Cessation of menses due to ovarian failure best defines
   A. secondary amenorrhea
   B. climacteric
   C. menopause
   D. none of the above
8. A 48-year-old woman with irregular menstrual cycles and menorrhagia
   A. should be evaluated for possible neoplasia
   B. is demonstrating normal menopausal changes in menstrual function
   C. should be treated with hormones
   D. should keep a basal body temperature chart to see if she is ovulating.
9. Ovarian failure causes atrophic changes in all of the following *EXCEPT:*
   A. uterus
   B. vagina
   C. bladder trigone and urethra
   D. pituitary
10. If in doubt about the diagnosis of menopause, the best laboratory test to confirm the diagnosis is
    A. FSH level
    B. LH level
    C. hCG level
    D. T_4 level
11. Commonly encountered environmental changes that may play a role in the symptomatology of menopausal patients include
    A. changes in family content
    B. goal realization or failure

C. diminished self-esteem and loss of sexual confidence
D. all of the above

12. Treatment of menopausal women with estrogens gives best results in managing
 A. hot flashes and vaginal atrophy
 B. insomnia and depression
 C. osteoporosis and coronary artery disease
 D. abnormal bleeding
13. The diagnosis of hydatidiform mole may be confused with
 A. twins
 B. intrauterine fetal death
 C. hyperemesis gravidarum
 D. any of the above
14. After evacuation of a hydatidiform mole, pathologic examination of the tissue will accurately predict the potential for malignancy.
 TRUE FALSE
15. Trophoblastic neoplasia is the first carcinoma that can be diagnosed and monitored biochemically. TRUE FALSE
16. Trophoblastic neoplasia is the first carcinoma where chemotherapy is the treatment of choice and is curative.
 TRUE FALSE

ANSWERS

1. menarche
2. oligomenorrhea
3. polymenorrhea
4. hypermenorrhea or menorrhagia
5. inflammation, neoplasia, pregnancy, hormonal problems
6. B
7. C
8. A
9. D
10. A
11. D
12. A
13. D
14. FALSE
15. TRUE
16. TRUE

part two
gynecologic pathophysiology

chapter seven
infections in gynecology

OBJECTIVES

1. Know causative organisms of and appropriate diagnostic modalities for sexually transmitted diseases.
2. Know the best therapy for each disease.
3. Understand the pathogenesis and treatment of pelvic inflammatory disease.
4. Describe symptoms, findings, and etiology of vaginitis.
5. Understand pathogenesis of septic abortion.

VENEREAL DISEASE

The term "venereal" is derived from the Greek goddess of love, Venus. Venereal diseases are transmitted most often by penile-vaginal sexual contact but occasionally by oral-genital and genital-anal contact.

The common venereal diseases are gonorrhea and syphilis. The rarer venereal diseases include chancroid, lymphopathia venereum, and granuloma inguinale. In addition, there has been a recent rise in the incidence of herpes genitalis caused by herpes simplex virus, type II.

Gonorrhea

Gonorrhea is now the most commonly reported communicable disease in the United States. More than 2.5 million cases are treated annually.

Since it is estimated that only 25% of cases are reported by physicians, the true incidence is probably much higher. From 0.2% to 7% of a population have gonorrhea. These variations depend on socioeconomic status and previous venereal disease exposure. Sixty to 70% of patients are under 25 years of age.

The causative agent of gonorrhea is *Neisseria gonorrhoeae*, a Gram negative diplococcus. It is an oxidase-positive facultative anaerobe.

The vast majority of women with gonorrhea, approximately 80%, are asymptomatic carriers. In males with gonorrhea, 15–20% are asymptomatic carriers. Symptomatic female patients may have a variety of syndromes: lower genital tract infection with vaginal and cervical discharge and dysuria, salpingitis with abdominal pain and fever, or disseminated gonococcal disease, including septicemia, arthritis, and endocarditis. Complicated forms of gonorrhea are discussed.

If inadequately treated, gonorrhea can be devasting, particularly with regard to fertility. In the female, unimpeded gonorrheal infection leads to necrosis, fibrosis, and scarring of the Fallopian tube, with subsequent tubal obstruction.

The gonococcus in clinical material usually appears in an intracellular position. In females, a Gram stain of the endocervical secretions may be falsely negative, as many patients have a positive culture but a negative smear. In addition, the Gram stain may also be falsely positive, as organisms other than *Neisseria gonorrhoeae* appear as Gram negative diplococci. A patient should never be told that she has gonorrhea solely on the basis of a Gram stain.

The cornerstone of the diagnosis of gonorrhea is made by cultures on selective culture media, such as Thayer-Martin medium. This medium contains antibiotics to suppress other genital flora. A 90–93% accuracy in diagnosis is achieved with this medium. With nonselective media, less reliable results are obtained since other organisms may overgrow and obscure the gonococcus. The most rewarding culture sites are the anterior urethra in the male and the endocervix in the female. Approximately 2–5% additional positive cultures will be obtained if the anal canal and urethra are also cultured. (These latter two sites should generally be reserved for culturing as a test of cure.) The gonococcus may also be isolated from the oropharynx where it is transmitted by oral-genital relations.

Uncomplicated gonorrhea (asymptomatic carriers or lower genital tract infection only) is best treated with 1 gm of probenecid orally, followed by intramuscular injection of 4.8 million units of aqueous procaine penicillin. This regimen is highly efficacious (98% cure rate) and preferred since it also protects against incubating syphilis. The following are acceptable alternative regimens:

1. ampicillin—3.5 gm as a single oral dose with 1 gm of probenecid orally. This regimen is somewhat less effective than the intramuscular penicillin.
2. tetracycline—1.5 gm initially and then 0.5 gm orally four times daily, for a total of 9.5 gm (4 days). This regimen may be used in patients with penicillin allergies, but it is contraindicated during pregnancy.
3. erythromycin—same schedule as for tetracycline. This regimen is preferred for the pregnant patient who is allergic to penicillin.
4. spectinomycin—2 gm intramuscularly as a single injection.

Injectible therapy is preferred to assure administration of the drug. None of the alternative therapies are protective against incubating syphilis. Serial serologic tests for syphilis should be performed in all patients, especially those treated with alternative therapies. Patients treated with gonorrhea should be seen 7–14 days after therapy is completed and recultured at two or three sites as a test of cure. In 1976, the first strains of penicillin-resistant gonococci appeared. The mechanism of resistance is by the elaboration of penicillinase by the organisms. These strains probably originated in the Far East, and, to date, only a few isolates have been reported in the United States. The Center for Disease Control (CDC) recommends that patients with a positive culture after initial treatment with penicillin be treated with spectinomycin. If the frequency of the penicillin-resistant gonococci increases, it may be necessary to abandon penicillin as initial therapy.

Syphilis

In the United States, the number of new cases of syphilis diagnosed each year is over 100,000. The ratio of unreported to reported cases of syphilis is estimated at 2:1 (lower than with gonorrhea). The causative agent of syphilis is a spirochete, *Treponema pallidum*.

The natural history of the disease is characterized by three stages. Primary syphilis has an incubation of 9–90 days after contact. During that time, a chancre appears. It is a characteristic lesion with a single painless, punched-out ulcer and a firm border. The lesion may appear in the vagina or on the cervix in the female and thus is often hidden. In the male, the lesion is most frequently at the corona or along the shaft of the penis. It will heal in one to five weeks even without specific therapy. Secondary syphilis makes its appearance in 9–90 days after the initial infection. It appears as a generalized papular eruption that persists for two to six weeks without therapy but will be ablated rapidly with adequate treatment. The secondary lesions characteristically appear on the palms and on the soles. During the primary and secondary stages of syphilis, the disease is highly infectious, as the lesions contain numerous spirochetes. Tertiary (late) syphilis occurs in approximately 25% of untreated cases and progresses to multiple organ involvement. Cardiovascular disease in the form of aortic insufficiency is most common. In addition, the tunica media of the aorta may be attacked, producing an aneurysm. Tabes dorsalis is seen due to spinal cord involvement, and paresis may develop because of cerebral involvement. Arthropathies with destruction of joints, most particularly the knees, are another common late effect.

Syphilis may be diagnosed by direct identification of treponemes in scrapings from a chancre. More commonly, syphilis is diagnosed by serologic tests. Serologic tests for syphilis may be classified as nontreponemal or treponemal. Nontreponemal tests react to cardiolipin, a nonspecific antigen. These tests include the Kline-Kolmer test, Venereal Disease Research Laboratory test (VDRL), and the Rapid Plasma Reagin (RPR) card test. Nontreponemal tests may yield false positive results in patients with collagen vascular disease such as systemic lupus erythematosis. The nontreponemal tests should be used for screening, and positive diagnoses should be confirmed with a specific treponemal test. The FTA-ABS (fluorescent treponemal antibody absorption) test is utilized for specific diagnosis of syphilis. Syphilis is insidious in its onset, frequently showing no initial clinical manifestations. The disease is commonly diagnosed by a positive serology obtained for other reasons, such as premarital examination or hospitalization.

Penicillin is the drug of choice treating syphilis in all stages. (See Table 7–1.)

Chancroid

Chancroid (soft chancre) is an uncommon disease, caused by *Hemophilus ducreyi*, a Gram negative, short, thick bacillus. The incubation period of chancroid is three to six days, leading to genital ulceration. These ulcers are different from the syphilitic chancre. They

Table 7–1. Treatment of Syphilis

Stage		*Treatment of Choice*	*Alternate Treatment (for patients allergic to penicillin)*
Early syphilis	Primary Secondary Latent less than 1 year duration	Benzathine penicillin G, 2.4 million units IM, single dose	1. Tetracycline 500 mg orally four times daily for 15 days.[a] 2. Erythromycin, same dosage as tetracycline
Late syphilis	Latent syphilis of indeterminant or more than 1 year duration Cardiovascular Late benign Neurosyphilis	Benzathine penicillin G, 2.4 million units IM, weekly for 3 successive weeks (7.2 million units total)	1. Tetracycline 500 mg orally, four times daily, for 30 days.[a,b]

[a]Tetracycline should be given one hour before or two hours after meals.

[b]Efficacy of these regimens for neurosyphilis is undetermined. Pre- and posttreatment examination of cerebrospinal fluid is highly recommended.

have a soft border (thus, the name soft chancre), are frequently multiple rather than single, and are painful as opposed to the painless chancre of syphilis.

Diagnosis may be confirmed by culture of the specific organism or by use of specific fluorescent antibodies. The usual treatment for chancroid is sulfisoxazole, 1 gm four times a day for a 10-day course.

Lymphopathia Venereum (lymphogranuloma venereum, LGV)

Lymphopathia is a rare disease. Its causative agent is a member of the Bedsoniae group (Chlamydia), an obligate intracellular parasite. The clinical manifestations of early disease are a three- to five-day incubation period followed by a small genital ulceration that has an irregular border, thus differentiating it from chancroid and syphilis. In addition, a coexistent marked local lymphadenitis is present. The untreated disease leads to marked fibrosis and scarring of the lymphatics, resulting in rectal and vaginal strictures and, in its late stages, severe vulvar lymphedema and fibrosis. The disease is diagnosed by a Frei intradermal test or, more reliably, the newer complement fixation test. Therapy is with tetracycline 250 mg four times a day for a 21-day course.

Granuloma Inguinale

Granuloma inguinale is caused by a Gram negative bacteria, *Calymmatobacterium granulomatis* (*Donovania granulomatis*). It manifests itself by a small papular lesion in the genital area, leading to inguinal abscess, then ulceration and scarring. The disease is primarily of the subcutaneous tissue, differing from the lymphatic involvement of LGV. Diagnosis is by the identification of "Donovan bodies" in a scraping or biopsy of the lesions. These are unique intracytoplasmic bodies. Granuloma inguinale is treated with tetracycline or ampicillin, 500 mg four times a day for 21 days. Frequently prolonged therapy of four to six weeks is necessary.

Herpes Genitalis

This disease is caused by *Herpes virus hominis* (herpes simplex) type II (HVH II). (See also chapter on infections in pregnancy.) It may be diagnosed clinically if typical in presentation. The patient complains of pain and burning of the vulva. Early in the disease, one finds small, multiple, exquisitely tender vesicles that later ulcerate and may become secondarily infected. Laboratory diagnosis can often be achieved by cytologic study of the lesion. The epithelial cells frequently show giant cell inclusions and multinucleated giant cells. An antibody neutralization test is available for herpes type II, but it is unreliable. The definitive diagnostic method is viral culture. Herpes virus grows readily in culture, and reports are usually available within 48 hours.

Treatment is symptomatic. Hygiene must be stressed, and Burow's solution compresses and topical analgesics may help. Secondary bacterial or monilial infection should be treated, also. In severe cases, hospitalization with potent analgesia and bladder drainage with an indwelling catheter may be necessary. Suggested specific treatments with heterocyclic dyes and light or with povidone-iodine have not as yet been proved efficacious.

PELVIC INFLAMMATORY DISEASE

Pelvic inflammatory disease (PID) is a common and potentially serious infection. Familiarity with this condition is essential, as it must be considered in every female with lower abdominal pain. It is estimated that 600,000 cases occur in the United States each year.

Acute salpingitis usually develops as an ascending infection from the cervix. Late effects of salpingitis depend on the severity of the initial attack and the effectiveness of therapy. The insult may be mild, without subsequent tubal obstruction. More severe cases result in pyosalpinges, hydrosalpinges, and peritubal adhesions with tuboovarian masses and extensive adhesions.

The bacterial etiology of pelvic inflammatory disease has recently been carefully investigated. Acute pelvic inflammatory disease may be either gonococcal or nongonococcal. In gonococcal PID, *Neisseria gonorrhoeae* is isolated from the cervix and commonly from the peritoneal exudate as well. In nongonococcal pelvic inflammatory disease, *Neisseria gonorrhoeae* is not found in the cervix or in the peritoneal exudate. Nongonococcal acute salpingitis is caused by a mixture of organisms, including a variety of aerobic and anaerobic bacteria found in the pelvis. Cases of gonococcal and nongonococcal PID occur in roughly equal numbers.

Chronic salpingitis is the result of mixed bacterial invaders even when the initial infection was caused by gonococcus. The major pathogens in these chronic infections are *Bacteroides fragilis*, aerobic and anaerobic streptococci, and *E. coli*. Early reports of a large number of sterile tubal abscesses reflect poor anaerobic culture techniques.

Complications of gonococcal salpingitis include perihepatic adhesions and inflammation (the Fitzhugh-Curtis syndrome). Disseminated infection, including gonococcal bacteremia, endocarditis, and arthropathy, may develop. Any joint may be involved with arthropathy, but large joints are more commonly involved. Salpingitis is usually not present when disseminated infection occurs. From indirect statistical studies, it appears that IUD users are at an increased relative risk for pelvic inflammatory disease.

The onset of symptoms of acute salpingitis commonly coincides with a menstrual period. Bilateral or diffuse pelvic pain are the most common initial complaints. Increased vaginal discharge and irregular vaginal bleeding are other common symptoms, while nausea, vomiting, and dysuria occur less frequently. Most patients who develop acute salpingitis are young, but chronic infections have been reported in patients well into their sixties.

Even during the acute attack, the patient with salpingitis rarely appears toxic. Mild degrees of pyrexia (100–102°F) are the rule; however, even with a temperature of 103°F, bacteremia is rare. The pulse rate parallels the temperature, and respirations are rarely embarrassed. Indeed, deterioration of vital signs and respiratory difficulty suggest either a ruptured tubo-ovarian abscess or another diagnosis, such as intraperitoneal hemorrhage from an ectopic pregnancy. The abdomen may be distended. Bowel sounds are often hypoactive, and there is generally diffuse guarding over the lower abdominal quadrants. On examination of the patient with more extensive disease, rebound tenderness may be found in the lower quadrants or in the upper quadrants. Pelvic examination may reveal inflammation of the Bartholin's gland or the urethra and vaginal and cervical discharge. These features, though, are nonspecific findings. Motion of the cervix and uterus is generally painful. Adnexal enlargement or an appreciable fullness may be discerned in 25% of patients. These enlargements may represent edema and induration of the tubes, hydrosalpinx, pyosalpinx, tubo-ovarian abscess, or adherent bowel. Most of these enlargements disappear with proper therapy.

No laboratory test is diagnostic of pelvic inflammatory disease. The white blood count is often mildly elevated (12,000–20,000) but may be normal. The erythrocyte sedimentation rate is elevated frequently, but this is nonspecific. As mentioned, a Gram stain of the cervix frequently yields false positive or false negative results. Even a positive endocervical culture for *Neisseria gonorrhoeae* is not diagnostic of gonococcal PID, as the patient may be an asymptomatic carrier and actually have another cause of her pain. (See differential diagnosis.) In addition, many patients have nongonococcal pelvic inflammatory disease in which there is an exceedingly poor correlation between the organisms found in the cervix and the peritoneal fluid. Culdocentesis may be performed if the diagnosis is unclear. If purulent fluid is obtained, pelvic inflammatory disease is likely, but this finding may occur with other conditions, such as appendicitis.

The differential diagnosis of pelvic inflammatory disease is extensive, and the clinical diagnosis may be difficult to make with certainty. Before the patient is diagnosed as having pelvic inflammatory disease, the following conditions must be ruled out:

Gynecologic:

1. Ectopic pregnancy
2. Hemorrhagic, ruptured, twisted, or infarcted ovarian cyst
3. Abortion, especially septic
4. Mittelschmerz (midcycle pain of ovulation)
5. Endometriosis

Gastrointestinal:

1. Acute appendicitis
2. Regional enteritis
3. Meckel's diverticulitis
4. Diverticulitis
5. Other gastrointestinal conditions

Urinary Disorder:

1. Urinary tract infections, especially pyelonephritis
2. Calculus of the kidney, ureter, or bladder

Scandinavian investigators have routinely employed laparoscopy as an aid in the diagnosis of acute pelvic inflammatory disease. These studies report 30% error in the clinical diagnosis of PID. Because of the need for general anesthesia, however, the use of laparoscopy has been reserved for patients in whom the diagnosis is uncertain.

Although surgical removal of acutely infected tubes is unnecessary, a sufficient number have been submitted to pathologic examination to allow description of the morbid anatomy. The tube in acute gonococcal salpingitis is reddened, swollen, and edematous, usually with a purulent exudate dripping out of the patent fimbriated end. Histologic examination reveals vascular engorgement and edema of the plicae. A marked inflammatory infiltrate involves the epithelium and often the lumen. The leukocytes are chiefly polymorphonuclear, although plasma cells may be present. The infiltrate is generally confined to the mucosa but sometimes involves the stroma and muscular layers. Small patches of epithelial desquamation may be seen.

Traditionally, acute nongonococcal salpingitis has been said to appear with large, grossly edematous tubes, with absence of luminal exudate and minimal insult to the mucosa. Perisalpingitis and involvement of the mesosalpinx, with superficial inflammatory infiltrate into the edematous musculature, are common findings in such cases. These correlations between infecting organism and histology are subject to qualification, as many exceptions occur.

Chronic salpingitis may appear in several forms. A pyosalpinx results from an acute infection superimposed on a chronic salpingitis. The occluded ostium prevents drainage of the purulent exudate that then accumulates in the tubal lumen. The distended "pus tube" becomes adherent to the surrounding structures due to associated perisalpingitis. If the tube becomes adherent to the ovary, direct invasion of the "usually resistant" ovarian tissue may take place with the formation of a tubo-ovarian abscess.

Microscopic examination reveals purulent exudate in the lumen. The mucosal plicae are flattened and often agglutinated. The lamina propria is infiltrated with plasma cells, lymphocytes, and some polymorphonuclear leukocytes. This infiltrate usually penetrates the markedly thickened tubal musculature with predominant perivascular concentrations. The peritoneum participates in this reaction with proliferation of the mesothelial cells, sometimes with giant cell and adenomatous formations. A hydrosalpinx is usually secondary to a pyosalpinx, with gradual substitution of a watery transudate for the purulent exudate after the contents of the tubal lumen become sterile. The walls of the hydrosalpinx are thin, and the luminal fluid is clear so that the cystic mass is translucent. The tubal ostium is closed.

The following criteria are offered as guidelines for admitting a patient with pelvic inflammatory disease to the hospital instead of treating her on an outpatient basis:

1. Pelvic inflammatory disease with an adnexal mass
2. Pelvic inflammatory disease in a moderately to severely ill patient,

including those with high fevers, marked nausea, and vomiting or rebound tenderness, especially if it involves the upper abdomen
3. An unclear diagnosis

If the patient is judged to be mildly ill and a candidate for outpatient treatment, her treatment regimen should consist of bed rest, abstinence from coitus, and antibiotics. Acute gonococcal salpingitis may be treated with procaine penicillin 4.8 million units intramuscularly with 1.0 gm probenecid orally, followed by ampicillin 500 mg by mouth four times a day for 10 days. For penicillin-allergic patients, tetracycline 1.5 gm orally initially, followed by 500 mg orally four times a day for 10 days.

Outpatient therapy is less well established for nongonococcal salpingitis. Since gonococcal salpingitis can never be ruled out with certainty in an emergency room setting (i.e. based on Gram stain), initial therapy ought to be effective against the gonococcus as well. Thus, the recommended antibiotic treatment for nongonococcal pelvic inflammatory disease is the same as for gonococcal salpingitis. It is necessary to see the patient within 7–10 days to ascertain response. If there is any worsening, the patient should return immediately.

For more severely ill patients, inpatient treatment is indicated. Most patients will respond to medical management consisting of bed rest, in semi-Fowler's position, correction of fluid and electrolyte imbalances, and antibiotics. Because of the wide spectrum of microorganisms that may cause salpingitis, the initial antibiotic selection should be a combination of antibiotics with efficacy against the gonococcus as well as the other pelvic pathogens. Various regimens include ampicillin (6–12 gm/day), penicillin (20 million units/day) and an aminoglycoside, clindamycin and an aminoglycoside, ampicillin and clindamycin, or penicillin and chloramphenicol. Results with most of these treatment regimens have been good. Even if the endocervical culture is reported positive for the gonococcus, it may not be appropriate to treat the patient only with penicillin. Recent reports indicate that in the cul-de-sac of such patients there may also be coliform and anaerobic organisms resistant to penicillin. Thus, we continue broad spectrum therapy.

Most infections respond to this therapy, although several days may elapse prior to satisfactory control. The best prognostic criteria are temperature, pulse rate, and abdominal findings. After the patient becomes afebrile and asymptomatic, it is important to perform a bimanual examination to evaluate residual adnexal disease.

Not all cases of chronic pelvic inflammatory disease respond to antibiotic therapy. Because of resistant organisms, failure to obtain adequate levels of antibiotics, or failure of the antibiotics to reach the organisms, some patients develop persistent parametritis and tubo-ovarian or pelvic abscesses. Such abscesses occasionally rupture, and rupture is catastrophic. Impending rupture may be suspected if the abdominal findings persist or worsen and if the patient's condition deteriorates or she becomes toxic. Vigorous antibiotic therapy and a radical change in surgical therapy have improved the prognosis. Prior to 1947, the treatment for ruptured adnexal abscesses with generalized peritonitis consisted of simple cul-de-sac or peritoneal drainage, and the mortality in the United States was 50–100%. Since the 1960s, therapy has included total extirpation of the uterus, tubes, and ovaries, and the mortality rate has fallen strikingly to 10–15%.

Some patients fail to respond to medical therapy. Any situation in which the patient's condition appears to be deteriorating under medical therapy warrants surgical intervention. Hysterectomy and bilateral salpingo-oophorectomy are often performed when medical therapy has failed. However, an occasional abscess may point in the cul-de-sac and be drained by colpotomy alone.

Clinical judgment is most critically challenged by the group of patients who complain of incapacitating chronic pelvic pain, dysfunctional bleeding, and/or repeated acute episodes of PID. These patients pose three questions for the gynecologist: 1) Who will be benefited by surgical intervention? 2) When is the optimum time for the operative procedure? 3) What procedure should be performed?

Nonemergent surgical therapy for chronic salpingitis should be offered to women who have persistent residual masses, are incapacitated by pain, or who have repeated exacerbations of acute symptoms. The decision is not always easy, but careful observation and thoughtful counseling will afford the best immediate and long-term results.

The prognosis for life with acute salpingitis is excellent. The prognosis for future well-being is less optimistic. Because many such patients are exposed repeatedly to venereal infection, it is difficult to decide whether subsequent exacerbations are due to recurrence or to reinfection. Acute salpingitis compromises future fertility. Swedish investigation has recently found tubal obstruction in 12% of patients with one infection, 35% of patients with two infections, and 75% of patients with three or more infections. It has previously been believed that gonococcal pelvic inflammatory disease was more devasting to fertility than was nongonococcal disease. However, in Sweden, results to the contrary have been found in patients in whom the diagnosis was confirmed laparoscopically and microbiologically.

As noted above, some patients with acute salpingitis develop chronic salpingitis, often complicated by recurrent acute exacerbations. These women endure several acute episodes of pelvic inflammatory disease yearly, generally in relationship to menstruation and the proliferative phase of the cycle. While most of them respond satisfactorily to conservative measures and content themselves with intermittent freedom from acute pain, a small percentage are incapacited. They are troubled with infertility and nonrhythmic uterine bleeding, and some develop life-threatening complications that warrant surgical intervention.

VAGINITIS

Vaginitis is an extremely common condition, experienced by most females at one time or another. Though specific infections often are the cause of vaginitis, it is also caused by noninfectious agents. Thus, each patient should be properly evaluated and appropriately treated. Vulvovaginitis may be due to *Monila*, *Trichomonas vaginalis*, or "nonspecific bacteria."

Monilial Vulvovaginitis

The agent for this common infection is *Candida albicans*. Predisposing causes are pregnancy, use of broad spectrum antibiotics, use of steroid medication, and diabetes mellitus. In its most classic form, the vulva is reddened and often excoriated, and the vagina contains white, curdy, nonfoul ("cottage cheese") discharge. The diagnosis should be confirmed by microscopic examination of the secretions, mixed with a drop or two of 10% potassium hydroxide. *Candida* appears as typical branching and budding hyphae and pseudomycelia. Cultures for *C. albicans* are not needed routinely to make the diagnosis.

The standard treatment for monilia remains nystatin vaginal suppositories, administered twice daily for 14 days. Oral nystatin tablets need not be used, as there is poor gastrointestinal absorption. Dramatic transient, symptomatic relief can be given in the office by painting the vulva and vagina with a 1% solution of gentian violet. Undergarments may be stained with the dye, however, and repeated use can cause drying and burning. For vulvar involvement, nystatin cream may be used. A combination of nystatin, neomycin, gramicidin, and triamcinolone is also available, but the antibacterial agents in this preparation are unnecessary. Recently other antimonilial agents have become available, including miconazole and candicidin. Patients need not be asked to douche.

Trichomonas Vaginitis

The agent for this frequent infection is *Trichomonas vaginalis*, a flagellated parasite. Typically, the patient will complain of marked vaginal discharge and itching, often with dysuria. Trichomonas infections are commonly passed back and forth between sexual partners. The discharge is often frothy, yellow-green, and malodorous. On occasion, the cervix may have punctate hemorrhages, producing the so-called "flea-bitten" appearance. The diagnosis should be made promptly by microscopic examination of a drop of the discharge with a drop or two of normal saline. The field usually contains many white cells. On close inspection, highly motile trichomonads can be seen.

The standard treatment in the nonpregnant female is metronidazole, 250 mg orally, three times a day for seven days. However, this agent is contraindicated in the first trimester of pregnancy. In the second and third trimesters, metronidazole therapy should be reserved for patients who have persistent marked symptoms despite treatment with topical agents such as sulfonamide vaginal cream. Recent investigations have suggested lower doses may be equally effective, but these regimens have not yet received approval from the FDA.

Nonspecific Bacterial Vaginitis

In practical terms, the diagnosis of a nonspecific bacterial infection usually is made following exclusion of trichomonas and monilia as the cause of the infectious vaginitis. Some authors have suggested that many (up to 90%) cases of nonspecific vaginitis are caused by *Hemophilus* (*Cornyebacterium*) *vaginalis*. However, this statement is controversial, as many cases of "nonspecific vaginitis" are undoubtedly mixed infections involving a variety of vaginal bacteria. Treatment of these infections consists of various sulfonamide vaginal creams, ointments, and suppositories. Douching may offer transient relief.

Foreign bodies are a rare cause of vaginal infections and produce heavy, perhaps bloody, often foul vaginal discharge. Rapid clearing can be expected on removal of the foreign body.

Gonorrheal vaginitis is discussed as a symptom in the section concerning PID.

Contact vulvovaginitis may be caused by irritating powders, douches, laundry detergents, or fabrics. Treatment consists of detecting and eliminating these agents. If pruritus is severe, topical steroid creams can be very helpful. Patients may also complain of vaginal discharge and have no discernible cause. These patients should be reassured and advised to employ perineal hygiene and absorbent undergarments and to avoid pantyhose and panty girdles. Occasionally, a complaint of a vaginal discharge is offered by the patient to seek the physician's advice regarding more sensitive problems such as sexual difficulties.

SEPTIC ABORTION

Even though legal abortion has been readily available in the United States, infected abortion is still a problem. Indeed, according to data collected by the CDC covering 1972–1974, infection was the leading cause of death from legal abortion and the leading cause of nonlethal complications following legal abortion. In addition, some women still become infected after illegal abortions, and others after spontaneous abortion.

Infection following abortion is an ascending process. It is most likely to occur in the presence of retained placental fragments or operative trauma, both of which provide an ideal environment for bacterial proliferation.

Septic abortion must be considered in every woman with lower abdominal pain, especially in the presence of fever. Patients who have had illegal abortions often vigorously deny the operation. Symptoms usually include fever, chills, malaise, abdominal pain, vaginal bleeding, and perhaps passage of placental tissue from the vagina.

On physical examination, there is usually lower abdominal tenderness, vaginal bleeding (often profuse), and a foul odor in the vagina. The cervix is open, and placental tissue may be present in the cervix or vagina. Vaginal or cervical lacerations may also be evident. The uterus is tender, and tenderness may extend laterally. The temperature is often markedly elevated, and sepsis and hypotension are not rare.

Treatment consists of vigorous antibiotic and supportive therapy, but an equally important part of therapy is drainage of the infected site. Most often, this is accomplished by curettage, but, on occasion, hysterectomy may be necessary.

SUGGESTED READING

Eschenbach DA, Holmes KK: Acute PID: Current concepts of pathogenesis, etiology and management. *Clin Obstet Gynecol* 18:35, 1975.

Evans TN: Sexually transmitted diseases. *Am J Obstet Gynecol* 125:116, 1976.

Monif GRG (ed): *Infectious Diseases in Obstetrics and Gynecology,* Hagerstown, Md. Harper & Row, Inc, 1974.

QUESTIONS

Circle the correct answers.

1. Gonorrhea is:
   A. the most commonly reported communicable disease
   B. of epidemic proportions due to large asymptomatic carrier population
   C. usually difficult to treat
   D. all the above
   E. A and B only
   F. B and C only
2. The diagnosis of syphilis can be made reliably by a VDRL serologic test. TRUE FALSE
3. Match diagnostic test or findings to disease.

A. Gonorrhea (G.C.)	——1. Donovan bodies
B. Syphilis	——2. FTA-ABS test
C. Lymphopathia venereum	——3. Cytology of lesion
D. Granuloma inguinale	——4. Frei test
E. Herpes genitalis	——5. Thayer-Martin culture medium

4. Match disease with appropriate treatment.

A. G.C. carrier	——1. Procaine penicillin, single injection
B. Syphilis	
C. PID	——2. Benzathine penicillin

D. G.C. carrier with penicillin allergy, not pregnant ——— 3. Penicillin and aminoglycoside

——— 4. Tetracycline × four days

E. Herpes genitalis ——— 5. Burow's solution

5. Gonorrhea is best treated by aqueous procaine penicillin, whereas syphilis identified serologically is best treated with benazthine penicillin. TRUE FALSE
6. Clinical PID may be either gonococcal or nongonococcal. TRUE FALSE
7. It is always possible to distinguish gonococcal from nongonococcal PID by histologic examination of the tube. TRUE FALSE
8. Removal of the infected tubes, ovaries, and uterus should be used in treating PID when:
   A. there is failure of fever or pain to respond to medical therapy
   B. there is persistent, incapacitating pain from adhesions
   C. there is persistent, large adnexal mass
   D. there is a ruptured abscess
   E. all of the above
9. Which of the following cause vaginitis?
   A. *Candida albicans*
   B. *Trichomonas vaginalis*
   C. *Trichuris vaginitis*
   D. various vaginal bacteria
   E. all of the above.

ANSWERS

1. E
2. FALSE
3. A/5, B/2, C/4, D/1, E/3
4. A/1, B/2, C/3, D/4, E/5
5. TRUE
6. TRUE
7. FALSE
8. E
9. A, B, D

chapter eight
the vulva

OBJECTIVES

1. Know the structures that constitute the vulva.
2. Describe the histology of the different structures of the vulva.
3. Know the terms commonly used to describe skin lesions.
4. Know the common benign inflammatory diseases of the vulva.
5. Know the common ulcerative lesions of the vulva.
6. Know the common benign cysts and tumors of the vulva.
7. Classify common white lesions seen on the vulva.
8. Know the "in situ" and malignant lesions found in the vulva.
9. Know the epidemiology of vulvar malignancy.
10. Know the treatment for vulvar malignancy.

The vulva and its diseases have often been a source of confusion to the student. The words used to describe structural alterations and pathologic entities are often imprecise and based on outmoded ideas concerning the pathophysiology of vulvar disease. This chapter will attempt to define the structures comprising the vulva, the histology of these structures, the terms commonly used in describing vulvar disease, and, finally, the common diseases of the vulva.

Structures in the Vulva

The vulva is comprised of skin and its appendages with modifications in certain areas. The mons pubis is essentially a fatty cushion covering the symphysis pubis. The skin of the mons is ordinary skin containing hair follicles, sebaceous glands, and sweat glands. The labia majora are longitudinal folds of skin blending into the mons pubis. They develop pigmentation and fatty tissue at puberty. The skin of the labia majora is normal skin like that of the mons pubis and contains skin appendages, including apocrine sweat glands. Apocrine glands are also found in the axillae, nipples, and periumbilical and perianal regions. The breast itself is a specialized apocrine gland. Apocrine glands are simple tubular glands usually located near a hair follicle. They secrete in response to stress. Apocrine gland secretions are sterile and odorless but are rapidly acted on by bacteria to produce distinctive odors. Apocrine glands do not function until after puberty and are, therefore, often considered to be a secondary sex characteristic.

The labia minora are small firm vascular folds of skin medial to the labia majora. They fuse anteriorly over the clitoris and blend together in the region of the posterior fourchette. The skin of the labia minora contains no hair follicles, and its sebaceous glands secrete directly onto the skin surface. The squamous epithelium covering the labia minora contains no granular cell layer.

The clitoris is a midline structure of vascular erectile tissue homologous to the penis in the male. It is richly supplied with nerve endings but contains no urethra or corpus spongiosum.

The urethral orifice opens in the vulva between the labia minora about 1.5 cm below the clitoris. The epithelium about the meatus is stratified squamous, but the urethral epithelium becomes stratified transitional closer to the bladder. Skene's ducts open to the vulva just below the urethral orifice and carry secretions from the paraurethral glands.

The hymen or its remnants, which partially occlude the vaginal opening, consist of a double layer of stratified squamous epithelium covering vascular connective tissue.

Bartholin's glands open on either side of the vulva, just outside the hymen approximately one-third of the distance from the fourchette to the clitoris. The glands are racemose and located deep beneath the fascia. The ducts of the glands are lined with squamous epithelium at the orifice.

Histology of Epidermis

The skin of the vulva is a physiologic and anatomic barrier, separating the external environment from the internal environment. Normally, cornified epidermis consists of two layers. The lower layer is the living layer, consisting of a basal cell layer, an intermediate "prickle" cell layer in which individual cells are connected by microscopic fibrils, and a superficial granular cell layer in which the cells contain keratohyalin granules.

The upper layer is the stratum corneum or horny layer in which there are no living cells, only the dead end products of the lower layers. Ordinarily, there are no nucleated cells in this layer.

Terminology of Vulvar Disease

When describing pathologic processes in skin, certain terms are used that need definition:

1. acanthosis—hyperplasia of the living layer of the epidermis, especially of the prickle cell layer.
2. hyperkeratosis—an increase in the horny layer of the skin generally associated with a prominent granular cell layer.
3. lichenification—a term meaning thickening of the skin with exaggeration of the normal fissures.
4. kraurosis—a term denoting a dry shriveled condition or a constricting change.
5. leukoplakia—a term identifying a white patch—no reference to a premalignant condition implied.
6. parakeratosis—a histologic abnormality of the horny layer of the epidermis with persistence of nucleated cells into the horny layer, often a diminution of the granular layer.

Inflammatory and Ulcerative Lesions of the Vulva

The vulva can be the site of virtually any dermatologic condition. However, a host of local irritants such as vaginal discharge, urine, and menstrual blood acting in conjunction with tight-fitting clothing of synthetic materials can produce an acute vulvar dermatitis. Atopic or neurodermatitis is often associated with the use of hygiene sprays, perfumed soaps and powders, and local medications or strong detergents. Frequently, vulvar dermatitis can be successfully managed by keeping the vulva clean and dry and by the avoidance of tight-fitting synthetic clothing. Several manufacturers are now producing panties and pantyhose with cotton or ventilated crotches.

Intertrigo, which is dermatitis between two folds of skin, may be seen around the vulva between the labia majora and thighs or along the inguinal regions. The skin may be reddened with fissures, or if it is a chronic process, lichenification may be present. Treatment is aimed at keeping the affected region clean and dry.

Seborrheic dermatitis can occur on the vulva as well as other hairy areas of the body. It is a chronic scaling condition that may be associated with erythema. Often there is lowered skin resistance to bacterial infection. Treatment should consist of scrupulous cleanliness and thorough rinsing to remove all soap. Nonmedicated talc may be used to reduce friction and keep the area dry. Steroid lotions may be of value in an acute situation, though the disease seems to be chronic and marked by remissions and exacerbations.

Inflammation around hair follicles may be due to folliculitis or superficial bacterial infection of the follicle. One should also be aware that hidradenitis suppurativa may appear as a similar condition but is a much more significant disease. Hidradenitis is a deep-seated infection of the apocrine glands and may also occur in the axilla. Multiple reddish nodules may appear and suppurate, forming sinus tracts. Antibiotics are the first line of treatment, but débridement of the labia may be necessary to resolve the infection. Modification of the cyclic activity of the apocrine glands by oral contraceptives or estrogens has been reported to be of value as an adjunct in the treatment.

Acute Bartholin's adenitis is often due to a gonococcal infection in the gland even though other bacteria may cause it. The gland becomes acutely swollen with erythema and edema of the vulva. The patient experiences acute pain. An abscess may form and spontaneously drain, usually incompletely, by rupturing. Treatment consists of bed rest, warm or cold compresses, antibiotics, and surgical drainage if an abscess is present. This can usually be accomplished by incision under local anethesthia. An attempt to establish permanent drainage of the infected glands, by marsupialization, may be made in the acute stage but is usually more successful after the inflammation has resolved. Chronic bartholinitis may appear as an indurated asymptomatic enlargement that needs no specific treatment. Should exacerbations of infection occur in such a patient, marsupialization should be performed.

Ulcerative lesions of the vulva caused by herpes virus, syphilis, chancroid, granuloma inguinale, or lymphopathia venereum have been covered in Chapter 7. It should also be remembered that vulvar cancer, which will be discussed later, can appear as an ulcerative lesion. The only other significant ulcerative lesion of the vulva is the acute simple ulcer (Lipschütz ulcer), which is usually multiple and may involve the lower vagina also. These ulcers are usually self-limited and cause little discomfort. The etiologic agent is unknown but may be viral. The importance of the lesion lies in differentiating it from herpetic ulcers, vulvar carcinoma, and the granulomatous venereal diseases.

Cystic and Solid Benign Lesions

Bartholin gland enlargement as a cause of cystic benign vulvar lesions has been discussed. Sebaceous cysts, resulting from inflammatory blockage of sebaceous glands, may be found on the labia majora and minora and are usually asymptomatic and require no treatment. Wolffian duct remnant cysts and endometriosis may be found on the vulva and cysts of the canal of Nuck, or hydroceles may be found in the labia majora—all of these lesions are rare.

The solid benign lesion of the vulva most commonly encountered is the viral lesion of condyloma acuminata. The lesions are treelike structures of hyperplastic epithelium with acanthosis, hyperkeratosis, and parakeratosis on a trunk of connective tissue. The virus responsible is antigenically different from the common wart virus. The lesions of condyloma acuminata are located on the vulva but may also be found in the vagina, on the cervix, on the perineum, and in the natal cleft. Condyloma frequently increase greatly in size during pregnancy and may become so extensive as to require abdominal delivery. There is often an accompanying vaginal infection that should be diagnosed and treated as an adjunct to therapy of the condyloma. For treatment, podophyllin 20% in tincture of benzoin may be painted sparingly on the lesions, taking care to protect normal skin, and then washed off in several hours to avoid severe burns. This therapy may be repeated in a week or two if necessary. Removal of condyloma by surgical excision may be desirable for some lesions; fulguration or cryotherapy may also be of value. An immunologic treatment of condyloma is currently under investigation with promising results.

Condyloma are not ordinarily considered a premalignant lesion, but at times malignancies have been misdiagnosed as condyloma. If there is any question about the diagnosis, a biopsy should be obtained. In addition, malignancies have developed in condyloma acuminata but usually only after the lesions were left untreated for many years.

Molluscum contagiosum is another viral disease producing benign solid lesions. The lesions are rounded, less than 1 cm in size, and slightly umbilicated. They may produce mild local irritation and are usually treated by evacuation with a curette.

Fibromas of the vulva are uncommon tumors that often contain some fat or smooth muscle. As they enlarge, they tend to become pedunculated. Treatment is by surgical excision.

Pigmented moles of the vulva are important because the vulva contains approximately 1% of the skin of the body, but about 5% of the malignant melanomas of the body originate in the vulva. If a nevus looks suspicious, it should be removed.

Hidradenoma is a small benign tumor arising in the apocrine sweat glands. Its importance is in knowing that it is benign. The appearance can be mistaken for an adenocarcinoma on histologic examination.

White Lesions of the Vulva

There is much concern expressed over white lesions of the vulva. Some of this concern revolves about their being called leukoplakia. The term leukoplakia is often used to connote a premalignant lesion on other parts of the body, but in reference to vulvar disease it has no such meaning. Perhaps it would be best to avoid the term altogether and simply term white lesions of the vulva as white patches.

White patches on the vulva ordinarily arise from either absence of pigment or from an increase in keratin formation.

Leukoderma refers to congenital absence of pigment from the skin of the vulva. This may not be evident until puberty, however, when the skin of the vulva normally darkens somewhat. Leukoderma is of no pathologic significance.

Vitiligo refers to an acquired loss of pigment. The etiology of this is unknown, but it may be associated with chronic inflammation or radiation. Both leukoderma and vitiligo may involve other parts of the body.

White patches of the vulva due to an increase in keratin formation can be caused by a variety of pathologic conditions ranging from chronic dermatitis to vulvar carcinoma.

Chronic dermatitis can produce redness and thickening of the skin with exaggeration of the normal skin folds—lichenification. It may also produce epithelial hyperplasia with hyperkeratosis and acanthosis but without atypical maturation of the cells of the epithelium. There may be an inflammatory infiltrate in the corium. This may be termed "typical hyperplasia" and has virtually no malignant potential.

Lichen sclerosis may occur at any age and on many parts of the body, but when concerned with the vulva, it is most often seen in the early postmenopausal years. The involved area of skin becomes thin, white, and parchmentlike. This atropic appearance led to the name lichen sclerosis et atrophicus. Metabolic studies, however, have revealed that this is not an atrophic epithelium, for while it is thinned, the cells are more active metabolically than normal skin.

As the epithelium thins, it becomes more easily irritated and pruritic, which may lead to excoriations of the skin. Histologically, the epithelium is thinned with loss of rete pegs. Hyperkeratosis may be present. It is thought that the epidermal changes are secondary to changes in the dermis, which include edema and collagenization in the upper third with an inflammatory infiltrate of lymphocytes below it. There seems to be a loss of elastic tissue. These processes often lead to a shrinking of vulvar structures with a narrowing of the vaginal inlet. The descriptive term for this process is kraurosis.

Treatment of lichen sclerosis is often unsatisfactory. Estrogen creams may be of value in prepubertal children and postmenopausal women. An attempt should be made to eliminate scratching, and hydrocortisone creams and antihistamines may be of value. Any vaginitis present should also be treated to decrease vulvar irritation. Plastic procedures to increase the size of the vaginal orifice may be necessary in some patients, but surgical removal of the involved areas is often disappointing because of the multifocal origin of the process and its frequent recurrence after removal.

Biopsies of the vulva may be of value in making the diagnosis of lichen sclerosis or investigating any hyperplastic areas.

Areas of atypical hyperplasia may be found in white patches. These hyperplasias often have multiple foci and can complicate any hyperplastic lesion. Punch biopsies under local anesthesia are usually well tolerated by patients as an office procedure. Atypical hyperplasias have a malignant potential and should be followed closely. Microscopically atypical hyperplasia shows abnormal maturation of cells in the lower half of the epithelium with or without acanthosis. Severe atypicality is noted by finding "pearl" formation in the tips of the rete pegs and/or extensive cellular abnormalities.

Senile atrophy of the vulva can, at times, become confused with lichen sclerosis clinically. The skin of the vulva can become thin and glossy with diminished elasticity. Kraurosis may be a feature of senile atrophy due to these changes. Microscopically, there is atrophy of both dermis and epidermis, but even though the rete pegs disappear, the dermal changes of lichen sclerosis are not seen.

Carcinoma of the vulva and carcinoma in situ may appear as white patches on the vulva. These entities will be discussed in the next section.

Vulvar Malignancies and Carcinoma in Situ

Three vulvar lesions are considered to be forms of intraepithelial carcinoma. They are carcinoma in situ, Bowen's disease, and Paget's disease. All are rare.

Epidermoid carcinoma in situ may present with a variety of appearances from small ulcers to small or diffuse white patches that may be on a reddened background. The disease tends to be multifocal on the vulva and may be present also in the vagina, cervix, or perineum. It may develop in an area of atypical hyperplasia and progress to invasive cancer, or it may be present on the periphery of invasive disease.

Selection of a site or sites for biopsy is facilitated by the use of toluidine blue. With this staining technique, toluidine blue is applied to the vulva and left for three minutes. Following this, a 1% solution of acetic acid is applied to remove the stain. The dye is retained in areas where nucleated cells are present, and these areas may then be selected for biopsy purposes.

Histologically, carcinoma in situ appears as a full thickness replacement of the epithelium by abnormal undifferentiated cells with increased mitotic activity. There is acanthosis with hyperkeratosis, which appears white grossly, and/or parakeratosis, which appears red grossly.

Treatment of epidermoid carcinoma in situ is by wide local excision if there is a single lesion or by simple vulvectomy if the lesion is multicentric. The value of topical chemotherapy for the disease may be promising.

Bowen's disease occurs at a relatively early age—mean age, 44 years. The lesion is circumscribed, slightly raised, discrete or confluent, dull or pale red, and in some areas crusted. The majority of patients have pruritus. Microscopically, the transition from normal to abnormal epithelium is sharply defined. The involved epithelium is several times the normal thickness with broadening and deepening of the rete pegs. Nuclei are large and hyperchromatic, with reduced amounts of cytoplasm in the cells. The cytoplasm appears to form doubly contoured halos about the hyperchromatic nuclei—these are "corps ronds." Giant cells may also be seen. Dermal changes are minimal. There seems to be an increased incidence of other malignancies in patients with Bowen's disease.

Paget's disease was originally described in the breast. The same type of cellular abnormalities have also been described in the vulva, and their relationship to underlying apocrine glands is a subject of conjecture. The vulvar disease usually presents with burning pain and pruritus. Gross inspection reveals sharply demarcated moist red lesions with occasional crusting. There are often white epithelial islands present. The disease is usually seen in the later decades of life and may involve any portion of the vulva.

Microscopically, the epithelium shows hyperkeratosis and acanthosis, but the most striking feature is the presence of Paget's cells in the epithelium. These are large, clear, polyhedral cells with vacuolated cytoplasm that contains mucin. By using differential staining techniques, Paget cells can be shown to contain acid and neutral mucopolysaccharides. Paget cells may contain melanin.

Paget's disease is thought to begin in the epidermis and spread along the epidermal appendages: hair follicles, sebaceous glands, and sweat glands.

Vulvar malignancies comprise 3–5% of all genital malignancies. The vast majority (80–90%) are epidermoid or squamous cell cancers. Malignant melanomas comprise 5%; undifferentiated carcinomas, 3%; sarcomas, 2%; basal cell carcinomas, 1–2%; and Bartholin gland carcinomas or adenocarcinomas, 1%.

Vulvar malignancies are usually seen in the postmenopausal years—three-fourths of the patients are age 55 or older, and the highest incidence is in the seventh decade of life. Precursors of vulvar malignancy include atypical hyperplasia as well as chronic irritative lesions. Granulomatous vulvar diseases such as syphilis, lymphopathia venereum, chancroid, and granuloma inguinale are often noted to be precursors of vulvar cancer, and in this group of patients, vulvar malignancy usually develops at an earlier age—before 50 years.

Vulvar malignancy often starts inconspiculously with a small nodular ulceration, but it can also appear as a whitish area or granulomatous area. Because it is so insidious in onset, there is often a delay of up to several years in seeking medical attention. Frequently, there is another delay of several months on the part of the physician in making the diagnosis. The results of therapy are often poor in vulvar malignancy because of these delays in diagnosis. The spread of this skin cancer follows predictable routes. The inguinal and femoral nodes are usually the earliest extravulvar tissues involved, and their involvement correlates very well with the size of the primary lesion. When the lesion is less than 1 cm in diameter, nodal metastases are unlikely; when 2 cm in diameter, 25% of patients have nodal spread, and if 3 cm, over 50% have nodal involvement. While vulvar cancer is not prone to widespread metastases, it can spread deep to involve vagina, urethra, bladder, rectum, and rectal sphincter.

Evaluation of the patient prior to treatment (staging the disease) is based on the size of the tumor, the presence of nodes, and the evidence of metastatic disease (tumor, nodes, metastasis—TNM). A *summary* of clinical staging is presented below.

- Stage I: Tumor confined to the vulva, 2 cm or less in diameter, no nodes clinically suspicious for cancer palpated, no evidence of metastatic disease
- Stage II: Tumor confined to the vulva, over 2 cm in diameter, no nodes clinically suspicious for cancer palpated, no evidence of metastatic disease
- Stage III: Tumor of any size, with spread to the urethra and/or perineum and/or anus, or a tumor of any size with nodes clinically suspicious for cancer palpated and no evidence of metastatic disease
- Stage IV: Tumor infiltrating urethra or bladder mucosa and/or rectal mucosa and/or fixed to bone; any lesion with fixed or ulcerated nodes or evidence of metastatic disease

Treatment is almost exclusively surgical, with radical vulvectomy and bilateral groin dissection giving the best results. If the node-bearing tissue contains no tumor, the five-year survival approaches 90%. If the nodes contain tumor, the survival falls to 50%, and if the deep femoral nodes (Cloquet's or Rosenmuller's node) contain tumor, it usually indicates spread of the disease to the deep pelvic nodes and a further reduction in survival to 20%.

Malignant melanomas usually originate in a previous mole or nevus and are relatively more common on the vulva where irritation may be more common than on other areas of the skin. There is thought to be a stimulating effect on their growth by steroid hormones, and the overall five-year survival is around 35%. Other vulvar malignancies are so rare as to deserve no mention in a text of this type.

SUGGESTED READING

Novak ER, Woodruff JD: *Novak's Gynecologic and Obstetric Pathology*. Philadelphia, WB Saunders Co, 1974, pp 1–51, chap 1.

QUESTIONS

1. Structures in the vulva include all of the following *EXCEPT*:
   A. labia majora

B. mons pubis
C. labia minora
D. inguinal nodes
E. clitoris

2. The skin of the labia majora contains:
A. sebaceous glands
B. apocrine glands
C. hair follicles
D. all of the above
E. none of the above

3. An abnormality of the epidermis causing thickening due especially to a thickening of the prickle cell layer best defines:
A. acanthosis
B. parakeratosis
C. hyperkeratosis
D. carcinoma in situ
E. kraurosis

4. Hidradenitis suppurativa is:
A. similar to folliculitis
B. a precursor of neoplasia
C. one of the sequelae of gonorrhea
D. only found on the vulva
E. an infection of the apocrine glands

5. Ulcerative lesions of the vulva may be caused by:
A. syphilis
B. herpes virus
C. granuloma inguinale
D. carcinoma
E. all of the above

6. Condyloma acuminata:
A. is caused by the common wart virus
B. is a premalignant lesion
C. may enlarge greatly during pregnancy
D. should not be removed surgically
E. is an atrophic lesion

7. Lichen sclerosis:
A. occurs only in the vulva
B. occurs only after the menopause
C. can cause kraurosis
D. should be surgically removed
E. is best treated with podophyllin

8. Vulvar lesions considered to be forms of carcinoma in situ include:
   A. leukoplakia
   B. Bowen's disease
   C. kraurosis
   D. atypical hyperplasia
   E. typical hyperplasia
9. Vulvar malignancies:
   A. are usually seen in patients over 55 years
   B. may be preceded by granulomatous disease
   C. are usually squamous cell tumors
   D. are often not diagnosed for several years after their onset
   E. all of the above
10. The treatment for vulvar malignancy is:
   A. simple vulvectomy
   B. radical vulvectomy and groin dissection
   C. radiotherapy
   D. topical chemotherapy
   E. systemic chemotherapy

ANSWERS

1. D
2. D
3. A
4. E
5. E
6. C
7. C
8. B
9. E
10. B

chapter nine

the vagina and cervix

OBJECTIVES

1. Describe the vagina.
2. Identify the embryonic origin of the vagina.
3. Identify inflammatory conditions of the vagina.
4. Know the common benign cystic and solid lesions of the vagina.
5. Know the types of malignancy found in the vagina.
6. Relate the evidence linking diethylstilbestrol to vaginal adenocarcinoma.
7. Describe the cervix—its epithelium, glands, and stroma.
8. Know the benign lesions of the cervix.
9. Identify risk factors for cervical carcinoma.
10. Relate the finding of precursor lesions to the development of cervical cancer.
11. Relate the use of diagnostic modalities to the screening of asymptomatic women for cervical malignancy.
12. Relate the treatment modalities to the disease state for cervical malignancies and precursor lesions.

The vagina is a tubular structure lined by a mucous membrane. It extends from the vulva to the lower portion of the uterus. It lies between the bladder and urethra anteriorly and the rectum posteriorly. In the usual state, the vagina is collapsed, and its anterior and posterior walls are in contact. There is some free space at its lateral margins, and if a cross-section were viewed, the vagina would be shaped like the letter H. The origin of the vaginal epithelium has been the subject of debate. At present, it seems that the upper vagina is formed from the Müllerian duct, while the lower vagina is derived from the urogenital sinus. If these two systems fail to meet and canalize, a transverse vaginal septum can result.

The vaginal mucous membrane contains no glands. The epithelium is noncornified stratified squamous that is loosely attached to the underlying connective tissue. Glycogen is present in the epithelial cells and is metabolized to lactic acid by the bacteria normally present in the vagina, thus creating a slightly acidic condition. The pH of the vagina is normally between 4 and 5.

The connective tissue beneath the epithelium is quite dense and contains elastic fibers, lymphoid nodules, and many blood vessels. The blood supply and lymphatic drainage of the vagina differ in different areas. The upper third receives blood from the cervicovaginal branches of the uterine arteries, and lymphatic drainage is to the iliac nodes; the middle third receives blood from the inferior vesical arteries, and lymphatic drainage is to the hypogastric nodes; the lower vagina receives blood from the middle hemorrhoidal and internal pudendal arteries, and lymphatic drainage is to the inguinal nodes.

The muscular coat of the vagina has two layers that are not sharply defined. There is an inner circular layer and an outer longitudinal layer. Closure of the vagina, however, is mainly accomplished by the levator ani muscles.

The anterior and posterior walls of the vagina contain prominent longitudinal ridges or columns from which extend transverse ridges or rugae at right angles. The rugae almost reach the lateral walls and give the vagina a corrugated appearance in women who have never borne children (nulliparous). After repeated childbirth, the rugae become less prominent, and after the menopause, they disappear entirely, and the vagina becomes smooth.

The vagina is separated from the bladder and urethra by the connective tissue of the vesicovaginal septum. Posteriorly, the rectovaginal septum separates the vagina from the rectum. The upper part of the posterior vagina is not adjacent to the rectum, however, but is adjacent to the cul-de-sac and is covered with peritoneum. The vagina is capable of great distention during childbirth, but stretching of the adjacent connective tissue can produce pelvic relaxation in later years. This topic will be covered in Chapter 12.

The vaginas of the guinea pig and mouse keratinize at the time of estrus due to high concentrations of estrogen in the blood stream. The epithelium of the human vagina never becomes keratinized unless exposed to air or some other unusual environmental factor but does show cyclic changes in response to hormonal changes. By looking at human vaginal epithelial cells, a trained cytologist can make a reasonable assessment of a woman's hormonal milieu. Smears obtained from the lateral vaginal wall may show mainly parabasal cells, indicating limited maturation and corresponding to the absence of estrogen and progesterone. Estrogen causes maturation of the epithelium to the point that mostly superficial cells are desquamated. Progesterone produces a pattern of maturation between these two—intermediate cells predominate. Much of the utility of this maturation index for assessing a woman's hormonal status has now been replaced by specific estrogen and progesterone determination in blood, but it can still be of value in some patients.

Inflammatory Diseases of the Vagina

The common forms of vaginitis have been discussed in Chapter 7 and will not be repeated here.

Vaginitis associated with an atrophic epithelium such as may be seen in the prepubertal or postmenopausal woman can be quite difficult to cure even though specific therapy is used. Often the addition of topical estrogen cream to the vaginal mucosa rapidly thickens the mucosa, which aids in controlling the infection.

Benign Solid and Cystic Lesions of the Vagina

Inclusion cysts are the most common tumors of the vagina. They are located on the posterior wall of the vagina near the outlet. They are rarely over a few centimeters in diameter and may be single or multiple. They occur when small bits of vaginal epithelium are buried beneath the surface following childbirth (laceration or episiotomy) or surgery (perineorrhaphy). The cysts are lined by stratified squamous epithelium, and continued desquamation of the epithelium produces the encysted material. They rarely cause symptoms and are easily removed if they do.

Gartner's duct cysts are formed by vestigial Wolffian duct remnants. These remnants may be found in the broad ligament, along the lateral margin of the uterus, and in the anterolateral vaginal wall. Cystic dilatation of the duct remnants in the vagina produce Gartner's duct cysts. They may be single or multiple, and they may be tiny or so large they protrude through the introitus. They often cause no symptoms but can be removed surgically if they do. Histologically, the epithelium may show a variety of cell types—cuboidal, columnar, or stratified squamous.

The cysts of endometriosis can be found anywhere in the vagina. They are usually deep and posterior in the vagina and appear as reddish lesions. Biopsy provides the diagnosis, and treatment follows the patient's general plan for treatment.

The finding of reddened, pebbly appearing elevations in the vagina of an asymptomatic woman suggests adenosis. Biopsy reveals tall, columnar, mucus-secreting cells in the vagina. Adenosis is almost always benign, but its presence has taken on added significance in recent years, as discussed more fully in the section on vaginal malignancies.

Solid benign tumors of the vagina include the rare case of fibromyoma, neurofibroma, papilloma, or myxoma as well as the much more common condyloma acuminata. Vagina condyloma are usually associated with vulvar lesions, but treatment is more difficult. Podophyllin applications are not recommended because of the possibility of extensive sloughing of vaginal mucosa. Frequently, the use of a sulfa vaginal cream to control associated infection will cause a diminution in size of the lesions. Surgical removal or cryotherapy may be useful.

Malignant Tumors of the Vagina

Malignant tumors of the vagina comprise 0.5% of genital malignancies in women. The disease most frequently occurs in the fifth and sixth decades of life and is usually asymptomatic in early stages. Bloody vaginal discharge is often the first symptom. Late symptoms include pain and a mass protruding from the vagina. Early invasion of the bladder and/or rectum occurs because the tumor is usually located on the anterior or posterior wall. The resultant fistula formation can be quite distressing. Distant spread is by lymphatics to the iliac, hypogastric, or inguinal nodes.

Almost all primary vaginal malignancies are squamous cell carcinomas. In situ carcinomas have also been described. Vaginal carcinoma in situ may be adjacent to a cervical cancer, or it may be one focus of a multicentric malignancy involving the vulva, vagina, and cervix. Secondary vaginal malignancies are much more common than the primary variety, and the usual primary source is the cervix or endometrium. Any tumor involving both cervix and vagina is assumed to be a cervical cancer with spread to the vagina.

The usual treatment of vaginal cancer is radiotherapy because the surgery necessary to remove the tumor would be so extensive and would involve the uterus, vagina, bladder, and rectum often with urinary and bowel diversions. Vaginectomy with a skin graft may be utilized in the treatment of carcinoma in situ in order to preserve function. Local chemotherapy may prove of value in the treatment of in situ disease.

Adenocarcinoma of the vagina received little attention until Herbst, Ulfelder, and Poskanzer reported eight cases at their hospital during a four-year period. Retrospective analysis revealed that the mothers of seven of these young women had taken the synthetic estrogen diethylstilbestrol (DES) during the first trimester of pregnancy. In a matched control group of 32 patients, none of the mothers had taken DES ($p < 0.00001$). DES was utilized to treat a variety of obstetric ills from the late 1940s into the 1960s. An unknown number of young women were exposed to DES in utero and so would seem to be at increased risk for developing vaginal adenocarcinoma. The magnitude of the risk seems quite small but is relatively greatly increased when one considers the rarity of the lesion.

When large numbers of young women who were exposed to DES in utero were examined, it was found that they had a high incidence of vaginal abnormalities. Vaginal adenosis has been reported in 30–90% of DES-exposed offspring, and gross cervical and/or vaginal abnormalities occur in about 20%. Ectropion, which is the finding of endocervical epithelium on the vaginal portion of the cervix, has been another very common association.

It is currently thought that the areas of vaginal adenosis could develop into adenocarcinoma since over 90% of patients with adenocarcinoma have had associated adenosis. No such transformation has yet been observed. It has also been stated that DES-exposed women with adnosis may be at high risk for the development of squamous cell carcinoma. This has not been reported yet, but epidermoid carcinoma in situ has been observed.

All DES-exposed women should be examined at least once a year and cervical and vaginal cytology obtained. Suspicious areas of epithelium identified by iodine staining or colposcopic examination should be biopsied. No definite statements can yet be made regarding the best method of patient follow-up or treatment.

Sarcomas of the vagina in adults are usually diffuse infiltrating lesions that may occur in any part of the vagina. The vaginal epithelium is usually intact until late in the disease.

Sarcoma botryoides is an unusual tumor that arises from the upper vagina and cervix. It occurs in young infants. Bloody vaginal discharge is usually the first symptom. Grossly the tumor is a mass of grayish red edematous polyps that are thought to originate from the growing tip of the Müllerian duct. Histologically, the tumor may appear innocuous with an edematous stroma. Rhabdomyoblasts can be found in the tumor.

The tumor commonly extends to the uterus, parametrium, and regional nodes by the time of diagnosis. The results of treatment are poor, and treatment modalities have included radical surgery, radiotherapy, and chemotherapy.

The Cervix

The cervix is that portion of the uterus that lies below the internal os; the lowermost portion of the cervix projects into the upper vagina and contains the external os. The endocervical canal runs between the internal and external os and contains an epithelium composed of a single layer of tall, mucus-secreting columnar cells with basal nuclei. Within the endocervical canal is an intricate system of ridges, tunnels, and clefts. These look like glands on microscopic examination. Technically speaking, they are not glands, but they are commonly called endocervical glands. The epithelium of the endocervix displays a limited response to varying concentrations of sex hormones in the circulation. At midcycle, when much estrogen and little progesterone are produced, the cervical mucus is thin, watery, and clear and can be stretched for several centimeters (spinnbarkeit). When progesterone is present in high concentrations, the mucus becomes quite viscid and tenacious—this is seen after ovulation or during pregnancy. Examination of the cervical mucus is an important part of an infertility evaluation, as discussed in Chapter 4.

The portion of the cervix that projects into the vagina is covered with a stratified squamous epithelium similar to that of the vagina. The point where the columnar and squamous epithelium meet is referred to as the squamocolumnar junction (SCJ), or transition zone. The anatomic location of the SCJ may vary from the lower endocervical canal to the outer part of the portiovaginalis of the cervix. The SCJ may be a sharp line clearly visible to the unaided eye where the grayish-pink squamous epithelium abuts the reddish-appearing columnar epithelium. More commonly, the exact location of the SCJ varies during a woman's life, and there is no sharp line, rather a transition zone where columnar epithelium on the portiovaginalis is replaced by squamous metaplasia to become a squamous epithelium. The SCJ is important because it is thought that most cervical malignancies arise in this area. Some physicians believe that DES creates vaginal adenosis by displacing the location of the SCJ from the cervix into the vagina.

The stroma of the cervix contains little smooth muscle and is mostly composed of collagen. During pregnancy, the cervix functions as a sphincter to hold the products of conception within the uterus. In late pregnancy, the collagen of the cervix seems to dissociate so that the cervix can dilate during labor. During pregnancy, the cervix becomes extremely vascular and edematous. The endocervical glands increase in number, size, and tortuosity, and by the end of pregnancy, the glands comprise about half the bulk of the cervix and may assume a hyperplastic appearance.

Benign Lesions of the Cervix

Cervicitis is an inflammatory lesion of the cervix that results from bacterial infection of the endocervical glands. Acute cervicitis is commonly due to gonorrhea and may cause little systemic symptomatology. The cervix is usually boggy and edematous. Chronic cervicitis is an extremely common finding of little significance. Chronic cervicitis may be associated with cervical lacerations and with eversion of the endocervical glands and epithelium out on to the portiovaginalis. It may be associated with a purulent discharge from the cervical os and is a common cause of leukorrhea.

Granulomatous and viral lesions of the cervix were discussed in Chapter 7.

Squamous metaplasia, as has been mentioned, occurs at the transition zone when endocervical glands and epithelium have come to occupy part of the portiovaginalis. The term denotes the presence of stratified epithelium in the tunnels or clefts of the endocervix. It is unknown if the squamous epithelium "invades" the columnar epithelium, if the columnar cells themselves become squamous cells, or if reserve cells in this area mature along squamous lines. Squamous metaplasia is extremely common and can be mistaken for neoplasia.

Cervical polyps are usually of endocervical origin. They are localized proliferations of the mucosa and are usually red in color and quite vascular. The symptoms they cause are usually from contact bleeding. They are often pedunculated and can be removed by twisting; if sessile, they may be removed by curette or biopsy. They have a very low malignant potential.

Premalignant and Malignant Lesions of the Cervix

Carcinoma of the uterine cervix is the most common genital malignancy, accounting for 50% of genital malignancies. It is the second or third most common malignancy is women. Breast is the most common, and cervical or colon-rectal alternate between second and third. It is estimated that 2% of untreated women will develop cervical carcinoma by age 40. Approximately 19,000 new cases of cervical carcinoma are diagnosed annually in the United States, and deaths are estimated at 7,800 annually. The median survival time for women with cervical cancer is 7.3 years, and the overall five-year survival is 52.2%. The average age at diagnosis of cervical cancer is 48 years.

Despite those bleak statistics, deaths from cervical cancer have decreased dramatically in the last generation. This decrease has been atttributable to public education and routine cytologic screening of asymptomatic women. By the time symptoms appear with cervical cancer, the disease is often far advanced with limited chance for successful treatment. Cytologic screening of asymptomatic women has led to detection of the disease at an earlier state as well as to the detection of precursor lesions that can be successfully managed. Cervical cancer should be a virtually preventable disease.

Epidemiologic Factors Associated with Cervical Cancer

The etiology of cervical cancer is unknown, but its development seems related to multiple insults sustained by the cervix. Cervical carcinoma is virtually nonexistent in a celibate population; only one case was recorded in a series of 13,000 nuns. It is more prevalent in women of lower socioeconomic groups and is correlated with early age of first coitus and with multiple sexual partners. It is no longer believed that male circumcision protects the female sexual partner from cervical carcinoma. Happily, cervical carcinoma does not seem to be related to the frequency of sexual intercourse.

Recently, herpes virus II has received attention as a possible etiologic agent for cervical cancer. When women with cytologic evidence of herpetic infection were followed prospectively, 23.7% developed cervical anaplasia, compared with 1.6% of a control group. Other studies have shown a higher incidence of antibodies to herpes virus II in the blood of cervical cancer patients than in a control group. Whether the virus is oncogenic, a cocarcinogen, or simply another variable associated with a multiplicity of sexual partners is uncertain. Another explanation might be that anaplastic cervical epithelium is more susceptible to herpetic infection.

Precursor Lesions of Cervical Carcinoma

A historical summary of the development of the concept of precursor lesions for cervical cancer is given below:

- 1886—Sir John Williams first noted changes in the epithelium similar to carcinoma but without invasive disease.
- 1900—Cullen identified similar epithelial changes at the margin of invasive cervical cancer.
- 1919—Rubin developed the hypothesis that these changes were a precursor of invasive disease.
- 1932—Broders introduced the term carcinoma in situ.
- 1932—Martzloff described epithelial cellular atypia (dysplasia) and suggested this as a precursor of carcinoma in situ.
- 1955—Reagan and 1959 MacKay supported dysplasia as a precursor of carcinoma in situ.
- 1963—Frick and 1968 Ullery presented the concept of a lesion designated "microinvasive carcinoma of the cervix." This lesion was intermediate between carcinoma in situ and invasive cancer and was defined as minimal stromal invasion—5 mm or less.

We now describe a spectrum of changes that have been well documented in prospective and retrospective studies as follows: normal cervical epithelium → dysplasia (atypia) → carcinoma in situ → microinvasive carcinoma → invasive cervical carcinoma.

Carcinoma in situ has also been called intraepithelial carcinoma or preinvasive carcinoma. In this condition the entire thickness of the epithelium has been replaced by cells indistinguishable from carcinoma but without stromal invasion. Not all cases progress to invasive disease. It has been found that without therapy approximately 25% of in situ carcinomas become invasive within five years, 35% become invasive within 10 years, and 60% become invasive within 20 years.

Dysplasia is a more difficult lesion to describe, and its fate is not so clearly defined as is carcinoma in situ. Dysplasia is defined as mild if there is a cellular atypicality or abnormal maturation extending through the lower 30% of the epithelium. Seventy-five percent of these lesions regress without treatment, 15% persist, and 10% progress to more serious conditions. Moderate dysplasia implies abnormal maturation through the bottom 60% of the epithelium. Of these lesions, 45% regress, 40% persist, and 15% progress to more serious lesions. Severe dysplasia is sometimes hard to differentiate from carcinoma in situ because the cellular atypicality extends through 90% of the epithelium. Only 25% of these lesions regress, whereas 50% persist, and 25% progress to a more serious lesion. In large populations with dysplasia, 7% of patients progress to invasive cancer. Forty percent persist as a dysplastic lesion and 13–23% progress to carcinoma in situ. Although dysplasia is a precursor lesion to invasive cancer, it does not represent as great a risk to the patient as carcinoma in situ.

Microinvasive cancer may be considered a precursor lesion to invasive cancer, although its definition varies. The term implies that the tumor has invaded the stroma but to a depth of less than 5 mm from the surface without evidence of lymphatic or vascular invasion. Others define it as invasion to 3 mm or less from the bottom of the epithelium without evidence of vascular or lymphatic invasion. It is very rarely associated with lymphatic spread of the disease.

Diagnosis

The diagnosis of malignant and premalignant lesions of the cervix is based on biopsy material. It is also complicated by the fact that dysplasia, carcinoma in situ, and microinvasive disease can be found adjacent to areas of frank invasion. Routine use of cervical biopsies is impractical for asymptomatic patients, so that examination of exfoliated cervical cells is usually utilized as a screening test for asymptomatic patients.

The results of cytologic screening may be reported in several ways, which are combined and summarized below:

- Class I = normal cytology = negative cytology
- Class II = benign atypia = minimal dysplasia
- Class III = suspicious for malignancy = moderate dysplasia
- Class IV = severe dysplasia or carcinoma in situ
- Class V = invasive carcinoma = positive cytology

Cytologic screening is over 85% accurate at finding abnormalities that are present, so a patient with negative cytology needs only to have cytologic evaluation repeated on a yearly basis. All patients with some cytologic abnormality detected need to have further evaluation. In the case of minimal dysplasia, this may be merely repeated cytologic evaluations every three to six months to see if the condition reverts to normal. Several diagnoses of minimal dysplasia would place the patient in an increased risk group and would necessitate evaluation similar to that received by the Class III and higher group.

In the further evaluation of abnormal cytology, several things need to be stressed; any gross lesion should be biopsied, and the only conclusive single biopsy is one that shows invasive carcinoma because all the other lesions can coexist with invasive disease. To completely define the extent of the abnormal process, a curettage of the endocervical canal should be performed. This endocervical curettage (ECC) will not truly define invasive disease in the canal but will determine whether or not there is any abnormality in the cervix that cannot be seen. Directed biopsies of the cervix should also be obtained. Selection of the site for biopsy can be done at random by making four quadrant biopsies to include the SCJ, but this is not as good a method as selection of a biopsy site by iodine staining. By applying Schiller's iodine solution to the cervix, the normal glycogen-containing epithelial cells take up the stain and become a dark color. Abnormal epithelium does not stain and is therefore selected for biopsy. Visualization of the cervix with a binocular microscope (colposcopy) has recently become popular as a means of identifying areas of abnormal epithelium by visualization of the vascular pattern. Colposcopic selection of biopsy sites can be very accurate. The most definitive biopsy technique for the evaluation of abnormal cytology is the cone biopsy. This procedure consists of removal of an annular wedge of tissue from the cervix. The base of the cone should include the SCJ, and the apex of the cone should lie near the internal os. With this specimen, the pathologist can make 32–64 slides, including the SCJ and the endocervical canal. Conization of the cervix is highly accurate at diagnosing the lesions and their extent but is not to be undertaken lightly. Conization must be carried out in the hospital under anesthesia and can cause cervical stenosis or incompetency during future pregnancies. If performed during pregnancy, it is a highly morbid procedure due to blood loss and can actually interrupt the pregnancy.

Usually, if the ECC is negative for abnormality, directed biopsy is adequate to define the extent of the abnormal process. This can be performed on an outpatient basis with no anesthesia and virtually no complications.

Treatment

After definitive diagnosis has been obtained, treatment follows. Patients with mild to moderate dysplasia do not have to be treated. They may be followed with cytologic and biopsy evaluations for years, but treatment is usually desirable. Such treatment often consists of destroying the abnormal epithelium by excisional biopsy or more often by cryotherapy (freezing). Freezing the cervix for three to six minutes at −80 F has been highly effective at destroying abnormal epithelium. The operator assumes that normal epithelium will regenerate, and this is usually the case.

Severe dysplasia and carcinoma in situ should be treated. Their management varies in approach, depending on the wishes of the patient. Should the woman desire to retain childbearing capacity, cryotherapy or conization may be utilized for treatment. These must be combined with close and prolonged follow-up. Should the patient so desire, hysterectomy may be utilized to treat severe dysplasia and carcinoma in situ. No matter which treatment is chosen, long-term follow-up has shown that 1–2% of such patients have recurrent carcinoma in situ, and some even develop invasive disease.

The diagnosis of microinvasive carcinoma can only be made after cone biopsy. Therapy is mandatory. It is generally believed that simple hysterectomy is adequate treatment for such lesions, although some authors prefer a modified radical hysterectomy since 1% of such cases have tumor spread to lymph nodes.

Invasive cervical carcinoma is epidermoid in 90% of cases, adenocarcinoma in 5%, and mixed adenoepidermoid in 5%. The cell type of the malignancy has little bearing on treatment. Invasive carcinoma produces irregular vaginal bleeding. Postcoital bleeding frequently occurs, and this often prompts the patient to see the physician. Infection and necrosis often produce a foul discharge. As the tumor spreads to adjacent organs, symptoms increase. Pain in the lower abdomen results from parametrial involvement with tumor. Leg pain and swelling are seen when tumor involvement of the deep pelvic nodes obstructs efferent lymphatics and veins draining the lower extremities. Finally, bleeding from the rectum or bladder signals the extension of tumor through the mucosa of those organs.

Cervical carcinoma spreads first by local extension, then through invasion of lymphatics, and finally via the vascular system. Tumor spread ultimately becomes visible to the naked eye on the cervix or in the vagina. If tumor reaches the lower third of the vagina, there is a great decrease in survival, presumably because lymphatics in that area drain to the groin. Tumor in the pelvis extends along the cardinal and uterosacral ligaments to the bony pelvis, and the tumor may also spread to the mucosa of the bladder and rectum. The ureters lie in a vulnerable position and may become obstructed by local tumor extension long before the disease has spread to the pelvic wall. Ninety-eight percent of untreated patients with cervical carcinoma are dead of the disease within five years. The three main causes of death are uremia due to ureteral obstruction; hemorrhage due to vascular invasion; and septicemia secondary to infection in the tumor.

Once invasive cancer is found, treatment must be instituted at once if cure is to be obtained. Prior to treatment, however, evaluation of the patient's general medical condition is performed as well as evaluation of likely sites for metastatic disease. A minimal evaluation would include: complete blood count, liver and kidney function studies, cystoscopic examination, sigmoidoscopic examination, intravenous pyelogram, and chest roentgenogram. The results of these examinations allow the physician to stage the extent of the disease clinically. Staging is an accurate means of clinical assessment and is used to define the mode of therapy. Its use also allows comparisons to be made between treatment modalities. A *summary* of staging for cervical carcinoma is given below:

- Stage I—Carcinoma is strictly confined to the cervix.
- Stage II—Carcinoma extends beyond the cervix but has not extended to the pelvic wall, and/or the carcinoma involves the vagina but not the lower third.
- Stage III—Carcinoma has extended to the pelvic wall (on rectal examination, there is no cancer-free space between the tumor and the pelvic wall), and/or the tumor involves the lower third of the vagina. Hydronephrosis places the patient in this stage.
- Stage IV—The carcinoma has extended beyond the true pelvis or has involved the mucosa of the rectum or bladder.

Seventy percent of all invasive cervical cancers in the United States are treated with radiotherapy. Such treatment consists of central application of radium and, in addition, treating the areas of nodal spread with approximately 4,000 rads of external beam therapy. Stage I carcinoma may be treated effectively with irradiation or radical surgery. The radical surgery includes removal of the upper vagina, uterus, parametrium, and pelvic lymph nodes in all cases other than microinvasive disease.

The five-year survival rates for cervical cancer are given below:

- CIS—100%
- Stage I—75%
- Stage II—50%
- Stage III—30%
- Stage IV—8%

The key to successful treatment lies in early diagnosis. The use of cytology in asymptomatic women reveals early anaplastic lesions with 85% accuracy. Endocervical curettage and directed biopsies of sites selected by iodine staining or colposcopy define the extent of the lesion in most patients. Cervical cone biopsy may be utilized for more difficult diagnoses or for therapy of preinvasive disease. Once invasion occurs and symptoms appear, the chance for cure is compromised.

SUGGESTED READING

ACOG Technical Bulletin: Management of abnormal cytology. 38: May 1976.

Gunning JE: The DES story. Obstet Gynecol Survey 31:827, 1976.

Herbst AL, Ulfelder H, Poskanzer DC: Adenocarcinoma of the vagina. N Engl J Med 284:878, 1971.

Kolstad P, Klem V: Long term follow up of 1121 cases of carcinoma in situ. Obstet Gynecol 48:125, 1976.

Novak ER, Woodruff JD: *Novak's Gynecologic and Obstetric Pathology*, Philadelphia, WB Saunders, 1974, pp 52–140.

QUESTIONS

1. The vagina originates from the:
   A. Müllerian ducts
   B. urogenital sinus
   C. both
   D. neither
2. The vagina is a mucous membrane because it contains glands.
   A. statement true, reason true
   B. statement true, reason false

C. statement false, reason true
D. statement false, reason false

3. The epithelium of the vagina is normally cornified because estrogen promotes the maturation of the vaginal epithelium.
   A. statement true, reason false
   B. statement true, reason false
   C. statement false, reason true
   D. statement false, reason false
4. The most common malignancy found in the vagina is:
   A. primary squamous cell carcinoma
   B. primary adenocarcinoma
   C. sarcoma botryoides
   D. secondary spread from cervical carcinoma
5. Women whose mothers took DES during early pregnancy:
   A. seem at increased risk for vaginal adenocarcinoma
   B. have a high incidence of vaginal adenosis
   C. both true
   D. neither true
6. Precursors of cervical carcinoma include:
   A. endocervical polyps
   B. chronic cervicitis
   C. squamous metaplasia
   D. none of the above
7. Most cervical malignancies arise in the:
   A. endocervical canal
   B. transition zone
   C. portiovaginalis
   D. none of the above
8. Factors that place patients at increased risk for the development of cervical carcinoma include all of the following *EXCEPT*:
   A. early age of first coitus
   B. frequent intercourse
   C. low socioeconomic status
   D. multiple sexual partners
9. The decline in deaths from cervical cancer has been attributed to:
   A. cytologic screening of asymptomatic women
   B. better surgical techniques
   C. better radiotherapy techniques
   D. more liberal use of cervical cone biopsy
10. Patients with abnormal cervical cytology should have:
    A. simple hysterectomy

B. radical hysterectomy
C. radiotherapy
D. ECC and directed biopsies

11. If invasive cancer is diagnosed on a cervical biopsy, the patient should have:
    A. cervical cone biopsy
    B. ECC
    C. additional biopsies directed by iodine staining or colposcopy
    D. none of the above
12. The usual treatment of cervical cancer is:
    A. simple hysterectomy
    B. radical hysterectomy
    C. radiotherapy
    D. hysterectomy plus radiotherapy

ANSWERS

1. C
2. B
3. C
4. D
5. C
6. D
7. B
8. B
9. A
10. D
11. D
12. C

chapter ten
the uterus

OBJECTIVES

1. Know the size, location, and support of the uterus.
2. Know blood supply to the endometrium and the endometrial cycle.
3. Describe uterine leiomyomas—gross and microscopic.
4. Identify the types of leiomyomas based on location and relate this to symptomatology.
5. Know types of degeneration that can occur in leiomyomas.
6. Know indications for surgical treatment of leiomyomas.
7. Define adenomyosis and know the symptoms produced.
8. Know the theories of the etiology of endometriosis.
9. Account for the symptoms produced by endometriosis.
10. Know three forms of management for patients with endometriosis.
11. Know the types of endometrial hyperplasia and relate this to the risk of developing endometrial adenocarcinoma.
12. Describe the phenotype of patients with endometrial adenocarcinoma.
13. Identify the factors that seem to increase the risk of endometrial adenocarcinoma.
14. Know the methods of detecting endometrial adenocarcinoma.
15. Know the routes of spread of endometrial adenocarcinoma.
16. Be familiar with the staging of endometrial adenocarcinoma and its relationship to patient survival.
17. Know the incidence of sarcomatous degeneration of uterine leiomyomas.

The uterus is a muscular organ formed by fusion of the Müllerian ducts. It is located between the bladder and rectum. Its posterior wall is covered by peritoneum. The upper portion of the anterior wall is also covered by peritoneum, but the lower portion is loosely attached to the posterior surface of the bladder. The uterus is composed of three layers: an outer serosal surface, a thick layer of smooth muscle (myometrium), and a mucous membrane (endometrium). The woman of reproductive age who has never been pregnant (nulliparous) has a uterus that weighs 45–70 gm, while the uterus of a parous woman weighs 80 gm or more. In the nonpregnant state, the smooth muscle fibers of the uterus are approximately 50 μ in length, whereas the uterus at term during pregnancy weighs 1,000 gm and has smooth muscle fibers of 500 μ in length. In the myometrium, the smooth muscle bundles are arranged in layers but are not sharply outlined. The myometrial layer also contains collagenous bundles, fibroblasts, numerous vascular channels, and some elastic fibers.

The uterus is a mobile organ but is supported by several ligaments. The broad ligaments, which are peritoneal reflections, sweep from either side of the uterus to the pelvic side wall and provide little support but do enclose vascular and lymphatic vessels as well as the Fallopian tubes and round ligaments. The round ligaments extend from the upper lateral aspect of the uterus to the inguinal canal and may be of some support to the uterus. The main supports of the uterus are the cardinal or transverse cervical ligaments located in the base of the broad ligament. They are condensations of pelvic fascia and serve to anchor the uterine cervix to the pelvic side walls. They also enclose the uterine arteries and ureters. The uterosacral ligaments are the other main support of the uterus. They are also condensations of pelvic fascia and anchor the upper portion of the uterine cervix posteriorly to the sacrum.

The mucous membrane of the uterus, the endometrium, is remarkably responsive to changes in the concentration of ovarian hormones. The endometrium varies in thickness from 0.5 mm at the end of a menstrual period to as much as 5 mm or more in thickness prior to a menstrual period. Its blood supply is via two sets of arterioles that enter at right angles to the endometrium. Straight arterioles serve a nutritive function and supply the basal layer of endometrium. This is the area that is not shed during menstruation and the layer from which the endometrium regenerates after menstruation. Coiled or spiral arterioles supply the upper functional zone of the endometrium. The functional zone is composed of a compact zone that includes the surface epithelium and stroma and a spongy zone that contains endometrial glands. Spiral arterioles, by virtue of their ability to respond to the concentration of ovarian hormones, are instrumental in producing cyclic sloughing of the functional layer of the endometrium during menstruation.

The response of the endometrium to ovarian hormone production has been termed the endometrial cycle, and the stages in this cycle have been divided and subdivided by many authors. At the end of menstruation, while the next follicle is developing for ovulation, estradiol is produced in increasing amounts. In response to estrogen production, the endometrium undergoes proliferation of the functional zone. The proliferative phase lasts until the time of ovulation and is characterized histologically by growth of the endometrium in thickness. The glands are straight and widely separated. Mitoses may be observed in the epithelium of the glands. The stroma is edematous in appearance, and there is an increase in ground substance.

After ovulation, estrogen and progesterone are both present in the circulation, and the endometrium progresses into the secretory phase. The earliest morphologic changes are recognized by the second day after ovulation. The basally placed nuclei of the epithelium are displaced toward the lumina of the glands by subnuclear vacuoles. With time, these secretory vacuoles slip past the nucleus to be secreted into the lumen, and the nuclei return to their basal position again. Secretory material becomes obvious in the glands, which become increasingly tortuous. A week prior to menstruation, secretory activity peaks. The endometrium is thick, the surface epithelium is tall, and the midportion of the glands is convoluted. The stromal cells become more noticeable as their cytoplasm thickens, and the basal and spiral arterioles also become more prominent.

If fertilization and implantation do not occur, the corpus luteum begins to regress, hormone production drops, and the endometrium enters the premenstrual phase. This phase lasts two or three days and is characterized by infiltration of the endometrial stroma by leukocytes—mononuclear as well as polymorphonuclear cells. The endometrium collapses somewhat, and the spiral arterioles coil more tightly.

The menstrual phase of the endometrial cycle follows and usually lasts three to five days. During this time, the functional zone of the endometrium is sloughed. Arteriolar and venous blood are both lost. Spasm of the spiral arterioles is thought to limit the blood loss to approximately 60 ml during the menstrual episode. It is thought that menstrual blood clots within the uterus and then liquefies. The passage of clots during menstruation signifies brisk bleeding. The endometrial surface is regenerated by epithelial proliferation from the edges of the glands in the basal zone. Regeneration takes place with amazing rapidity and is complete by the end of menstruation.

Should fertilization and implantation occur, hormone production from the corpus luteum continues, and the endometrium remains in the secretory phase of the cycle. Endometrial stromal cells become larger and polygonal and are now termed decidual cells. They are most marked in the upper compact zone. In the spongy zone, hypertrophy and convolution of the glands become even more pronounced.

In the absence of estrogen, such as occurs postmenopausally, the endometrium usually exhibits atrophy. It becomes quite thin with sparse glands, and the epithelium becomes low cuboidal. Stromal cells often regress and become fibrous in appearance.

BENIGN UTERINE DISEASES

Leiomyomas are the most frequent uterine neoplasms. These benign smooth muscle tumors are also known as myomas, fibromyomas, and most commonly as fibroids. They are usually multiple and range in size from microscopic lesions to over 100 lb in weight. It has been estimated that over 20% of women above the age of 30 harbor leiomyomas in their uteri. Myomas are rarely found before the age of 20. Most myomas that become symptomatic do so between the ages of 35 and 45. They are slow growing, but the growth may be accelerated by endogenous or exogenous estrogenic compounds. Myomas usually cease to grow and frequently regress after menopause. Black women are reported to have a higher incidence of leiomyomas than white women. Myomas are also associated with infertility. They may be a cause of infertility, but there is also a higher incidence of myomas in women who have never married.

Leiomyomas are firm, solid masses of muscle and connective tissue with the consistency of a hard rubber ball. They have no true capsule but are easily separated from the normal myometrium. Microscopically, they are composed of whorls of smooth muscle fibers. Bundles of muscle run in all directions, and varying amounts of fibrous tissue are present.

Physical findings of leiomyomas depend on their size and location. The location of myomas in relation to the uterus determines the type of myoma and the symptomatology produced. The types of myomas are listed and described below:

- Subserous—protrude from the serosal surface of the uterus and covered by peritoneum. They may be multiple, mobile, pedunculated masses that feel like a "bag of rocks" on pelvic examination. They may actually develop a blood supply from the omentum, lose their attachment to the uterus, and become "parasitic" myomas.
- Intramural—contained within the myometrium. They cause enlargement of the uterus and uterine cavity.
- Submucous—extend into the endometrial cavity and are covered only by endometrium. These have a tendency to cause abnormal bleeding through interference with menstruation or by actual ulceration. They may become pedunculated and be expelled from the uterus through the cervix—"aborting" myoma.
- Interligamentous—located within the leaves of the broad ligament. They may be mistaken for immobile ovarian tumors and cause technical difficulty at the time of surgical removal.
- Cervical—an uncommon type. Located in the cervical stroma, they may interfere with obstetric delivery and may be technically difficult to remove surgically.

Most leiomyomas are asymptomatic, but symptoms include abnormal or excessive bleeding, pelvic pressure, pelvic pain, and interference with pregnancy. Abnormal bleeding is usually associated with submucous myomas, and the mechanism of the abnormal bleeding is not fully defined. These myomas can usually be detected during a curettage for evaluation of abnormal bleeding and may require surgical therapy even though small. Pelvic pressure symptoms are often related to myomas pressing on the bladder or may be reflected as a heavy sensation in the pelvis. Pain may occur with abortion of a submucous myoma or with degeneration of, or hemorrhage into, a myoma. During pregnancy, carneous or red degeneration may occur. Its cause is not clear, but it produces acute symptomatology, with pain, tenderness, and possibly fever. No specific treatment is necessary. Other pregnancy-related problems include infertility, abortion, premature labor, malpresentation of the fetus, and obstruction of the birth canal during delivery.

Degenerative changes of many kinds can occur in myomas. Red degeneration produces the most symptoms and has been mentioned. Hyaline degeneration of the center of the myoma is the most common change; the center can also degenerate by liquefaction—cystic degeneration. Postmenopausally, calcification occurs as the blood supply to the uterus decreases. Torsion of a pedunculated myoma can cause necrosis; infection can occur in ulcerated submucous myomas, and fatty degeneration can occur in any myoma. The most significant degenerative change is sarcomatous degeneration, which occurs in less than 1% of leimyomas. It may be associated with rapid growth of the tumor and is more fully discussed later in the chapter.

Leiomyomas require only observation if small and asymptomatic. Surgical intervention may be necessary in the absence of symptoms if the uterus is larger than the size of a 12-week pregnancy, if myomas are growing rapidly, or if they exhibit growth after the menopause. The latter two conditions suggest the possibility of sarcomatous change. Recurrent symptoms or inability to exclude the possibility of an adnexal mass may also necessitate surgical therapy. Hysterectomy is the most common surgical treatment for symptomatic myomas, but myomectomy may be considered for the patient who wishes to retain the potential to become pregnant.

Adenomyosis is a condition characterized by finding benign endometrial tissue within the myometrium. By convention, it should be located more than one high-power field from normal endometrium on microscopic examination. The process is discovered most often in multiparous women near the end of their reproductive years. Dyspareunia, heavy menses, pelvic discomfort, and bowel or bladder pressure are common symptoms. Pelvic examination often reveals the uterus to be moderately enlarged and softer than normal. These signs and symptoms are thought due to the menstrual activity of the aberrant islands of endometrial tissue. Adenomyosis is an extremely common condition and may be found in over a fourth of hysterectomy specimens when the uterus is removed for some other reason. Even though adenomyosis can be suspected clinically, the diagnosis can only be made pathologically in an extirpated uterus. If adenomyosis is suspected and is of such significance as to warrant treatment, menstrual suppression with progestational agents may be tried before resorting to hysterectomy.

Endometriosis is not a uterine disease, for it is defined as the presence of ectopic rests of endometrium at sites outside the uterus. The topic is included in this chapter because it relates to the endometrium.

Endometriosis is a disease of women in the reproductive years. It does not occur before puberty, is uncommon before the age of 20, and usually regresses after menopause. Most cases are diagnosed in the fourth decade of life, but the classical presentation is in upper middle-class nulliparous women in the third decade. The disease is not uncommon and has been reported to be present in 18–30% of gynecologic operations.

There are several theories of the etiology of endometriosis. One theory contends that retrograde menstruation can occur through the Fallopian tubes; microscopic foci of endometrium then implant throughout the pelvis and grow. This theory is strengthened by animal experimentation in which retrograde menstruation in monkeys has been shown to produce typical endometriosis. It is also known that islands of endometrium can implant and grow on a traumatized cervix or vagina or in an abdominal incision. This theory is also supported by the fact that most lesions of endometriosis are located in the pelvis at sites such as the ovaries, uterosacral ligaments, Fallopian tubes, cul-de-sac, appendix, and pelvic peritoneum where retrograde menstruation might be expected to cause implantation. Another theory contends that since primitive peritoneum forms the Müllerian ducts that form the endometrium, some unknown stimulus could cause transformation of the coelomic epithelium into endometrium at other times during life. This theory accounts for the sites listed above as well as the more uncommon sites of endometriosis, such as the inguinal canal, ureter, renal pelvis, urethra, and umbilicus. A third theory, hematogenous dissemination of endometrial fragments, tries to account for such unusual sites of endometriosis as the lung and extremities.

The symptomatology of endometriosis is caused by the periodic bleeding from aberrant rests of endometrium with subsequent formation of adhesions. Fully 50% of women with endometriosis are asymptomatic, but when present, the symptoms can be quite severe and may have no constant relationship to the size or extent of the lesions. Common symptoms include dysmenorrhea, dyspareunia (usually during deep penetration), painful defecation, and infertility.

Pelvic examination may be completely normal, but common findings include the presence of tender nodules on the uterosacral ligaments, fixed retroflexion of the uterus, thickening of the adnexa, and ovarian enlargement. Visualization of the pelvis may reveal bluish-purple or reddish-brown "powder burn" lesions on the peritoneal surface. These lesions may be of variable size and location. There may be gross distortion of pelvic anatomy caused by cementlike adhesions between the uterus, adnexa, cul-de-sac, sigmoid colon, cecum, and omentum.

To make the diagnosis of endometriosis, one needs to identify endometrial glands, stroma, and evidence of hemorrhage microscopically. This is often impossible. No conclusive pathologic diagnosis can be made in approximately one-third of cases because there has been so much pressure or scarring that the original lesion is no longer visible.

The treatment of endometriosis is difficult. Different modalities are utilized, depending on the patient's symptomatology and plans for childbearing. Hormonal treatment to render the patient amenorrheic may be utilized for the patient with small lesions who has pain and does not desire childbearing in the near future. Good pain relief has been achieved with menstrual suppression by estrogens, progestins, estrogens and progestins, androgens, and, most recently, with danazol (17 alpha ethinyl testosterone), which inhibits hypothalamic-releasing hormones. None of the hormonal treatments cure endometriosis, and they often have severe side effects that make them unacceptable to patients for prolonged use. Danazol is also very expensive.

For the patient who sees fertility as the main goal of her treatment, a conservative surgical approach offers the best result. This involves lysis of adhesions to restore proper pelvic anatomy, removal of visible endometrial implants, removal of ovarian endometriomas, suspension of the uterus in an anterior position, and presacral neurectomy for relief of pain. This management offers the patient the best chance for pregnancy with some pain relief.

Definitive surgery involves removal of the uterus and usually the tubes and ovaries associated with lysis of adhesions. The main indication for this approach is for the relief of pain, and it is usually successful. Estrogens can be administered to these patients postoperatively without causing exacerbations of endometriosis.

ENDOMETRIAL HYPERPLASIA

Hyperplastic growth of the endometrium is somewhat analogous to dysplasia of the cervix. Some of the lesions may revert to normal spontaneously or with treatment, some lesions persist as hyperplasia, and a few of the lesions progress to endometrial adenocarcinoma. The diagnosis of endometrial hyperplasia can only be made on pathologic examination of the endometrium, and it embraces many different kinds of lesions. There is no reliable, commonly used method of screening asymptomatic women for endometrial hyperplasia, so most patients with endometrial hyperplasia are diagnosed because they presented for medical care with symptoms—usually abnormal bleeding—and had endometrial samples obtained.

True benign cystic endometrial hyperplasia is characterized by an increase in the epithelial and stromal components of the endometrium. Mitoses are present in the epithelium and the stroma. A marked disparity in gland size exists with some glands large and cystic and others quite small; this distinctive pattern has led to the descriptive phrase "Swiss cheese" hyperplasia. This lesion is apparently benign if the patient is in the reproductive years.

In addition to cystic hyperplasia, various patterns termed proliferative or adenomatous hyperplasia may be seen. Focal adenomatous hyperplasia is the mildest change and is probably the result of several months without ovulation. It is of little significance as a precursor of malignancy. Mild adenomatous hyperplasia denotes glandular crowding, pseudostratification of the glandular epithelium and minimal intraluminal infolding of the epithelium, and budding of the glands. Moderate adenomatous hyperplasia is an exaggeration of the same pattern with more pseudostratification, infolding, and budding, and the glands are more closely packed. In its most abnormal form, adenomatous hyperplasia may be termed atypical or severe adenomatous hyperplasia or carcinoma in situ of the endometrium. This lesion is characterized by extreme glandular proliferation with obliteration of the stroma. Glands are in a "back to back" pattern with no intervening stroma. Intraluminal tufting of the epithelium and glandular budding are marked, and the epithelium of the glands is composed of cells that vary in size, shape, and polarity.

Endometrial hyperplasias seem to be caused by sustained stimulation of the endometrium by estrogen. This occurs in several clinical situations. In the perimenopausal years, ovulation does not occur regularly, and estrogen stimulation may persist for several months without the presence of progesterone to oppose it by inducing secretory changes and menstruation. Patients with polycystic ovarian disease may also have prolonged, unopposed estrogen stimulation as well as patients with estrogen-producing ovarian tumors and patients taking estrogenic medications. Hyperplasias of pathologic significance can be produced by unopposed estrogen within as short a time as three to six months.

In the management of patients with endometrial hyperplasias, one needs to consider the type of hyperplasia as well as the age of the patient and her hormonal status. Thorough uterine curettage for diagnosis will also be therapeutic for most patients with endometrial hyperplasia. In one series of 500 women who underwent uterine curettage for abnormal bleeding, 70% were "cured" of the bleeding abnormality with one curettage regardless of whether the diagnosis was cystic or adenomatous hyperplasia.

Patients with a diagnosis of cystic or "Swiss cheese" hyperplasia after uterine curettage can simply be observed since only 1–2% of these lesions progress to carcinoma within 10 years. If abnormal bleeding recurs, the patient deserves repeat curettage.

Adenomatous hyperplasia is a more significant diagnosis, especially if there is cellular atypia or if the patient is postmenopausal. It is also a serious diagnosis when it occurs in young women as a result of anovulation. Five to 15% of patients with adenomatous hyperplasia can be expected to progress to endometrial adenocarcinoma within three to six years. Fortunately, this practically never occurs in the absence of recurrent abnormal bleeding. Young women with anovulation and adenomatous hyperplasia are at special risk to develop carcinoma unless progesterone is provided to oppose prolonged estrogenic stimulation of the endometrium. This may be achieved by inducing ovulation, by giving progestins in the second half of the cycle to induce secretory changes and menstruation, or by giving combination oral contraceptives. The perimenopausal patient may be observed for the recurrence of abnormal bleeding and curetted again if it occurs. Hysterectomy may ultimately be desirable for some women with recurrent adenomatous hyperplasia if the bleeding is severe or if there is cellular atypia.

UTERINE MALIGNANCIES

Endometrial adenocarcinoma has become more prevalent as our population has increased in affluence and longevity. It is now the second most common gynecologic malignancy and in some areas is more common than cervical carcinoma. Endometrial carcinoma is usually diagnosed in the sixth decade. Three-fourths of patients are postmenopausal. It is uncommon before the age of 40 and rare before the age of 30. Patients predisposed to endometrial adenocarcinoma are obese, hypertensive women with diabetes mellitus and low fertility. They often have a past history of irregular bleeding, anovulation, endometrial hyperplasia, and a family history of endometrial carcinoma. They are more often of higher socioeconomic status.

The etiology of endometrial adenocarcinoma is unknown, but estrogens seem to play a role in its development. The increased use of estrogens by postmenopausal women seems to parallel the increasing incidence of endometrial adenocarcinoma. It is estimated that there are 8 million prescriptions for estrogen filled annually in the United States; 80% of these are for conjugated estrogens, which are mostly estrone. Estrogen use and cancer risk has been recently reviewed by Lipsett, and there seems to be a three- to eight-fold increased risk of developing endometrial adenocarcinoma by women who take estrogens postmenopausally. An increase in breast cancer has not been associated, however.

It has been theorized that unopposed exposure of the endometrium to estrone may be a causal factor in the development of endometrial neoplasia. Androstenedione can be aromatized into estrone by the skin, fat, and liver. Androstenedione is produced mainly in the adrenal cortex. Increased androstenedione production can occur from the ovary in patients with ovarian tumors or polycystic ovary disease; this results in an increase in circulating estrone. An increase in the conversion rate of normal amounts of androstenedione to estrone occurs with obesity and also with age. Thus, an obese older woman has a 15–20-fold increase in estrone production. Liver disease decreases excretion of androstenedione so that more is available to convert into estrone. The estrone theory of Siiteri and MacDonald is unproven but quite attractive because it incorporates many known associations already established for endometrial adenocarcinoma. They also hypothesize that pregnancy exerts a beneficial effect by producing high levels of estriol, which compete with estrone for binding sites in the endometrium, thereby temporarily shutting off estrone stimulation.

There is no suitable screening device for the early detection of uterine cancer in asymptomatic patients. In a series of symptomatic patients with known endometrial adenocarcinoma, cervical cytology detected 15% and vaginal pool cytology detected 40–60% of cases. Endometrial biopsy has been reported to detect 78–86% of cases, and jet washings of the endometrial cavity provide a similar rate of detection. Endometrial carcinoma is most accurately diagnosed by cervical dilatation and fractional curettage of the endocervix and endometrium (fractional D and C). Patients with abnormal bleeding may be investigated by an office procedure such as endometrial biopsy, but if this does not reveal carcinoma, they should receive a fractional D and C, especially if they are near or after the time of menopause.

Endometrial adenocarcinoma may spread into the lower uterine segment and cervix, in which case, further metastases follow the pattern of cervical carcinoma. It may also invade the myometrium, whose lymphatics accompany the ovarian vessels, and subsequently spread to aortic, lumbar, and perirenal nodes.

Endometrial carcinomas may be of several cellular types. Adenocarcinoma is the most common. Should benign squamous metaplasia occur in the tumor, it is called an adenoacanthoma; this does not worsen the prognosis. If the adenocarcinoma contains malignant-appearing squamous elements, the tumor is called adenoepidermoid or adenosquamous, and the prognosis is worse. Clear cell tumors also occur.

The degree of differentiation of the cells in the tumor relate to patient survival. A grade 1 tumor is well differentiated, a grade 3 is undifferentiated or solid, and a grade 2 is intermediate. Other correlates of survival include cervical extension, uterine size, and myometrial invasion with tumor.

After a diagnosis of uterine adenocarcinoma is made, a general evaluation is performed similar to that described for cervical carcinoma. Staging of endometrial carcinoma and five-year survival are *summarized* below:

- Stage I: carcinoma confined to the body of the uterus—76% survival.
  IA—uterine cavity less than 8 cm in length.
  IB—uterine cavity greater than 8 cm in length.
  Stage I should also be subgrouped according to the degree of differentiation—grade 1, 2, or 3.
- Stage II: the carcinoma involves the body of the uterus and cervix—51% survival.
- Stage III: the carcinoma extends outside the uterus but is confined to the pelvis—26% survival.
- Stage IV: the carcinoma extends outside the true pelvis or involves the mucosa of the bladder or rectum—9% survival.

Patients with endometrial adenocarcinoma are usually treated with combinations of surgery and radiotherapy. Stage I disease is usually treated by hysterectomy and removal of the tubes and ovaries with pre- or postoperative irradiation to decrease the incidence of recurrent disease in the vaginal vault. Stage II is treated by radical hysterectomy or radiotherapy similar to cervical carcinoma, and stage III is usually treated by radiotherapy alone. Some patients with recurrent endometrial adenocarcinoma may be successfully treated by high doses of progestins. One-third of the patients exhibit good response, and one-third of responders may actually be cured.

Although there is no good screening test for endometrial carcinoma, the early appearance of abnormal bleeding usually brings the woman to medical attention, and it should be investigated promptly and completely to diagnose the disease as early as possible.

Endometrial sarcomas are uncommon tumors whose peak incidence is in the sixth decade of life. Patients often complain of abnormal bleeding. The tumors are pinkish polypoid masses that may be aborted through the cervix. They may contain cartilage, bone, and striated muscle and are often called mixed tumors or mixed mesodermal tumors. Treatment is by hysterectomy and irradiation, and the five-year survival rate is low—less than 25%.

Leiomyosarcomas are also uncommon uterine malignancies that may develop from within the myometrium or from a leiomyoma. It is estimated that two or three of each thousand women with myomas will develop leiomyosarcomas. Sarcomatous change in a myoma may be suspected by rapid growth or softening of a myoma. The diagnosis is rarely made prior to surgery, however. Microscopically, these tumors may be hard to distinguish from cellular leiomyomas. Patient survival correlates best with the average number of mitoses counted on microscopic high-power fields (HPF). Less than five mitoses per HPF correlates with 77% five-year survival, six to 10 mitoses per HPF correlates with 33% five-year survival, and over 10 mitoses per HPF rarely correlates with survival. Surgery is the only treatment for these tumors; adjuvant radiotherapy has been of no value.

SUGGESTED READING

Lipsett MB: Estrogen use and cancer risk. *JAMA* 237:1112, 1977.

MacDonald PC, Siiteri PK: The relationship between the extraglandular production of estrone and the occurrence of endometrial neoplasia. *Gynecol Oncol* 2:259, 1974.

Noyes RW, Hertig AT, Rock J: Dating the endometrial biopsy. *Fertil Steril* 1:3, 1950.

Rutledge F, Boronow RC, Wharton JT: *Gynecologic Oncology,* New York, John Wiley & Sons, Inc 1976, pp. 97–138.

QUESTIONS

1. The main supports of the uterus are:
   A. broad ligaments

B. infundibulopelvic ligaments
C. round ligaments
D. cardinal ligaments

2. The stage of the endometrial cycle characterized by tall, straight glands with mitosis and an edematous stroma is the:
   A. proliferative
   B. secretory
   C. premenstrual
   D. menstrual
3. A myoma that causes irregular uterine bleeding is probably:
   A. subserous
   B. intramural
   C. submucous
   D. interligamentous
4. The degenerative change of a myoma that causes pain during pregnancy is:
   A. red degeneration
   B. hyaline degeneration
   C. cystic degeneration
   D. sarcomatous degeneration
5. Indications for the surgical treatment of leiomyomas include:
   A. inability to distinguish them from ovarian tumors
   B. irregular bleeding
   C. infertility
   D. all of the above
6. Adenomyosis is the finding of endometrial glands and stroma:
   A. on the pelvic peritoneum
   B. on the ovaries
   C. in the myometrium
   D. in the umbilicus
7. Which theory for the etiology of endometriosis has been supported by data in animals?
   A. retrograde menstruation
   B. coelomic metaplasia
   C. intravenous seeding
   D. none of the above
8. The symptoms of endometriosis are caused by:
   A. bleeding of the implants
   B. adhesions around the implants
   C. both A and B
   D. neither A nor B

9. The definitive management of endometriosis is:
   A. lysis of adhesions and removal of implants
   B. removal of uterus, tubes, and ovaries and lysis of adhesions.
   C. hormonal suppression of the implants
   D. none of the above
10. Endometrial hyperplasia is most often detected because of:
   A. excellent screening tools for asymptomatic patients
   B. abnormal bleeding
   C. pelvic pain
   D. none of the above
11. Patients with endometrial adenocarcinoma often relate a history of:
   A. abnormal bleeding
   B. diabetes mellitus and/or hypertension
   C. low fertility
   D. all of the above
12. Endometrial adenocarcinoma may be related to prolonged stimulation of the endometrium by:
   A. estrone
   B. estradiol
   C. estriol
   D. progesterone
13. Endometrial adenocarcinoma that involves the cervix is at least:
   A. Stage II with 90% survival
   B. Stage II with 50% survival
   C. Stage III with 25% survival
   D. Stage IV with 10% survival
14. Sarcomatous degeneration of a leiomyoma is suspected when any of the following occur *EXCEPT*:
   A. sudden growth of the myoma
   B. postmenopausal growth of a myoma
   C. calcification in a myoma
   D. softening of a myoma

ANSWERS

1. D
2. A
3. C

4. A
5. D
6. C
7. A
8. C
9. B
10. B
11. D
12. A
13. B
14. C

chapter eleven
the adnexa

OBJECTIVES

1. Identify and describe structures in the adnexa.
2. Describe the events necessary for reproduction that occur in the oviduct.
3. Recall the origin and hormonal function of the ovaries from previous chapters.
4. Know six things to be described in evaluating adnexal masses.
5. Know three categories of origin of adnexal masses.
6. Know origins of nonovarian adnexal masses.
7. Discuss tubal pregnancy from the standpoint of incidence, etiology, diagnosis, treatment, and impact on future reproductive performance.
8. Know the origins of nonneoplastic ovarian masses.
9. Discuss functional ovarian cysts from the standpoint of etiology, clinical findings, and management.
10. Discuss ovarian neoplasia from the standpoint of symptomatology, diagnosis, and spread of disease.
11. List five general categories of ovarian neoplasia.
12. List the epithelial tumors.
13. List the germ cell tumors.
14. List the gonadal stroma tumors.
15. List the mesenchymal tumors.
16. List the common types of nongenital tumors that metastasize to the ovary.

17. Discuss the preoperative evaluation of patients with suspected ovarian neoplasia.
18. Know the summarized staging of ovarian carcinoma.

The uterine adnexa are composed of the ovaries, oviducts, and their mesenteries. Each oviduct is approximately 12 cm in length and extends from the endometrial cavity to near the ovaries. Four distinct areas of the oviduct are recognized. The intramural portion extends from the endometrial cavity through the myometrium. The isthmus is 2–3 cm in length and contains a heavy muscular wall. The lumen of the oviduct is narrowest in the isthmus. The ampulla is adjacent to the isthmus, and the ampullary-isthmic junction (A-IJ) can be detected by palpating the oviduct and noting the change in consistency. The ampulla is 5–8 cm in length and the lumen varies from 1 mm in diameter near the A-IJ to a centimeter or more near the infundibulum. The infundibulum is the trumpet-shaped distal portion of the oviduct that terminates in the fingerlike fimbria that delineate the abdominal ostium of the tube.

The oviduct is enclosed in the leaves of the broad ligament, and this upper portion of the broad ligament is the tubal mesentery-mesosalpinx. This mesentery blends into the infundibulopelvic ligament that encloses the ovarian vessels. The blood supply of the oviduct is from ovarian and uterine arteries, and the venous drainage follows the arterial supply. Tubal lymphatics drain to the paraaortic and lumbar nodes.

The mucous membrane of the oviduct is a single layer of columnar cells. Some cells are secretory, others have cilia beating toward the uterus, and some cells seem to be exhausted secretory cells called "peg cells." The mucous membrane is located on extensive longitudinal folds that decrease in height and complexity when examined from the ampulla to the intramural portion of the tube. There is some change in the appearance of the epithelium during different phases of the ovarian cycle. Tubal fluid is actively secreted as well as transudated into the lumen.

The muscular coat of the tube seems to consist of two or three layers. The intramural portion of the tube has an inner longitudinal layer that virtually disappears in the isthmus, leaving only a circular layer and an outer longitudinal layer for the remainder of the tube.

Physiologic events necessary for reproduction that occur in the oviduct include sperm transport and capacitation, ovum pickup and capacitation, fertilization, zygote transport, and zygote nourishment. The numbers of sperm that are necessary for normal fertility are far in excess of the number that participate in fertilization. The chief sites of reduction in the number of sperm are the cervix, the uterotubal junction, and the isthmus. The uterotubal junction may serve as a local reservoir that replaces to the isthmus the sperm that leave the isthmus for the ampulla. Sperm have been recovered from human oviducts within five minutes after being placed on the cervix. The ascent of sperm to the oviduct thus occurs more rapidly than could be accounted for by the sperm's intrinsic motility. The mechanism by which the female genital tract facilitates sperm transport has not been elucidated.

Freshly ejaculated sperm have poor fertilizing ability. During the interval that spermatozoa await the arrival of ova in the tube, they acquire the ability to penetrate the ovum. The enhancement of penetrating ability is called capacitation. Most evidence suggests that capacitation involves the removal of a substance of high molecular weight from the sperm head. Capacitated sperm can be decapacitated by exposure to seminal or epididymal fluid and then recapacitated in the Fallopian tubes.

The mechanism of ovum pickup by the fimbria has been a subject of debate. One mechanism proposes that the infundibular portion of the tube is brought into opposition with the ovary at ovulation by action of the tubal musculature. The tubal fimbria thus holds the inferior pole of the ovary for several minutes, during which time the ovum is conducted into the ampulla by action of the fimbria on the cumulus mass. In contrast, it is also clear that pickup of the ovum can occur from the cul-de-sac; women with a single ovary on one side and an oviduct on the other can become pregnant. Conduct of the ovum into the ampulla is not due to local suction but by currents caused by cilia.

The egg must spend some period of time within the oviduct before becoming receptive to sperm penetration. Following this, the ovum has a finite fertilizable life span that is estimated to be less than 24 hours in the human. It has been suggested that the corona radiata plays a role in maintaining the fertilizable potential of ova. Denudation of the corona may limit the length of time during which an ovum can be fertilized.

Once fertilized, the ovum pauses for several days at the A-IJ. The duration of this pause is precisely timed, and successful nidation depends on this timing within fairly narrow limits. In the rabbit, zygote transport through the A-IJ can be accelerated by prior injection with progesterone and delayed by injection with estrogen. The mechanisms by which ovarian hormones alter ovum transport rates are currently under investigation.

Tubal secretions are absorbed into the developing zygote. The salutary effects of the oviductal environment on metabolism during the early cleavage stages is apparently not species specific, as normal blastocyst formation can occur after the heterospecific transfer of eggs into receptive oviducts. Tubal fluid is actively secreted, and its constituents vary with the endocrine status of the animal. Many of these constituents are necessary for the maintenance of the male and female gametes and of the developing ovum after fertilization. Thus, the oviduct, in addition to providing a thoroughfare for sperm and eggs, takes an active part in their maintenance, in preparation for fertilization and segmentation, and in supporting blastulation.

The ovaries are dull-white almond-shaped structures located in the uterine adnexa. Their embryology and hormonal production have been discussed in previous chapters. Ovarian size varies with age. During the reproductive years, the ovaries are approximately 3 cm in length, 2 cm in breadth, and 1.5 cm in thickness. Postmenopausally, the ovaries decrease in size to approximately .5 cm in each dimension. The ovaries are attached to the posterior aspect of the broad ligament. Laterally, they are attached to the pelvic side wall by the infundibulopelvic ligament, which is a vascular structure containing the ovarian vessels. The blood supply to the ovary enters via the mesovarium into the hilum of the ovary. Medially, the ovaries are attached to the uterus by the utero-ovarian ligament, which is posterior to the tubal junction with the uterus.

Histologically the ovary is quite complex. An epithelial covering, incorrectly called the germinal epithelium, is usually described. This is best seen as a single layer of cuboidal cells in neonates. In the reproductive age and in older women, the epithelium is quite thin and may be impossible to detect with certainty. The cortex of the ovary is the thickest and most functional layer. It is composed of ova surrounded by granulosa cells as an avascular covering. Surrounding this are the vascular theca interna and theca externa. The cellular origin of the granulosa and theca layers is in debate. They may develop from the ovarian mesenchyme or from embryonic invaginations of the germinal epithelium. The granulosa and theca interna produce the ovarian hormones associated with ovulation. After ovulation, a corpus luteum forms from the ruptured follicle, and the granulosa layer becomes vascularized. If pregnancy does not occur, the corpus luteum begins to regress after about 10 days and becomes a hylanized corpus albicans. If pregnancy and implantation occur, chorionic gonadotropin formed by the trophoblast stimulates the corpus luteum to remain functional. The corpus luteum does not seem necessary for the maintenance of pregnancy after eight weeks, when the placenta becomes capable of sufficient hormone production.

The medulla of the ovary is the central portion that is continuous with the mesovarium. It contains loose connective tissue, vascular structures, some smooth muscle, and remnants of embryonic structures. Hilus cells have also been described that seem histologically and functionally similar to the interstitial cells of the testicle.

In discussing adnexal diseases, it should be remembered that the normal oviduct and round ligament are not palpable. Likewise, the postmenopausal ovary cannot be felt, and the ovaries of a reproductive aged woman are not always palpable, depending on their location, the obesity of the patient, and the skill and persistence of the examiner. Most adnexal diseases are first detected on pelvic examination when an abnormal mass is felt. In evaluating adnexal disease, the patient's age, hormonal status, and symptomatology should be considered. In the remainder of this chapter, adnexal masses will be discussed from the standpoint of what they could represent and from the point of view of a physician who has encountered an adnexal mass and is now attempting to evaluate and treat it. Three broad categories will be discussed: nonovarian disease, ovarian non-neoplasia, and ovarian neoplasia.

In evaluating any adnexal mass, the examiner should carefully describe the mass in terms of its size, shape, consistency (cystic or solid) mobility (fixed or freely mobile), tenderness, and location (anterior or posterior to the broad ligament and unilateral, bilateral or midline). Careful observation plays an important role in determining the further evaluation of the patient.

NONOVARIAN ADNEXAL MASS

Sources of nonovarian adnexal masses include structures that are or may be located in the pelvis. A partial list includes:

- *Bowel*—feces, diverticulitis, regional enteritis, carcinoma, appendicial abscess
- *Urinary tract*—full bladder, pelvic kidney
- *Uterus*—leiomyoma, uterine anomaly
- *Retroperitoneal mass*—neoplasm or anterior meningocele
- *Broad ligament*—parovarian cyst
- *Oviduct*—tubal pregnancy; sequellae of salpingitis, such as pyosalpinx or hydrosalpinx and carcinoma

Bowel lesions of an inflammatory nature may be detected on pelvic examination as adnexal masses because of the adhesions present between loops of bowel, omentum, and possibly the adnexa. These do not generally cause a problem in differential diagnosis with possible ovarian neoplasia. The possibility of an occult bowel carcinoma, detected as an adnexal mass, should always be kept in mind, and appropriate questions relating to bowel function should be asked. Barium enema, colonoscopy, or protoscopy should be utilized if needed to make a diagnosis. Feces in the sigmoid colon can be detected as an adnexal mass on the left. Characteristically, indentation of the mass on examination identifies it as feces, and repeat examination after evacuation, with an enema, if necessary, proves the etiology of the mass.

A full bladder may be mistaken for an anterior pelvic mass. The patient should always void prior to pelvic examination to eliminate this source of confusion, but to be sure one is not dealing with a full bladder, catheterization may be required. A pelvic kidney can be a source of confusion in dealing with adnexal masses, and an intravenous pyelogram prior to surgical treatment of an adnexal mass is always appropriate. It locates the course of the ureters as well as the location of the kidneys.

The uterus is often the source of nonovarian adnexal masses. This may be due to a uterine anomaly resulting in a bicornuate uterus or rudimentary horn or much more commonly to the presence of uterine leiomyomata. If myomas are multiple and subserous, one can be fairly certain of the diagnosis. A pedunculated subserous myoma or an interligamentous myoma can simulate an ovarian neoplasm and necessitate surgical removal. Whenever a pelvic mass reaches a size of 12–15 cm, it rises out of the true pelvis and can be palpated on abdominal examination. A uterus so enlarged with myomas can make evaluation of the adnexa impossible and may necessitate surgical removal.

Retroperitoneal neoplasms that appear as pelvic masses usually are not diagnosed until surgery.

The broad ligament contains vestigial remnants of the mesonephric duct that persist, especially in the mesosalpinx, between the tube and hilum of the ovary. Cystic dilatation of these remnants can produce a parovarian cyst that is clinically indistinguishable from an ovarian cyst. At surgery, the oviduct is seen stretched across the cyst, which is separate from the ovary. Parovarian cysts are thin walled, unilocular, and translucent. They are lined by a single layer of cuboidal cells and have no malignant potential. Treatment consists simply of excision of the cyst.

The oviduct is the most common site of nonovarian adnexal masses. This is often because of the sequellae of salpingitis. In the acute phase of salpingitis, a pyosalpinx can form and adhesions result with bowel and omentum, which may be detected later as a mass. More commonly, the purulent contents are replaced by fluid forming a hydrosalpinx. These may persist for years and be identified as sausage-shaped cystic adnexal masses. Surgery is often necessary to reveal their true identity. (The topic of salpingitis was covered in Chapter 7.)

Tubal carcinoma is very rare, comprising less than 1% of genital malignancies. The mean age of the patient at the time of diagnosis is 52, and the disease is almost never diagnosed prior to operation. It is usually thought to be an ovarian tumor or the sequellae of endometriosis or salpingitis.

Tubal pregnancy may be found as an adnexal mass. All forms of eccyesis will not be considered in this book, but tubal pregnancy will be covered in detail because it accounts for 95% of ectopic gestations and is often a consideration in the differential diagnosis of abdominal pain and adnexal mass.

The incidence of ectopic pregnancy varies from 1 in 30 to 1 in 300 live births, depending on the population surveyed, and is increasing. During the last decade, the death rate from ectopic pregnancy was 2.1 per 100,000 live births, accounting for 6% of all maternal deaths.

The causes of tubal pregnancy have been divided into conditions retarding the passage of the ovum, conditions increasing tubal receptivity, and factors intrinsic in the conceptus. This subject has been extensively reviewed, and only selected problems need be discussed here.

Although pelvic inflammatory disease is usually thought of as an antecedent of eccyesis, no gross or microscopic evidence of salpingitis is found in a majority of women who suffer ectopic pregnancy. Recently, tubal diverticulae of the isthmus (salpingitis isthmica nodosa) and proximal ampulla were identified in 49% of women with ectopic pregnancies, as compared to 11% of controls. Such diverticulae are acquired and may be the result of salpingitis. Changing sexual mores and early antibiotic treatment of salpingitis may contribute to the increased incidence of ectopic pregnancy but are not the sole explanations.

An association of ectopic pregnancy and the intrauterine contraceptive device (IUD) has recently become apparent. The IUD does not increase the absolute incidence of eccyesis but actually reduces it. For example, in a population in which 150 tubal pregnancies per 30,000 women would be expected, only 45 ectopic pregnancies were seen. However, the IUD increases the relative incidence of tubal pregnancy. When women wearing IUDs become pregnant, the ratio of ectopic pregnancy to intrauterine pregnancy is greatly increased. Even more striking is the increase in ovarian pregnancies among IUD users. These data have been interpreted as demonstrating that the IUD reduces uterine implantation by 99.5%, tubal implantation by 95%, and does not affect ovarian implantation at all, thus accounting for the striking relative increase in tubal and ovarian pregnancy.

Increased utilization of induced abortion may raise the incidence of eccyesis. A recent study concluded that a previous induced abortion may increase the risk of a subsequent ectopic pregnancy 10-fold. More information is needed on this important question. Finally, the use of large doses of estrogen to prevent implantation may increase the relative incidence of ectopic pregnancy. Among 29 pregnancies resulting from failure of this method, 3 were ectopic. This 10% ectopic rate is about a 10-fold increase over the expected rate. Unfortunately, we cannot test contraceptive methods for their likelihood of inducing ectopic pregnancy. Experimental arrest of the fertilized ovum in the oviducts of infraprimates does not cause eccyesis, but rather ends in blastolysis. Although extremely rare instances of ectopic pregnancy may occur in infraprimates, there is no embryogenis, but only placental growth. Subhuman primates very rarely develop eccyesis. Only 3 ectopic pregnancies have been noted among nearly 3,000 primate pregnancies. The usual explanation for the relatively frequent occurrence of tubal pregnancy in humans has been that man is subject to diseases that do not affect other species. Although this might account for the more frequent arrest of zygotes in the human oviduct, it does not explain why nidation, rather than degeneration, results. Current investigators suspect that the early human embryo may be less fastidious than that of other species.

Ectopic pregnancy is sometimes clinically obvious but may also be quite difficult to diagnose. In the past, about 20% of patients operated on with a diagnosis of ectopic pregnancy have had some other condition, and about 20% of patients in whom an ectopic has been found come to surgery with some other diagnosis. Although the clinical diagnosis of ectopic gestation remains a problem, the error can be held to less than 10% by maintaining a high awareness of this condition and by using some diagnostic aids. The clinician must be aware that many patients afflicted by tubal pregnancy do not experience the classic symptom complex of pain, amenorrhea, bleeding, and fainting. A large number of patients present with what has been called "chronic" tubal pregnancy. These patients usually experience a bout of pain with relatively little bleeding at time of rupture. The process then becomes "walled off," leading to recurrent episodes of pelvic pain. The average duration of symptoms prior to diagnosis is much longer in the chronic group. The symptoms of chronic eccyesis vary greatly. Such peculiar presentations as hemothorax, hernia, ureteral obstruction, and icterus have been reported.

In over 4,000 cases of eccyesis reviewed from the literature, the symptom frequency was as follows: abdominal pain, 96%; shoulder pain, 17%; amenorrhea, 76%; abnormal vaginal bleeding, 83%; fainting, 21%. It is possible that meticulous questioning would reveal some abnormality or delay of the menses in a higher percentage of cases. The vaginal bleeding may vary from spotting to severe hemorrhage. One of the commonest errors made by young house officers is to eliminate ectopic pregnancy as a possibility because vaginal bleeding is profuse.

The physical findings in ectopic are also quite variable. In the acute case, the diagnosis is usually apparent, with obvious signs of intraperitoneal hemorrhage and associated vascular collapse. In the more occult variety, careful physical examination is important. The temperature rarely rises above 101.5 F. The patients are usually normotensive or slightly hypotensive because their youth and antecedent good health enhances homeostasis. Tachycardia or bradycardia may be the only manifestations of decreased blood volume. The breasts may be engorged or occasionally lactating. Inspection of the abdomen with the patient supine may reveal infraumbilical distention, with a relatively flat hypochondrium. Such distention is evidence of intraperitoneal hemorrhage. Auscultation of the abdomen often will reveal active peristaltic sounds, whereas with the same degree of distention and acute tenderness in the inflammatory diseases, diminished or absent peristalsis is the rule. Abdominal tenderness to palpation is often but not invariably present. Rebound tenderness may be elicited due to the generalized peritoneal reaction in the pelvis. The uterus may or may not be enlarged. Added to the difficulty in accurately evaluating uterine size is the presence of pain and muscle guarding; the uterine size varies with the pathophysiology at the time of examination, and it is not uncommon to encounter a normal-sized uterus at operation for ectopic pregnancy. A uterus that feels larger than the size of a 3½-month gestation is usually not compatible with the diagnosis of ectopic pregnancy.

A pelvic mass is present in a significant number of patients with ectopic pregnancy, especially of the "chronic" variety. The mass may be noted in either adnexal region or as a bulging in the cul-de-sac. Adnexal tenderness is generally present on palpation and may be exquisite. Most patients with ectopic pregnancy complain of pain on movement of the cervix, probably due to the tension on the involved tube resulting from the manual displacement of the corpus. Unilateral adnexal pain upon cervical motion is a valuable sign if present; however, the pain is commonly elicited by motion of the cervix to either side.

Other clinical procedures have been employed to enhance the accuracy of the clinical diagnosis of tubal gestation. These procedures include culdocentesis, diagnostic colpotomy, and endoscopy.

Although there has been some controversy about the use of diagnostic culdocentesis in the past, most gynecologists are now convinced of its value. Culdocentesis is classed as negative if clear peritoneal fluid is obtained, positive if nonclotting blood is obtained, and unsatisfactory if neither peritoneal fluid nor blood is obtained. There is an incidence of both false negative and false positive culdocentesis.

Colpotomy has been employed in the investigation of suspected ectopic pregnancy for nearly three decades. The procedure can be accomplished under local anesthesia and will often allow direct visualization of the adnexa. Colpotomy can be performed on nearly all patients provided a cul-de-sac mass is not present. Both false positive and false negative examinations occur, although the latter are infrequent.

Laparoscopy is also utilized in the diagnosis of eccyesis. Although this procedure was advocated in America more than 25 years ago, enthusiasm was lacking until the past decade. Laparoscopy is reasonably accurate in experienced hands. The patient can be discharged quickly if ectopic gestation is not present. Reported complications of the procedure include hemorrhage into the abdominal wall, intestinal perforations, gas embolism, parietal and mediastinal emphysema, and accidental pneumothorax.

The clinical diagnosis of ectopic pregnancy, even in the most experienced hands, continues to suffer from serious diagnostic errors. Procedures such as culdocentesis, colpotomy, and laparoscopy, while neither without complication nor infallible, clearly facilitate the diagnosis of suspected ectopic pregnancy. Every procedure in medicine has some cost and some risk, and the art of medicine consists in great part of deciding which procedure is indicated in a given case.

Immunologic tests for pregnancy are of limited value. Some tests, particularly the slide latex agglutination varieties, may give false negative tests in as many as 63% of ectopics. In other cases, entities such as tubo-ovarian abscess may give a false positive immunologic test apparently due to a nonspecific interference with agglutination-inhibition. Overall, the pregnancy test is positive in about 50% of patients with eccyesis. The high percentage of negative tests is not surprising because a positive test depends on adequate circulating levels of chorionic gonadotropin, which, in turn, requires the presence of viable trophoblast in proximity to maternal vascular channels. On the other hand, a positive test, even if accurate, demonstrates viable trophoblast but does not indicate its location. There is little indication for ordering a pregnancy test as an aid to the diagnosis of tubal gestation.

The finding of intrauterine decidua without villi on endometrial biopsy is suggestive of extrauterine pregnancy. The frequency of this finding varies from 15 to 62%. The longer the history of bleeding, the less the likelihood of finding decidua. In addition to finding decidua without villi, the pathologist suspects ectopic pregnancy if the endometrium contains the histologic features first described by Javier Arias-Stella. These changes, which are also seen in association with abortion, recent delivery, and trophoblastic tumors, reflect the presence of active chorionic tissue. The endometrial changes are characterized by foci of abundant vacuolated foamy cells, with hyperchromatic nuclei and bizzare variations in nuclear size and shape. Syncitial-like changes of the gland cells with definite loss of polarity may also be seen.

Fertilization normally takes place in the ampulla, and any delay in zygote migration can result in tubal implantation. The incidence of ampullary implantation is roughly twice that of isthmic implantation. Together, these locations account for nearly 90% of tubal pregnancies, the rest being divided nearly equally between the fimbriated extremity and the interstitial portion. Tubal implantation is similar to uterine implantation in that a placenta and intervillous space are formed. However, the decidua basalis is, at best, poorly developed, and the decidua vera and capsularis are nonexistent. The deep penetration of the trophoblast into and through the myosalpinx is usually thought to be a consequence of defective decidualization. Tubal rupture is thought to be contingent upon trophoblastic penetration of the myosalpinx, coupled with distention by the growing conceptus.

The treatment of ectopic pregnancy is surgical. The surgeon must first decide whether to utilize an abdominal or a vaginal approach. Once within the peritoneal cavity, he must decide which operation to perform on the oviduct, whether to include cornual resection if tubal excision is elected, and whether to remove or conserve the ovary on the involved side.

A reasonably experienced gynecologic surgeon can usually make an accurate judgment concerning the patency of the opposite tube from its appearance. If he is uncertain, the patency of the opposite tube can be determined by injecting warmed methylene blue in saline into the uterus and observing fimbrial spill. If the opposite tube is judged normal, the involved tube should be excised. In this situation, conservative operations on the pregnant tube do not improve the pregnancy rate but markedly increase the frequency of future ectopic pregnancies.

The surgeon may elect to preserve the pregnant tube if it is the patient's only tube, she desires more children, and is willing to accept the risk of another tubal pregnancy. Sometimes the pregnancy can simply be "milked out" of the abdominal ostium. In other unruptured tubes, linear salpingotomy is performed, and the gestational sac is shelled out. If the tube is ruptured, repair of the rupture or partial salpingectomy with anastomosis or reimplantation of the oviduct may be performed. If the opposite tube is present but abnormal, many gynecologists try to preserve the pregnant tube and also to repair the opposite tube if fertility is important to the patient. In this situation, many patients would be better served by sterilization. Although reparative surgery allows the chance for future pregnancy, it carries a greatly increased risk of recurrent eccyesis. If reparative operation is done, many clinics add medical therapy designed to maintain patency and reduce adhesions.

The risk of death from ectopic pregnancy is about 0.12%, or 1 in 800. In addition to this small but real threat to life, ectopic pregnancy is associated with compromised future fertility, and increased risk of recurrent eccyesis. After tubal pregnancy, only about a third of patients will ever deliver a living child. Although more than this become pregnant, the incidence of abortion as well as eccyesis is increased. However, the association of an ectopic pregnancy with decreased future reproductive success is not necessarily causal. Approximately 25% of these women were involuntarily infertile prior to their initial ectopic pregnancy. Some of them offer a history of pyogenic or granulomatous salpingitis, but in most, no provisional cause can be identified.

In addition to the likelihood that eccyesis and infertility are both symptoms of an underlying process in many women, it is suspected that the tubal pregnancy may itself contribute to future infertility. One obvious possibility is that intraperitoneal blood causes an inflammatory reaction that, in turn, causes adhesions. This sequence is unlikely. Ruptured corpus luteum hematomas sometimes cause massive intraperitoneal hemorrhage. Women suffering from such ruptured hemorrhagic ovarian cysts enjoy normal fertility, with no increased incidence of eccyesis, after the original operation.

In a recent series, about 50% of women suffering ectopic pregnancy conceived afterward. The conception rate was 52.8% for patients under age 30 at the time of the ectopic and 38.1% for those over age 30. Women who had never been pregnant prior to the tubal pregnancy had the lowest reproductive success after the ectopic (34.6%), whereas those who had four or more children had the highest conception rate (81.8%). The repeat ectopic rate was 16.2% for the entire group. The incidence of repeat ectopic pregnancy was not affected by age but was lowest in multiparous women (9.1%). Conservative surgery was done in nine women, resulting in four pregnancies, of which three were ectopic.

In summary, women who develop tubal pregnancy seem to fall into two groups; those of normal fertility in whom the ectopic is an accident and those in whom eccyesis is a symptom of compromised fertility. More of them fall into the latter category; thus, overall future reproductive success is low. Conservative tubal operations significantly increase the risk of repeat eccyesis.

NONNEOPLASTIC OVARIAN MASSES

Ovarian masses of a nonneoplastic nature are often encountered in the adnexa. These masses are generally asymptomatic, and a clue to their benign nature is usually available on obtaining a history and performing a physical examination. A partial list of nonneoplastic ovarian enlargements includes:

- Functional ovarian cysts
- Polycystic ovaries
- Theca lutein cysts
- Pregnancy luteomas
- Endometriomas

The term functional ovarian cyst has often been misused to cover a multitude of pathologic entities. Here it applies to only two: follicle cyst and corpus luteum cyst. A follicle is designated as a cyst when it reaches a diameter of 2.5 cm; and it may even attain a size of up to 10 cm. Follicle cysts are thought to originate from atretic ovarian follicles after death of the ovum. The cyst fluid may be clear or chocolate in appearance if hemorrhage has occurred into it. The cyst wall may be lined by granulosa or theca cells. Usually, there is little hormone production by the cyst, but menstrual alterations may be noted by some patients. Corpus luteum cysts are usually less than 6 cm in diameter. They may orginate because of persistent luteinizing hormone stimulation or from hemorrhage into a normal corpus luteum. They are often associated with intraperitoneal bleeding of varying degree. When they occur with an early intrauterine pregnancy, the picture of amenorrhea, lower abdominal pain, tender adnexal mass, and a soft, slightly enlarged uterus is often interpreted as a tubal pregnancy. Even if the patient is not pregnant, a persistent corpus luteum may be associated with anovulatory bleeding and abdominal pain so that ectopic pregnancy is suspected. Intraperitoneal hemorrhage from a corpus luteum cyst may require surgical exploration to control the bleeding; in this case, treatment consists of cystectomy.

Functional cysts are often noted in young women, and the main concern is in differentiating them from ovarian neoplasms. Several criteria may be kept in mind to avoid needless surgery as well as harmful procrastination. First, functional cysts usually have a short life span. If the patient is reexamined in a month or six weeks, the mass is usually gone or much smaller. Second, they essentially occur only in women of the reproductive age who are not receiving oral contraceptives. Finally, they are unilateral and mobile, have a cystic consistency, and are less than 10 cm in diameter. It is generally thought that administration of combination-type oral contraceptives will hasten their regression by shutting off secretion of pituitary gonadotropins.

In a series of patients with adnexal masses reported by Spanos, 286 women in the reproductive age were thought to have a functional cyst because of the finding of a unilateral, cystic, mobile adnexal mass less than 10 cm in diameter. They were given oral contraceptives and re-examined in six weeks. The cyst had disappeared in 205 patients. Eighty-one patients had a persistent mass and were subjected to surgery; 35 had nonneoplastic masses such as endometriosis, parovarian cyst, or hydrosalpinx, and 46 had ovarian neoplasms, of which five were early malignancies. It can be concluded from this that the risk of malignancy is low if all criteria for a functional cyst are present and that if the cyst persists for six weeks, surgical intervention is warranted.

It should be mentioned that any ovarian cyst can undergo "accidents" that may warrant surgical intervention; these include torsion of the cyst or adnexa, hemorrhage into the cyst or peritoneal cavity, rupture of the cyst, and, occassionally infection.

Polycystic ovaries of patients with Stein-Leventhal syndrome may be palpated as bilateral ovarian enlargements. In this situation, the ovaries may be two to three times normal size. The patient's history of oligomenorrhea and menstrual irregularities usually gives sufficient clues to the etiology of the adnexal masses so that surgical intervention is not necessary.

Theca lutein cysts are due to multiple follicle cysts with luteinization of the theca cells. They are bilateral and are associated with high levels of chorionic gonadotropin such as occurs with hydatidiform mole, choriocarcinoma, or normal pregnancy, especially with twins. The cysts regress when the source of gonadotropin stimulation is removed. Unless torsion or hemorrhage occurs, no treatment is necessary even though they initially may be greater than 25 cm in diameter. These cysts may also be produced iatrogenically by attempting to induce ovulation in anovulatory patients.

Pregnancy luteomas result from hyperplasia of luteinized ovarian cells during pregnancy. They are usually bilateral and regress during the postpartum period. Generally asymptomatic, they are usually incidental findings at the time of cesarean section or puerperal tubal ligation. Their large size, up to 15 cm, may prompt unnecessary surgical removal. If doubt exists as to the cause of the lesion, biopsy and pathologic examination should be performed.

Endometriomas are usually found in conjunction with other symptomatology of endometriosis but rarely may be the only finding. In the latter situation they are diagnosed at the time of surgery. (Endometriosis was discussed in Chapter 10.)

NEOPLASTIC OVARIAN MASSES

Not all ovarian neoplasms are carcinomas, but the possibility of ovarian carcinoma must always be considered whenever an adnexal mass is palpated. Ovarian carcinoma is the third most common genital neoplasm but is the leading cause of death from genital malignancies. It is estimated that there are 17,000 new cases diagnosed and 10,800 deaths annually for ovarian carcinoma. Median survival for all stages of ovarian carcinoma is 1.4 years, and the five-year survival rate for all stages is 34%. Over half the patients are between 45 and 65 years of age at the time of diagnosis.

There is no early warning of ovarian carcinoma. Cervical cancer can be detected early by cytologic screening of asymptomatic women, and endometrial adenocarcinoma usually produces abnormal bleeding that leads to early detection, but ovarian carcinoma usually does not produce symptoms until the disease is widespread. There is no satisfactory screening tool available for asymptomatic patients. The best chance for cure is obtained when an asymptomatic adnexal mass is detected on pelvic examination.

Ovarian carcinoma spreads by topical implantation of cells exfoliated from the primary tumor as well as by lymphatic and hematogenous routes. Intraperitoneal metastases are often found at the time of surgery, and retroperitoneal nodes may be found as well. Metastases to peritoneal surfaces may not penetrate deeply but often spread along the peritoneal surface and cause intraperitoneal adhesions. The omentum is often the site of early metastases, and "cakes" of omental tumor are common in advanced disease. These metastatic masses are frequently easier to palpate abdominally than the primary ovarian tumor.

The symptoms of ovarian cancer occur late and are related to metastatic disease or the size of the primary tumor. Increasing abdominal girth is the most common symptom and may be related to the size of the tumor or to ascites. Ascitic fluid is not produced by the peritoneal implants but originates from the normal-appearing peritoneum. Malignant pleural effusions are commonly found with metastatic pleural disease resulting from either hematogenous spread or from transdiaphragmatic lymphatics. Chronic low abdominal or pelvic pain or pressure is the second most common symptom and is thought to be related to the presence of the tumor. Some type of abnormal uterine bleeding is the third most common symptom and is related either to hormone production by the tumor or to anovulation caused by destruction of ovarian tissue by the cancer. Gastrointestinal symptoms such as nausea, vomiting, and weight loss are sometimes the symptoms produced by ovarian cancer, and urinary tract symptoms predominate in some patients. Acute pain caused by torsion, hemorrhage, or rupture of the tumor may also cause the woman to seek medical attention. The symptoms and signs of ovarian carcinoma vary widely, and the clinician must be ever alert to the possibility of ovarian carcinoma. Adnexal masses are assumed to be neoplastic whenever the patient is not in the reproductive years or if the mass is solid, immobile, bilateral, or greater than 10 cm in size. Such findings should prompt rapid evaluation and treatment.

Until now, ovarian carcinoma has been discussed as if it were one disease. It is not. Ovarian neoplasms can occur from the many different cells of origin present in the ovary and exhibit many different types of biologic behavior. The student is often bewildered by the complexity of ovarian neoplasia, so an attempt will be made to discuss ovarian neoplasms according to the cell types of origin in hope of providing information without oversimplification. By asking what in the ovary can become neoplastic, we can develop the following list:

- Epithelium
- Germ cells
- Gonadal stroma
- Mesenchyme
- Metastatic disease to the ovary

Epithelial Tumors

These tumors are believed to originate from the germinal epithelium of the ovary that is of Müllerian derivation. This may or may not be true, but the epithelial tumors usually appear similar to the epithelium of the oviduct (serous), endometrium (endometroid), or endocervix (mucinous). These neoplasms may be benign, malignant, or of low potential malignancy, and the survival rates for patients vary with the histologic appearance. Numerically, epithelial tumors account for 70% of ovarian neoplasms and 85% of ovarian cancers. The six types of epithelial tumors are discussed below.

Serous cystadenoma accounts for 25% of all benign ovarian neoplasms and usually occurs in women between 20 and 50 years of age. It is bilateral in 20–25% of cases. The cyst may be unilocular or multilocular. Cyst fluid is usually clear yellow, and the microscopic appearance of the epithelium is so reminiscent of Fallopian tube that the term "endosalpingoma" has been applied to these cysts. It is often difficult to differentiate benign from malignant lesions at surgery, so conservative surgery is advocated for young patients. Peritoneal implants, ascites, external papillations, and bilaterality suggest malignancy. Serous cystadenocarcinoma is the most frequent ovarian malignancy, representing about 40% of all ovarian cancers. It is bilateral two-thirds of the time. Five-year survival is only 25% and correlates with histologic grading of the tumor. Psammoma bodies found in these malignancies probably represent either degeneration with calcification or a tissue response to the neoplasm. The presence of psammoma bodies is associated with a better prognosis.

Mucinous cystadenomas account for 20% of all benign ovarian neoplasms, and they may reach tremendous dimensions. Microscopically, one notes a typical mucin-producing epithelium that is one or two cell layers deep. The cells resemble endocervical epithelium with clear cytoplasm and basal nuclei that gives a "picket fence" appearance to the epithelium. Mucinous cystadenocarcinomas are associated with a 50% five-year survival but represent only 10–15% of ovarian malignancies. They are bilateral in 20% of cases and tend to be large, multilocular, and filled with gelatinous fluid. They may be associated with peritoneal implantation and an outpouring of mucinous fluid in the peritoneal cavity, designated "pseudomyxoma peritonei."

Endometroid tumors, which resemble endometrium, are rarely benign. Endometrial carcinomas comprise 20% of ovarian malignancies and resemble endometrial adenocarcinoma. They may rarely develop from the lesions of endometriosis. They are slow-growing tumors which metastasize late. Bilaterality occurs 30% of the time, and they have a 46% five-year survival.

Clear cell or mesonephroid epithelial tumors are rarely benign, and the malignant variety accounts for 5% of ovarian cancers. The prognosis is poor, with a five-year survival of 25%. The tumors are often large and may be cystic or solid. Microscopically, they are composed of clear glycogen-containing cells or a hobnail type of cell that may represent a clear cell devoid of glycogen.

Brenner tumors account for 2% of ovarian neoplasms and only rarely are malignant. They are composed of nests of cells resembling transitional epithelium embedded in whorls of fibrous stroma. The cell of origin for Brenner tumors is uncertain, and not all authorities classify them as epithelial tumors.

The final type of epithelial tumor is the undifferentiated or solid adenocarcinoma. There is no benign form, and the malignant variety composes 5% of ovarian cancers. It has the lowest five-year survival, 18%. Some of these tumors may be of nonepithelial origin, but they are included here because no other subdivision has been established.

Germ Cell Tumors

Neoplasms of germ cell origin can be understood more easily if we recall the things that germ cells produce. They can make other germ cells (dysgerminoma and gonadoblastoma); they can make babies (teratoma); they can make placentas (choriocarcinomas), and they can make yolk sacs (endodermal sinus tumors).

Germ cell tumors account for 20% of all ovarian neoplasms and for 5% of ovarian malignancies. The most common germ cell tumor is the benign cystic teratoma or dermoid, which contains all three embryonic germ layers. The cysts are unilocular and usually contain fat, hair, teeth, and cartilage. Abdominal roentgenograms may reveal calcification in areas of teeth or bone. Microscopic examination reveals skin and its appendages as well as intestinal and respiratory epithelium, brain, cartilage, and bone. Functioning areas of thyroid tissue may produce hyperthyroidism, struma ovarii. These tumors represent approximately 18% of all ovarian neoplasms and are bilateral 15% of the time. They are usually less than 15 cm in diameter, and if they rupture, the contents are quite irritating and can produce aseptic chemical peritonitis. The treatment of benign cystic teratomas is by removal of the cyst, not removal of the ovary. Dermoids rarely undergo malignant change (1%). If malignancy does develop, it is usually a squamous carcinoma. Malignant teratomas can also develop *de novo* but are quite rare.

Dysgerminomas are the ovarian homologue of the seminoma of the testes. They account for 1–2% of ovarian malignancies and occur mainly in children and young women. They may occur in a pure form or may be mixed with other germ cell tumors. Characteristically, the tumors show nests of germ cells with connective tissue septa and a lymphocytic infiltration. The five-year survival is 25% for dysgerminomas, but the tumors are quite radiosensitive.

Gonadoblastomas are rare tumors that consist of germ cells and immature-appearing granulosa cells. They may coexist with dysgerminomas and are found almost exclusively in genetically abnormal people, such as those with gonadal dysgenesis, mixed gonadal dysgenesis, or male pseudohermaphroditism.

Primary choriocarcinoma of the ovary is a rare germ cell tumor. It is histologically identical to gestational choriocarcinoma and occurs primarily in young women. It does not respond to chemotherapy as well as gestational choriocarcinoma, however, and the prognosis is poor.

Endodermal sinus or yolk sac tumors are so named because they resemble the endodermal sinus of the rat placenta. This tumor is rare but is frequently found in association with other germ cell malignancies. The tumor is very aggressive, but vigorous chemotherapy has apparently cured some patients.

Gonadal Stroma Tumors

These interesting ovarian neoplasms comprise 5% of ovarian tumors but have attracted much attention because of their propensity to be hormonally active.

Granulosa cell tumors are the most common functioning ovarian neoplasms. They are usually unilateral (95%) and are generally estrogen producing. They usually occur in women between the ages of 40 and 55, but their occurrence in children can cause a picture similar to precocious puberty. Postmenopausally, they cause uterine bleeding and endometrial carcinoma; during the reproductive years, they cause menstrual aberrations. Granulosa cell tumors are highly vascular and may be discovered when laparotomy is performed for intraperitoneal hemorrhage. Many histologic patterns are recognized, but the finding of rosettelike Call-Exner bodies is diagnostic. Granulosa tumors are low-grade malignancies. Survival rates of 70–90% have been reported in some series, but they have a tendency to late recurrence. They tend to respond well to radiotherapy.

Theca cell tumors are generally estrogen-producing tumors of theca cells. Pure thecomas are benign. When they are mixed with granulosa cells as granulosa-theca tumors, they are managed and classified like granulosa cell tumors.

When the ovarian stroma differentiates along the lines of testicular structures, it is called a Sertoli-Leydig cell tumor or arrhenoblastoma. These are very unusual tumors but are always remembered because of the striking changes they can produce in the female body habitus. In over three-fourths of cases, they produce androgenic hormones that produce virilization of the female with amenorrhea, breast atrophy, acne, clitoromegaly, beard growth, and temporal balding. They are often small tumors and yield a five-year survival of approximately 65%.

Lipid or hilus cell tumors are usually small unilateral tumors found in the hilar region of the ovary. They are usually virilizing, and this is what brings them to medical attention, for they are usually not palpable. They behave in a benign fashion.

It might be concluded from this section that all gonadal stroma tumors are hormonally active; this is not true. Most are hormonally active, as described, but some are not, and some tumors that characteristically produce estrogens may produce androgens. Additionally, almost any ovarian tumor can be associated with peculiar hormonal production by the ovarian stroma.

Mesenchyme Tumors

Tumors of the ovarian mesenchyme account for 5% of ovarian neoplasms. Lymphomas, sarcomas, hemangiomas, and lipomas have been reported, but for practical purposes, almost all of the mesenchyme tumors are fibromas. Fibromas occur most often between ages 35 and 50, are bilateral in 15%, and are usually asymptomatic. The neoplasm arises from the connective tissue of the ovarian stroma and is firm, white, and whorled on cut surface. Microscopically, the tumor consists only of fibroblasts and may be difficult to distinguish from a thecoma. Treatment consists of excision. The fibroma has achieved fame because of its association with Meig's syndrome, which consists of ascites and pleural effusion associated with a benign ovarian tumor, usually a fibroma, although the syndrome can occur with thecomas or Brenner tumors.

Metastatic Disease to the Ovary

The possibility of tumor metastatic to the ovary should always be considered when an adnexal mass is felt. Five percent of ovarian cancers are a result of metastases. Classically, this is a "Krukenberg" tumor with a primary lesion in the stomach. Metastases from other gastrointestinal primary sites as well as from the breast are frequently seen.

When it has been decided that an adnexal mass warrants surgical therapy, a preoperative evaluation should include a minimum of:

- *Abdominal roentgenogram*— This can identify perhaps 50% of benign teratomas by their pattern of calcification or a "halo" of fat around the tumor. Old calcified myomas that do not require surgical treatment as well as some serous tumors with dense psammoma bodies may be identified.
- *Intravenous pyelogram*— This can identify a pelvic kidney but, more importantly, traces the course of the ureters, reveals anomalies, and detects obstruction.
- *Chest roentgenogram*— This is to identify metastatic lesions and pleural effusions.
- *Barium enema*— This is especially important if the patient has symptoms of colonic disease or a fixed pelvic mass.
- *Upper gastrointestinal series*— This should be performed to detect a primary tumor in the stomach if the patient has suspicious symptoms.
- *Liver function tests*— These should be performed, especially if there is suspicion of ascites.

The clinical staging for ovarian cancer allows one to utilize the findings at laparotomy in the staging procedure. The staging of ovarian carcinoma is *summarized* below:

- Stage I—growth limited to the ovaries
- Stage Ia—one ovary, no ascites
- Stage Ib—both ovaries, no ascites
- Stage Ic—one or both ovaries; malignant ascites present
- Stage II—pelvic extension
- Stage IIa—extension/metastasis to uterus/tubes/ovary
- Stage IIb—extension to other pelvic tissues
- Stage III—intraperitoneal metastases
- Stage IV—distant metastases or parenchymal liver disease
- Special Category—unexplored cases that are thought to be ovarian carcinoma

The standard treatment for ovarian carcinoma involves a total abdominal hysterectomy and bilateral removal of the tubes and ovaries. This may be followed by radiation or chemotherapy, depending on the stage of the disease. At the time of surgery, malignancy is suggested if the tumor involves both ovaries, if there is ascites, if the tumor is adherent to other structures or has excrescences from the surface, or if there are peritoneal implants. Malignancy is also suggested when the opened tumor reveals areas of hemorrhage or necrosis, solid or semisolid state, or papillations into a cyst cavity. There is generally no need to do a complete operation on a young woman in whom the diagnosis of malignancy is in doubt. She may best be treated by ovarian cystectomy and then reoperated if final pathologic diagnosis confirms malignancy. This is greatly preferred to a castrating operation for benign disease.

SUGGESTED READING

DiSaia RJ, Morrow CP, Townsend DE: *Synopsis of Gynecologic Oncology*, New York, John Wiley & Sons, Inc, 1975, pp 136–202.

Pauerstein CJ, Hodgson, BJ, Kramen MA: Anatomy and physiology of the oviduct, in Wynn RM (ed): *Obstetrics and Gynecology Annual 1974*. New York, Appleton-Century-Crofts, 1974, pp 137–201.

Spanos WJ: Preoperative hormonal therapy of cystic adnexal masses. *Am J Obstet Gynecol* 116:551, 1973.

QUESTIONS

1. Which of the following structures is *not* in the adnexa?
   A. mesosalpinx
   B. infundibulopelvic ligament
   C. oviduct
   D. ovary
2. The lumen of the oviduct is narrowest in:
   A. interstitial
   B. isthmus
   C. ampulla
   D. indundibulum
3. Events necessary for reproduction that occur in the oviduct include:
   A. sperm transport and capacitation
   B. development of the blastocyst
   C. nidation
   D. all of the above
4. In a 62-year-old woman, an adnexal mass detected that is unilateral, 6 cm in diameter, round, cystic, mobile, nontender, and posterior to the broad ligament is most likely:
   A. a functional cyst
   B. parovarian cyst
   C. uterine myoma
   D. ovarian neoplasia
5. IUDs:
   A. increase the relative number of tubal pregnancies
   B. increase the relative number of ovarian pregnancies
   C. both true
   D. neither true
6. The most common finding with tubal pregnancy is:
   A. positive pregnancy test
   B. abdominal pain
   C. fainting and shock
   D. amenorrhea
7. In evaluation for tubal pregnancy, a culdocentesis that obtains clear fluid from the peritoneal cavity is termed:
   A. negative
   B. positive
   C. nondiagnostic
   D. none of the above

8. Nonneoplastic ovarian cysts include:
   A. theca lutein cysts
   B. thecoma
   C. dermoid
   D. all of the above
9. Ovarian carcinoma:
   A. causes symptoms early in the disease
   B. is detectable by cervical cytology
   C. causes more fatalities than any other genital malignancy
   D. usually does not occur after menopause
10. The most common ovarian neoplasms are:
   A. epithelial tumors
   B. germ cell tumors
   C. gonadal stroma tumors
   D. mesenchymal tumors
11. Serous cystadenocarcinomas are:
   A. epithelial tumors
   B. germ cell tumors
   C. gonadal stroma tumors
   D. mesenchymal tumors
12. The most common germ cell tumor is:
   A. dysgerminoma
   B. gonadoblastoma
   C. choriocarcinoma
   D. cystic teratoma
13. The most common functioning ovarian tumor is:
   A. theca cell tumor
   B. arrhenoblastoma
   C. granulosa cell tumor
   D. hilus cell tumor
14. Ovarian carcinoma that involves the omentum is stage:
   A. I
   B. II
   C. III
   D. IV

ANSWERS

1. B
2. B
3. A
4. D
5. C
6. B
7. A
8. A
9. C
10. A
11. A
12. D
13. C
14. C

chapter twelve

pelvic support and pathophysiology of the pelvic floor

OBJECTIVES

1. Know factors that alter the pelvic floor.
2. List pathologic-anatomic changes of the pelvic floor.
3. Define urethrocele, cystocele, procidentia, rectocele, enterocele, and symptoms that are associated with them.
4. List symptoms of pelvic relaxation—congestion.
5. List the neurologic control of two major areas of the bladder.
6. List the causes of urgency incontinence.
7. Define stress urinary incontinence (SUI).
8. State the basic pathophysiologic problem in SUI.
9. List four necessary components of SUI evaluation.
10. List and define four methods of treating SUI.

Genital prolapse, pelvic relaxation-congestion, and female urinary dysfunction are related anatomically, etiologically, and clinically. They are anatomically related because of proximity and similarity of structural support. Etiologic factors that seem to alter the pelvic floor include: aging; heredity; trauma of pregnancy, childbirth and physical labor; scarring after inflammatory and surgical insults; psychosomatic control of smooth muscle and vascular components of the pelvis; sex steroid changes producing changes in smooth muscle response, submucosal support, and mucosal integrity; and inability of striated muscle to maintain support of the pelvic floor. Clinically, they are similar in their importance and in the lack of objective evidence to permit solid confirmation of their pathophysiology.

It is difficult to estimate the prevalence of these entities, for they are often relatively asymptomatic. When they are brought to the physician's attention, it is often because of such complaints as pelvic pain, vaginal discharge, protrusions from the vagina, low back pain, urinary tract infection, fecal obstipation, dull pressure in the pelvis, urinary dysfunction, or ureteral obstruction. We do know, however, that they comprise a large part of gynecologic disease. Surgical repair of pelvic floor entities accounted for 12–15% of major gynecologic surgery at the Bexar County Hospital in 1975–1976.

Diagnosis is made by clinical judgment through understanding the woman's history and physical findings. There is no consistent objective test to document either the initial problem or therapeutic result. Treatment modalities range from counseling and exercises to medication and surgical repair. Effective therapy largely depends on a clear understanding of the probable pathophysiologic mechanisms and the woman's specific clinical circumstances.

GENITAL PROLAPSE

The structural support that suspends the internal genitalia several centimeters above the pelvic floor is the endopelvic fascia. Subdivisions of the endopelvic fascia include the cardinal ligaments, pubourethral ligaments, and uterosacral ligaments. The cardinal ligaments sweep from the lateral pelvic walls to encircle the uterine cervix. The cardinal ligaments and their posterior condensations, the uterosacral ligaments, are the primary support of the uterus, broad ligaments, and adnexa (Fig. 12–1A and B). These ill-defined ligamentous structures contain fibrous tissue and an irregular amount of smooth muscle that is continuous with smooth muscle of the cervix, bladder floor, urethra, and rectum. The round ligaments may provide some positional support to the uterus.

The pelvic floor support of the female is divided into two triangular areas separated by a line drawn between the ischial tuberosities. The anterior triangle is composed of endopelvic fascia, smooth muscle, and fibers of striated muscle from the ischiocavernous and bulbocavernous

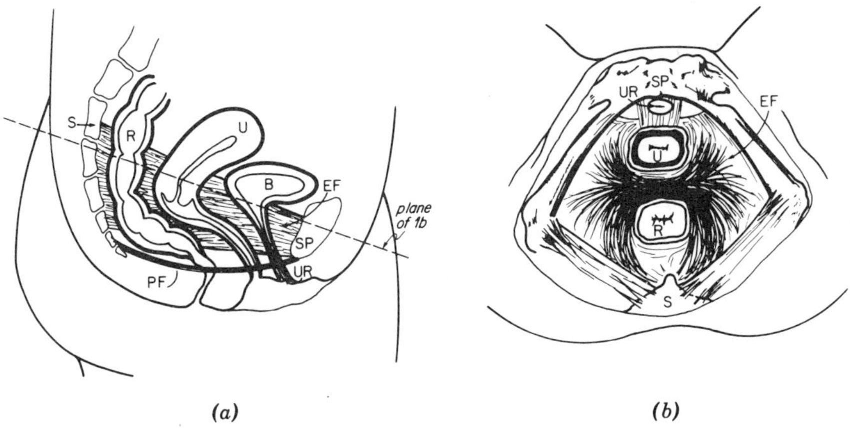

Figure 12–1. (*a*) Lateral view of structures and support in the female pelvis. SP = symphysis pubis, UR = urethra, B = bladder, U = uterus, R = rectum, S = sacrum, PF = pelvic floor, EF = endopelvic fascia. (*b*) Transverse view of structures and support in the female pelvis. The endopelvic fascia is given different names in certain areas (cardinal ligament, pubourethral ligament, uterosacral ligament) but is continuous throughout.

muscles and from the pubococcygeus portion of the levator ani. The anterior triangle of the pelvic floor supports the bladder and anterior vaginal wall and is most easily damaged by the previously mentioned etiologic factors, especially childbirth. The posterior triangle is composed of fascia as well as striated muscle from the levator ani.

Abnormalities of pelvic floor anatomy can result in prolapse of the structures supported by the pelvic floor and include urethrocele, cystocele, uterine prolapse, enterocele, and rectocele. These conditions may occur individually or together in various combinations. They may or may not be associated with symptoms.

The urethra is supported by the anterior triangle and the pubourethral ligaments beneath the pubic symphysis. When these support structures are destroyed, the urethra rotates beneath the symphysis into the anterior vagina and eventually through the introitus. Although often asymptomatic, symptoms that may be associated with this anatomic abnormality are stress urinary incontinence and prolapse when associated with a cystocele.

A cystocele is a herniation of the bladder floor into the vagina and possibly through the vaginal outlet. Weakness in the support structures of smooth muscle and fascia in the trigone area may be associated with ulceration of the prolapsed vaginal mucosa and/or recurrent urinary tract infections because of residual urine after ineffectual bladder emptying.

Procidentia or prolapse of the uterine cervix onto the pelvic floor and/or through the introitus may cause pelvic heaviness, low back pain, cervical ulceration, or ureteral obstruction. An enterocele is a true herniation of intestine into a diverticulum of peritoneum from the cul-de-sac of Douglas between the vagina and rectum. It is usually important only as it occurs with general genital prolapse, but it has the potential to compromise or obstruct the small bowel.

Herniation of the anterior wall of the rectum into the vagina is a rectocele. It may cause fecal obstipation, which necessities digital manipulation of the posterior vaginal wall to allow expulsion of feces through the external anal sphincter. Constipation is erroneously associated with rectocele; it is a dysfunction related to diet, autonomic nervous system activity, and/or physical inactivity.

Clinical grading of pelvic prolapse can be recorded in a standard fashion by use of the pelvic profile system. With the woman examined while "bearing down" in the dorsal lithotomy position, areas of abnormality are quantitatively described by a six-digit system. The first digit represents the urethra; the second, the bladder; third, cervix; fourth, enterocele; fifth, rectocele; and sixth, perineal body. The first five are scored in similar fashion. A score of 0 represents the anatomy of a nulliparous woman; 1 represents descensus from the normal to less than half the distance to the introitus. A score of 2 represents descensus over halfway to the introitus, and 3 represents descensus through the introitus. A score of 4 means vaginal inversion with the uterus out of the pelvis. The sixth digit, the perineal body, is evaluated differently. A score of 0 means an intact hymen; 1 means a thick perineal body; 2, an absent perineal body but intact anal sphincter; 3, loss of rectal sphincter; and 4, a cloaca. The pelvic profile system is shown in Figure 12–2.

Although isometric contractions of the skeletal muscle components of the pelvic floor (Kegel's exercises) several times a day may help prevent or minimize genital prolapse, the final therapeutic modality for symptoms of overt genital prolapse is surgical repair. Urethral prolapse, cystocele, enterocele, and rectocele may be repaired effectively without removal of the uterus. Future pregnancy with or without vaginal delivery may jeopardize the effectiveness of the repair, however.

For symptoms of uterine prolapse, the only nonsurgical treatment is a pessary. This is a plastic or rubber device worn in the vagina to

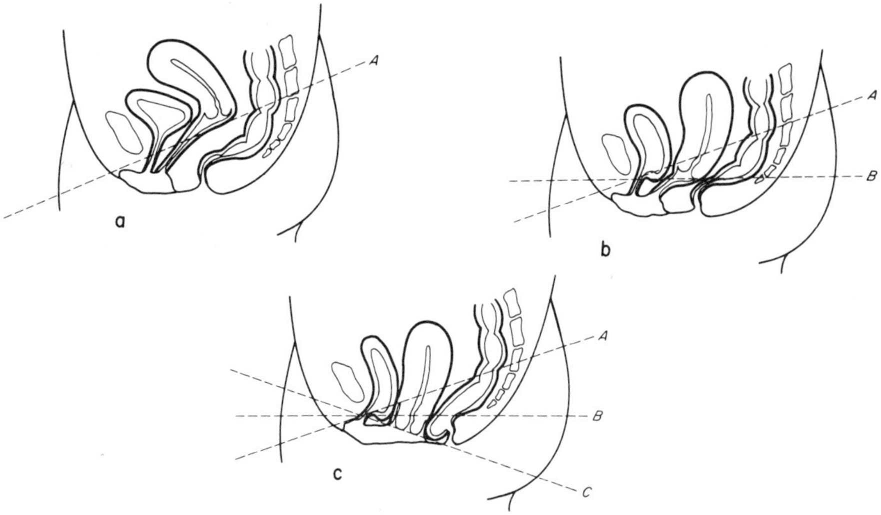

Figure 12–2. (*a*) Normal support of the female pelvis. The urethra, bladder, cervix and rectocele would all be given a pelvic profile score of 0. Enterocele and perineal body are not represented in the diagram. (*b*) Decensus from plane A to plane B represents a pelvic profile score of 1 for the four entities named above. (*c*) Decensus to plane C represents a pelvic profile score of 2 for the urethra, bladder, cervix and rectocele. Decensus below plane C represents a score of 3.

support the internal genitalia several centimeters above the pelvic floor. When reproductive function is not necessary, the most effective treatment of uterine prolapse is hysterectomy.

PELVIC RELAXATION-CONGESTION

A vague clinical syndrome called "pelvic relaxation," "pelvic congestion," "broad ligament laceration" (Allen-Masters), or "premenstrual tension" is often associated with mild degrees of relaxation of pelvic support. Symptoms include pelvic heaviness, low back pain, deep thrust and postcoital ache dyspareunia, and chronic lower abdominal aching or vague pelvic pain. These symptoms are diurnal in nature. They are minimal or absent when the woman awakens but increase when she is sitting or standing. Symptoms are usually accentuated near the menstrual period.

Although the nulliparous woman may develop this problem, most patients are multiparous. Pelvic examination may demonstrate uterine descensus and hypermobility of the cervix or simply vague uterine or adnexal tenderness, especially if performed near the end of the day.

The pathophysiologic basis of this controversial clinical entity has been described as either psychosomatic or organic, but as with most clinical entities, it probably involves both. The pathophysiologic process appears to be pelvic venous congestion. The psychosomatic component acts through the autonomic nervous system dysfunction related to psychic stress. This is similar to the parasympathetically and sympathetically mediated physiologic changes associated with sexual arousal. Venous congestion that occurs through the above mechanisms may progress to a persistent organic problem similar to that of varicose veins in other parts of the body. The organic component is related to sex steroid effects on pelvic edema and uterine venous engorgement. Uterine venous engorgement may also be exacerbated by gravitational effects and broad ligament scarring following childbirth.

Various therapeutic approaches from psychiatric care to hysterectomy have been suggested. The most important therapy is time investment by the woman and physician in order to clarify the diagnosis, exclude the possibility of life-threatening disease, and identify alterations in environment and life style that might diminish symptoms. Medications such as diuretics, hormones, and psychoactive drugs have helped some women. Acknowledging patient discomfort and educating her about the process is of primary importance to reduce anxiety. Many women will have the desire and ability to live with moderate symptoms of this entity once they understand its benign nature. Although psychiatric disease must always be considered, surgical therapy is effective in most cases. Hysterectomy, however, only removes the "end organ" and does nothing to relieve the psychosomatic stress aspects of the patient's life.

It is important to recognize and differentiate another entity from pelvic relaxation-congestion syndrome because the symptoms are so similar but the management is different. Iliopsoas muscle spasm produces low back pain, pelvic pain, and deep thrust dyspareunia. Iliopsoas spasm occurs as a result of pregnancy, when the abdominal musculature has lost its tone, the paraspinal muscles of the back are shortened, and the iliopsoas muscles are stretched in the upright position. The only effective treatment of this process is an understanding by the woman of muscle spasm and a therapy program of exercises (touching toes, sit-ups, leg lifts) to stretch the paraspinal muscles and increase the tone of abdominal muscles. Removal of the uterus will not affect this cause of pelvic pain.

FEMALE URINARY FUNCTION AND DYSFUNCTION

Gynecologic urology deals with diseases of the urethra, urethrovesicle neck, trigone and detrusor muscle of the bladder as they relate to urinary incontinence in women. The main complaints are stress urinary incontinence and urgency urinary incontinence; other diagnostic possibilities are contained in Table 12–1. Proper diagnosis is the major concern. Detailed medical history, physical examination, urine cultures, radiographic studies, cystourethroscopy, and constant flow carbon dioxide cystometric evaluation can differentiate the causes of incontinence in women.

A brief review of the functional anatomy of the bladder and the physiology of micturition is pertinent. Smooth muscle of the detrusor has a different embryologic origin from smooth muscle of the proximal urethra and urethrovesicle neck. Histologically inseparable after fetal life, the two sources of smooth muscle retain separate innervation. The proximal urethra and urethrovesicle neck are primarily innervated by

Table 12–1. Female Urinary Dysfunction

I. Congenital or acquired urinary tract abnormalities
   A. Ectopic ureter
   B. Fistula—urethrovaginal, vesicovaginal, ureterovaginal

II. Neurologic Bladder Dysfunctions
   A. Hyperreflexic (Spastic)—automatic uninhibited micturition (may be uncoordinated) at 50–100 ml urine volume.
   B. Hyporeflexic (flaccid)—large bladder volume (over 800 ml) and incomplete emptying.
   C. Combined—most bladder dysfunctions have components of both of the above and are abnormalities of coordination of filling and emptying.

III. Urethrovesical Dysfunction (gynecologic urinary dysfunction)
   A. Urgency incontinence—may be associated with funneling of the bladder neck, infection, and irritation or psychosomatic problems.
   B. Detrusor dyssynergia—may be associated with funneling of the bladder neck, drugs, or psychosomatic problems.
   C. SUI—urethrovesical prolapse
   D. Combined—SUI plus urgency incontinence or detrusor dyssynergia.

alpha adrenergic fibers, while the detrusor muscle is primarily innervated by parasympathetic fibers. There is no circular sphincter muscle in the urethral area; instead, integration of smooth muscle in the trigone and proximal urethra provide increased pressure in the proximal urethra, which functions as a sphincter. Because of additional urethral factors, such as mucosal venous pressure, peristaltic waves, and inherent tone, the entire urethra functions as a sphincter to some extent.

The neurologic control of micturition results from a fine balance of the autonomic nervous system influenced by central autonomic training, the ratio of adrenergic and cholinergic receptors, the integrity of central and peripheral nerve pathways, and physical characteristics of the urethra. Micturition is initiated by relaxation of the urethrovesicle neck coordinated with contraction of the detrusor. Contraction of the detrusor causes flattening of the bladder base with loss of the posterior urethrovesicle angle, elevation of the bladder floor, and funneling of the proximal portion of the urethra. The presence of urine in the proximal urethra stimulates afferent nerve endings that act via a reflex to produce parasympathetic discharge and continued detrusor contraction. Micturition is terminated by contraction of the urethrovesicle neck associated with reinstitution of the posterior urethrovesicle angle and relaxation of the detrusor. When the intraurethral pressure becomes greater than intravesicle pressure, continence is restored.

Urgency Incontinence and Detrusor Dyssynergia

Urgency incontinence and detrusor dyssynergia may be considered together. They are both manifestations of dysfunction or inappropriate coordination between the detrusor muscle of the bladder and the functional sphincter area, the urethrovesicle neck. They are also manifestations of uncontrolled or inappropriate micturition. Urgency incontinence means the inability to control urination when the urge to empty the bladder occurs. Detrusor dyssynergia is the initiation of micturition without any feeling of urge and is manifested by episodes of uncontrolled urinary loss.

Urgency incontinence is frequently a symptom of urinary tract infection, but other etiologic factors are similar for both conditions. Both conditions may be associated with psychosomatic dysfunction, early neurologic lesions, local mucosal and neuromuscular irritation, drugs (especially those affecting the autonomic nervous system), and anatomic funneling of the bladder neck. These etiologic factors appear to operate by causing inappropriate urethral "sphincter" contraction or relaxation, detrusor contraction, or both. The diagnosis of urgency incontinence is usually made by obtaining a careful history. The diagnosis of detrusor dyssynergia is suspected by history and confirmed by cystometric evaluation (Fig. 12–3B).

Once infection and local inflammation have cleared, remaining symptoms may be treated in several ways. Medications such as psychoactive drugs and alpha adrenergic agonists have helped some women. Biofeedback offers hope for the future. Surgical repair of bladder anatomy may also be of value. Urgency incontinence is often diminished following repair of the anterior vaginal wall to correct funneling of the trigone area. Proposed mechanisms for relief of some cases of urgency incontinence by surgery include removing the stimulus of urine in the urethra that causes detrusor contraction by anatomic reconstruction of the bladder base or possibly by partial denervation of the urethrovesicle neck by the operation.

Stress Urinary Incontinence (SUI)

SUI is defined as the involuntary loss of urine at the exact time of an increase in intraabdominal pressure. This may occur with such activities as lifting, coughing, running, laughing, intercourse, sitting up, or getting out of bed. Simplistically, SUI is the involuntary loss of urine that occurs when the intravesicle pressure exceeds intraurethral pressure during the time of increased intraabdominal pressure. This implies that intraabdominal pressure increase is applied to the bladder (an intraabdominal organ) but is not effectively transmitted to the paraurethral areas as well.

Pathophysiology of SUI

There are two components to the forces that provide a functional sphincter for the bladder. The first component is the intraurethral forces. The closing forces of the urethra are generated by smooth muscle of the urethrovesicle neck and striated muscle of the pelvic floor. Minor components to urethral closure are provided by venous pressure in the submucosal plexus and smooth muscle tone along the length of the urethra. The intraurethral forces may be decreased by hormonal changes affecting smooth muscle tone such as occur after menopause or during pregnancy, from psychosomatic disorders, and by decreased compliance of the urethra secondary to scarring.

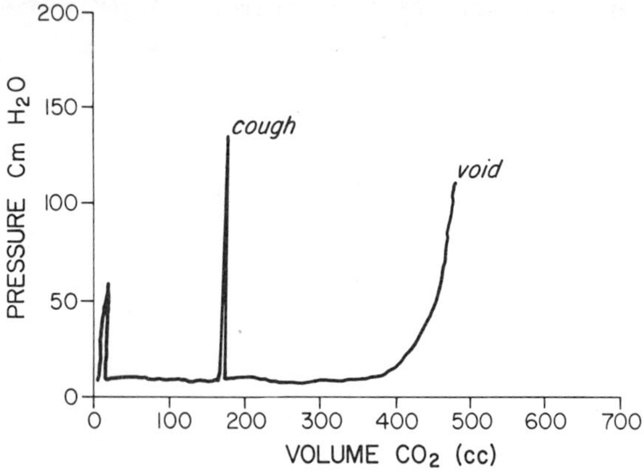

(a)

Figure 12–3*(a)*. Normal cystometric examination with no increase in intravesicle pressure until a volume of 300-600 ml is reached. Coughing transiently raises pressure but there is no detrusor contraction.

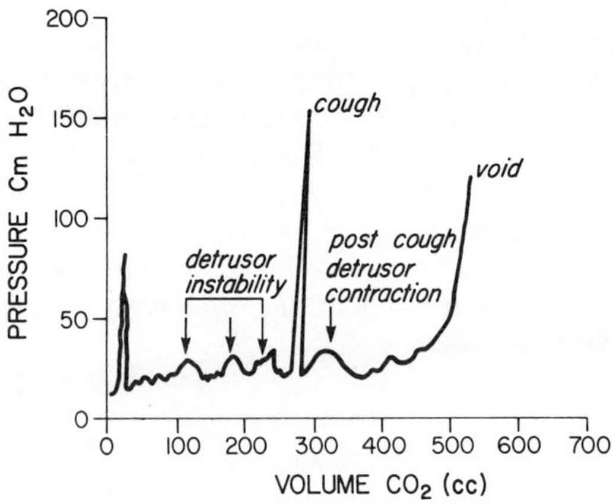

(b)

Figure 12–3*(b)*. Cystometric examination of a woman with detrusor dyssynergia. The detrusor contracts at low bladder volumes and again after a cough.

The second component of the functional bladder sphincter is the paraurethral forces. Paraurethral support assures the transmission of increased abdominal pressure to the urethra as well as to the bladder so that intraurethral pressure remains higher than intravesicle pressure during episodes of coughing or straining, thus maintaining continence of urine. The paraurethral forces can be diminished by pelvic floor trauma, such as pregnancy or childbirth, as well as by physical labor and aging. Illnesses such as pulmonary disease and obesity can cause frequent increases in intra-abdominal pressure that can weaken paraurethral support. The loss of transmission of intraabdominal pressure to the urethra by weakening of the paraurethral support is the major pathophysiology of stress urinary incontinence. If the urethra is not well supported, increased intraabdominal pressure will be transmitted only to the bladder, the urethra will rotate under the pubic symphysis, paraurethral forces will not be applied to it, and SUI will result. Surgical reattachment of paraurethral tissues to the pubis should correct the anatomic problem.

Correlation of Pathophysiology and Clinical Information

A number of other factors, such as urethral length, angle of the urethra to the symphysis, relationship of the urethrovesicle neck and the lowest region of the bladder, and loss of mobility of the dependent portion of the bladder, have also been proposed as important components in the pathophysiology of SUI. Urethral length is probably unrelated to SUI as long as the proximal 2–3 cm of urethra are intact and well supported.

Factors involving angles are stigmata of SUI rather than causes. A bead chain cystogram is often employed to identify anatomic abnormalities of urethral angles. In this test, a beaded chain is introduced into the bladder and left hanging out the urethra. The chain provides contrast for roentgenograms that are taken when the woman is relaxed and while straining. The outline of the urethra and lower bladder is thus obtained. Several authors believe this test can determine accurately the extent of anatomic abnormality present and the type of surgical repair that should be performed. Other studies have shown poor reproducibility of the test and poor correlation with the results of surgery. The problem with the bead chain cystogram test is that it measures a subjective effort by the patient. To obtain best results, the physician should be present to administer the test and evaluate the patient's effort at stress at the time roentgenograms are made; with good patient cooperation, the bead chain cystogram can provide accurate documentation of anatomic defects.

Another test used to assess the degree of anatomic abnormality is the applicator stick test, which is easily performed at the time of pelvic examination. This test is performed by introducing an applicator stick coated with local anesthetic jelly the length of the urethra. The woman is then asked to bear down maximally, and the angle the applicator stick makes with the horizontal is estimated. The more the urethra rotates due to loss of support, the greater the angle will be. The degree of pelvic relaxation should be documented by the pelvic profile system.

That the physician witnesses loss of urine by the patient at the time of straining (positive stress test) does not prove the diagnosis of SUI. The Bonney test can be helpful in confirming that incontinence is due to straining. It is performed after a positive stress test by elevating tissue lateral to the urethrovesicle neck to a position behind the pubis. The woman is asked again to strain or cough; the absence of incontinence now implies that surgical support of these tissues would correct the incontinence. Figure 12–4 depicts findings of bead chain cystogram, applicator stick test, and Bonney test.

Cystometric studies evaluate bladder capacity and the ability of the detrusor muscle to accommodate bladder filling without increasing bladder pressure (isotonic relaxation). The test is best performed by filling the bladder with a continuous flow of carbon dioxide and constant

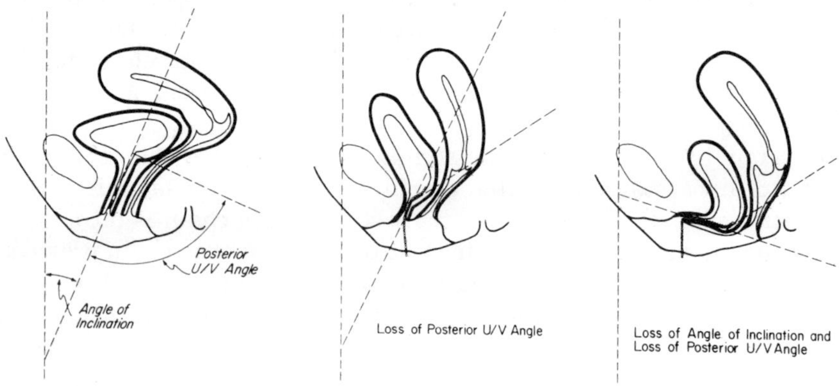

(a)

Figure 12–4*(a)*. Depicts findings of the bead chain cystogram. The first panel shows a normal angle of inclination between the urethra and the vertical axis and a normal posterior urethrovesicle (U/V) angle. The second panel depicts loss of the posterior U/V angle with funneling of the bladder neck. The third panel represents loss of the angle of inclination as well as loss of the posterior U/V angle.

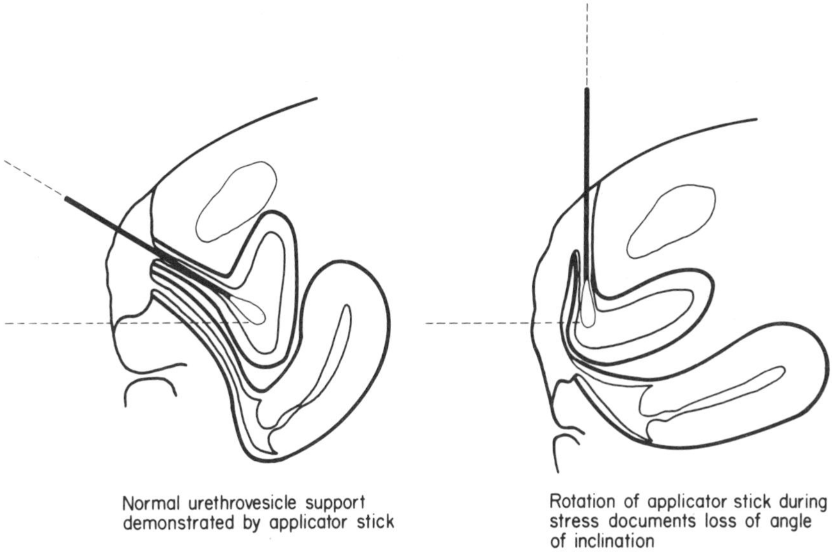

(b)

Figure 12–4*(b)*. The applicator stick test shows a normal angle of inclination in the first panel. The second panel depicts rotation of the applicator stick during stress. This indicates loss of the angle of inclination due to decensus of the urethrovesicle neck.

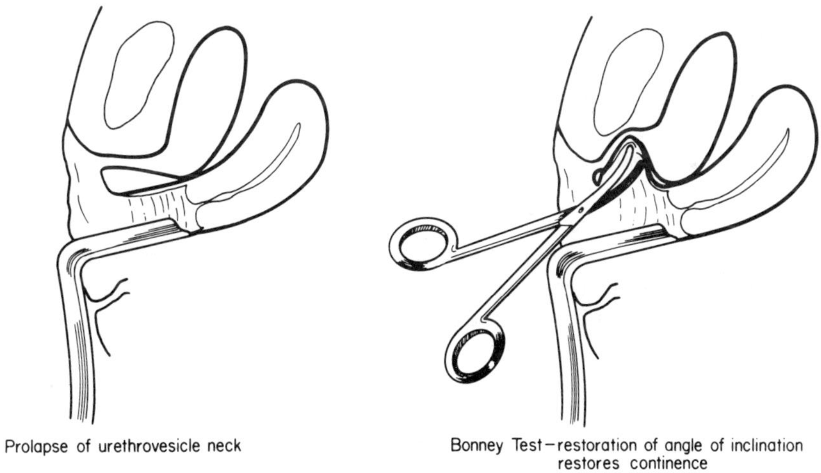

(c)

Figure 12–4*(c)*. The Bonney test. Prolapse of the urethrovesicle neck is shown in the first panel. Restoration of the angle of inclination corrects stress urinary incontinence.

pressure monitoring but can also be performed satisfactorily using saline to fill the bladder. Normal cystometric examination (Fig. 12–3A) shows no increase in pressure until a volume of 300–600 ml is reached. At this point, either a detrusor contraction associated with micturition occurs, or the woman complains of pain.

Other tests that may be performed to evaluate women with urinary dysfunction include urethroscopy to detect urethral disease, such as chronic urethritis or diverticula, urofluorimetry to detect obstruction or poor micturition function, and urethral pressure determinations to evaluate the functional sphincter mechanism. These tests, however, need further evaluation to ascertain their clinical value.

Clinical Evaluation of SUI

Several studies have shown that approximately 50% of normal nulliparous women without anatomic abnormalities experience occasional (even daily) loss of urine while straining. Even after successful surgical repair, most women continue to have occasional, mild urinary incontinence while straining. This is explained as an inevitable consequence of the functional anatomy of the female pelvis. Prior to therapy for SUI, it is necessary to obtain a thorough history to determine the amount of urine lost, the circumstances under which urine is lost, and the social significance of SUI. The severity of incontinence has been "graded" in many ways, but no system is superior to carefully recording all aspects of the problem. A careful history should also detect the possibility of urgency incontinence or neurologic dysfunction, which should be corrected before treatment for SUI. General physical examination to identify problems that might affect therapy as well as pelvic examination to record anatomic defects and to document the loss of urine with straining are necessary. Pelvic neurologic examination to elicit bulbocavernosus and pubococcygeus reflexes as well as perianal reflexes is necessary. Urinalysis and culture should be obtained from all patients prior to treatment and any infection cleared. Table 12–2 summarizes the clinical evaluation of SUI.

Treatment

Treatment for women with SUI includes counseling, Kegel exercises, medication, and surgery. The patient must be educated about the medical and social significance of her degree of stress incontinence. Once she understands the importance of an exact history, involving the activities associated with urine loss, the amount of urine lost, the exact

Table 12–2. Clinical Evaluation of SUI

A. History
   1. Does the incontinence correlate with stress?
   2. Severity—frequency of SUI
      volume of urine lost
      social significance—wear a pad? afraid to go in public?
   3. Are there other causes of incontinence?
      urgency incontinence
      detrusor dyssynergia
      hyporeflexia—overflow incontinence
      hyperreflexia—spastic bladder

B. Physical Examination
   1. General—identify other problems that may affect therapy
   2. Pelvic examination
      a. Anatomy—pelvic profile
      b. Estrogen deficiency?
      c. Applicator stick test
      d. Stress test
      e. Bonney test
      f. Residual urine after voiding
      g. Neurological—sensation and muscle tone of pelvic floor

C. Laboratory
   1. Urinalysis and culture
   2. Other tests as indicated by history and physical

time relationship between intraabdominal pressure and urine loss, and the relationship to psychic stress factors, a reasonable approach to her individual degree of urinary incontinence can be determined. It is important that the physician and woman both be aware that the problem is a common one for many healthy females, that it may be episodic and recurrent, and that it is primarily a disease of local discomfort and social significance. Evaluation and exact diagnosis of bladder function are paramount before any bladder repair is done since the best surgical procedure is the first surgical procedure.

The first treatment approach to SUI is counseling. The physician and patient must work together to determine the significance of the symptoms and the adequacy of simple measures such as preventive emptying of the bladder at high-risk times.

Textbooks state that approximately 15% of women with SUI improve with use of Kegel exercises. Better results have been obtained when

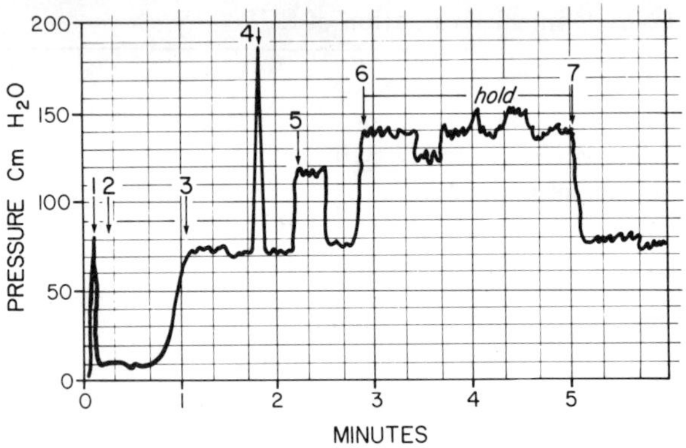

Figure 12–5. Recording of normal changes in urethral pressure. Number 1 represents passage of the pressure transducer through the urethra. The drop in pressure (2) indicates passage of the transducer into the bladder. At 3, the transducer is pulled back into the urethra where it remains. The rise in pressure (4) with a cough demonstrates that the increased abdominal pressure is transmitted to the urethra. Point 5 shows that Kegel's exercises can increase intraurethral pressure. Between points 6 and 7 Kegel's exercises are performed and held for several minutes. Relaxation occurs at point 7 and intraurethral pressure immediately returns to baseline level.

patient education is more extensive and patient performance has been documented by use of recording instruments. Women can be taught this exercise at the time of pelvic examination and should perform it several times a day. Objective recordings (Fig. 12–5) have shown that intraurethral pressure can be greatly raised by voluntary contraction of the pelvic floor. Although Kegel's exercise may improve symptoms of SUI, it is not curative because the voluntary muscle of the pelvic floor is distal to the functional bladder sphincter and urine that enters the upper urethra can trigger reflex contraction of the detrusor. Also, many episodes of stress occur before the woman is ready to contract the pelvic muscles. As with any form of exercise therapy, success is related to motivation and participation by the patient.

Drugs can effect SUI. Adequate estrogenic effect in the vaginal and urethral area is extremely important. Several studies have shown increased intraurethral pressure after estrogen usage by postmenopausal women. This occurs because of increased compliance and thickness of the urethral mucosa and an increased smooth muscle tone in the urethrovesicle neck area. An adequate estrogen effect can be obtained by applying small amounts of estrogen cream to the introitus and anterior vagina several times a week.

Alpha adrenergic stimulants and parasympatholytic drugs may act directly to increase intraurethral pressure or to reduce the urgency incontinence component often seen with SUI. Urethral dilations and long-term low-dose antibiotic therapy may reduce an infectious component of SUI. Conversely, drugs affecting the autonomic nervous system, such as antihypertensives, cold remedies, and psychotrophic drugs, can actually decrease the threshold for continence.

Once identified, studied, and confirmed, the anatomic abnormalities producing SUI can be surgically corrected by several techniques. The surgical approach should be individualized to the patient.

Anterior colporrhaphy (anterior repair) and its numerous modifications are often effective. The urethra and urethrovesical neck are dissected from the vaginal mucosa, and the pubourethral ligaments are then reapproximated and shortened to increase the paraurethral forces contributing to the functional sphincter mechanism (Fig. 12–6A). Abdominal urethropexy (Marshall-Marchetti-Krantz procedure) elevates the urethrovesical neck by approximating paraurethral tissues to the back of the pubis (Fig. 12–6B). The strap procedure and its modifications elevate the urethrovesical neck with a strip of fascia connected to the anterior abdominal wall so that straining by the patient results in tightening of the fascial strap and application of high paraurethral forces (Fig. 12–6C).

Surgical therapy for SUI has a 90% success rate two years after surgery. The success rate drops to between 50 and 85% by five years after surgery. This is not difficult to understand since the same factors related to tissue integrity that caused the initial problem often continue to exist after therapy. Patients also have recurrences of the urgency component of SUI, estrogen deficiency, infections, psychosomatic stress, and changes in bladder anatomy.

Prior to performing a second operative procedure for SUI, a greater preoperative evaluation must be performed to document fully the woman's anatomy, which has now been altered by previous surgery.

Frequently, a "little bladder repair" is performed at the time of other gynecologic surgery. This is a poor practice. The best opportunity for corrective pelvic surgery is the first opportunity, so the operation should be carefully considered and performed.

The long-term management of a woman who has mild SUI or has had a surgical procedure for SUI includes patient education about the condition, continuation of pelvic muscle exercises, adequate estrogen

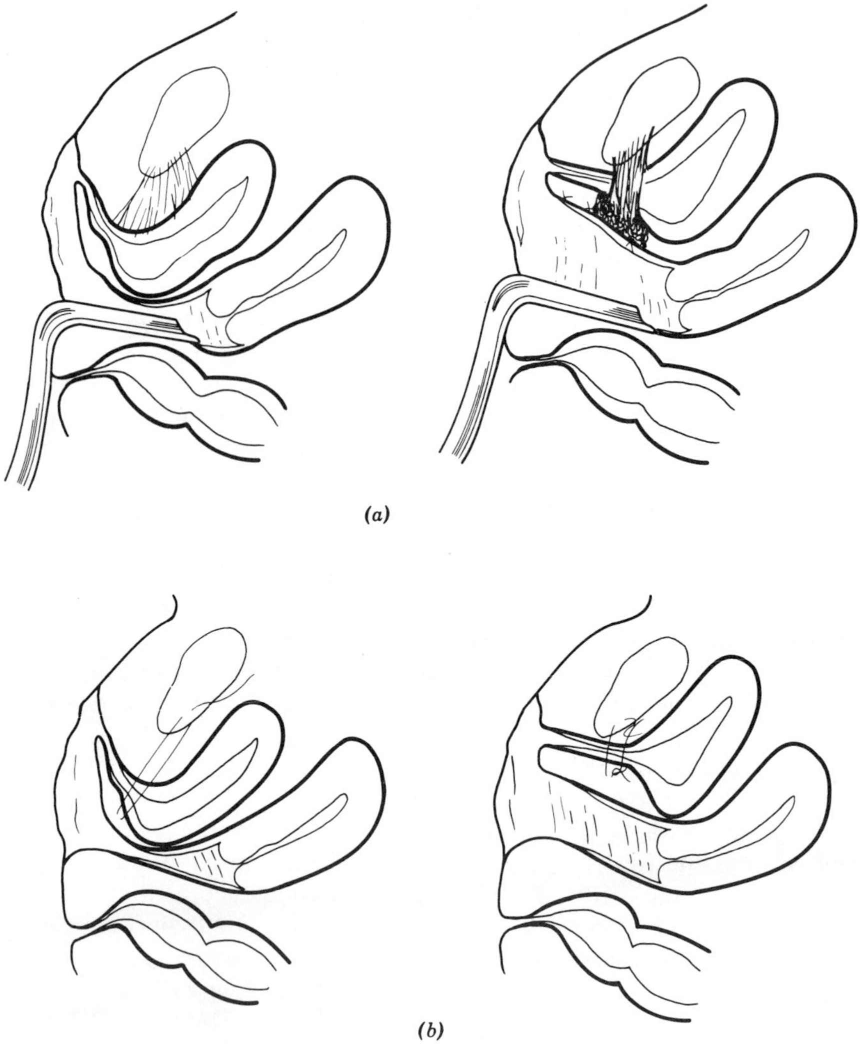

Figure 12–6. (*a*) The first panel shows pelvic relaxation which is corrected by surgical shortening of the pubourethral ligaments and development of a bridge of tissue beneath the urethrovesicle junction. (*b*) Abdominal urethropexy elevates the urethrovesicle neck by approximating the paraurethral tissues to the periosteum of the pubis.

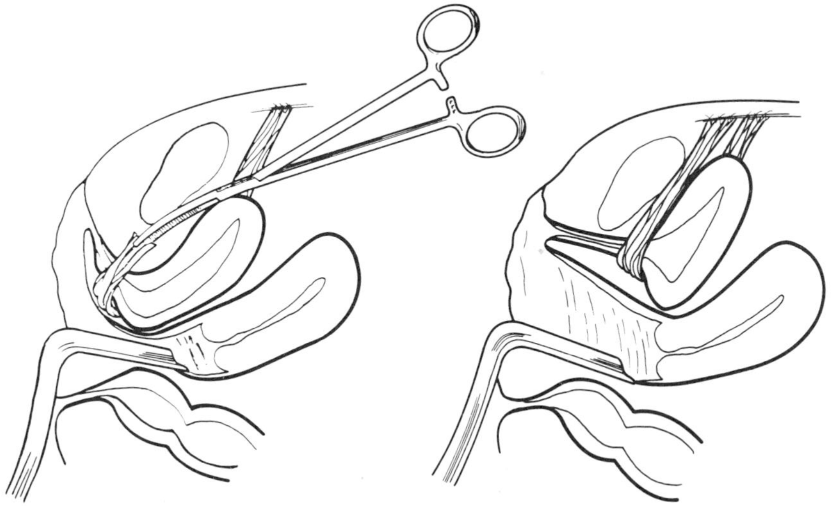

(c)

Figure 12–6*(c)*. Fascial strap urethropexy elevates the urethrovesicle neck with a strip of fascia which is sewn to fascia of the anterior abdominal wall.

support of the tissues, and prompt treatment of urinary tract infections.

SUGGESTED READING

Allen WM, Masters WH: Traumatic laceration of uterine support. The clinical syndrome and the operative treatment. *Am J Obstet Gynecol* 70:500, 1955.

Baden WF, Walker TA: Genesis of the vaginal profile: a correlated classification of vaginal relaxation. *Clin Obstet Gynecol* 15:1048, 1972.

Green TH Jr: *Gynecology: Essentials of Clinical Practice*, ed. 2. Boston, Little, Brown and Company, 1971, chap 18.

Ingelman-Sundberg A: Urge incontinence in women. *Acta Obstet Gynecol Scand* 54:153, 1975.

Marshall VF, Marchetti AA, Krantz KE: The correction of stress incontinence by simple vesicourethral suspension. *Surg Gynecol Obstet* 88:509, 1949.

Taylor HC Jr: Vascular congestion and hyperemia. Their effect on structure and function in the female reproductive system. *Am J Obstet Gynecol* 57:211, 637, 654, 1949.

QUESTIONS

1. Presenting symptoms of cystocele may include:
   A. protrusion from vagina
   B. low back pain
   C. stress incontinence
   D. pelvic pressure
   E. A, C, D
   F. All of above
2. The following are causes of urinary urgency incontinence:
   A. psychic stress
   B. infection
   C. local irritation
   D. neuropathy
   E. all of the above
3. A hernia of the cul-de-sac of Douglas into the vagina is:
   A. a rectocele
   B. a cystocele
   C. an enterocele
   D. procidentia
4. Incontinence may be a symptom of which of the following diseases:
   A. diabetes
   B. multiple sclerosis
   C. spinal cord injury
   D. cystourethrocele
   E. all of the above
5. The *sine qua non* of continence is:
   A. an intact pelvic diaphragm
   B. the sphincteric mechanism of the bladder
   C. an intraurethral pressure higher than intravesicle pressure
   D. none of the above

6. Which of the following are necessary in the evaluation of SUI:
   A. history
   B. physical
   C. cystometric evaluation
   D. urinalysis and culture
   E. all of the above
7. Which of the following primarily controls the bladder neck?
   A. sympathetic nervous system
   B. parasympathetic nervous system
8. Anticholinergic agents primarily affect the:
   A. trigone or bladder neck
   B. detrusor or bladder fundus
9. Ureteral obstruction is likely to be caused by:
   A. procidentia
   B. rectocele
   C. urethrocele
   D. pelvic relaxation-congestion
10. Cystocele may require surgical repair for:
   A. recurrent urinary tract infections
   B. neurogenic bladder

ANSWERS

1. F
2. E
3. C
4. E
5. C
6. E
7. A
8. B
9. A
10. A

part three
pregnancy

chapter thirteen

genetics in human reproduction

OBJECTIVES

1. Understand the mechanisms of cellular division and reproduction.
2. Describe the difference between male and female meiosis and gamete formation (the stages in which chromosomal anomalies may arise).
3. Review the abnormalities of chromosomal structure and number and the principles of Mendelian inheritance.
4. Know examples of sex chromosome abnormalities in order to correlate the effect of the genotype on the phenotype.
5. Review the prenatal diagnosis of hereditary diseases—methods and indications.
6. Discuss the role of cytogenetics in pregnancy wastage.

Genetics encompasses many different branches of medicine. In obstetrics, its main role is the evaluation of the fetus through amniotic fluid culture and cytogenetic examination of the conceptus in miscarriages and in abnormal newborns. In gynecology, chromosomal analysis is necessary in primary amenorrhea, male infertility, habitual abortion, sex anomalies, and hermaphroditism. At research levels, it is of interest in studies of teratogenic effects on preimplanting blastocysts and the evaluation of gynecologic tumors, such as teratomas. It is beyond the scope of this chapter to discuss all the above subjects. Cytogenetic and genetic principles will be outlined and a brief description of their clinical applications made. The extent of fetal wastage will be briefly discussed and prevention through prenatal diagnosis described.

Mitosis, Meiosis, and Gametogenesis

Mitosis

Mitosis is the mechanism by which somatic cells reproduce normally (skin, kidneys). Some of them are end-stage cells and will not reproduce again except under special conditions (lymphocytes). Four morphologic stages of cellular division can be seen: prophase, metaphase, anaphase, and telophase. The interphase nucleus is seen in the non-dividing stage of a cell. It represents a resting stage that is divided into three components (Fig. 13–1):

1. Gl period or gap 1 extends from the end of mitosis to the beginning of the S phase (duration: 12–24 hours up to years). During this period, nucleotides and enzymes are compiled in preparation for synthesis.

2. S phase or synthesis extends from the end of G1 to the beginning of G2 period. DNA synthesis, replication, and protein synthesis occur

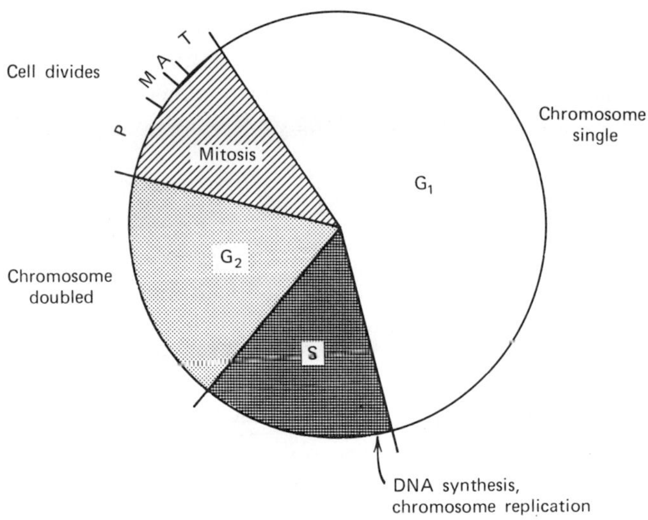

Figure 13–1. The cell cycle. (From Hamerton JL: *Human Cytogenetics*. New York, Academic Press, 1971, p. 41, Fig. 3–6.)

in this phase. The chromosomes are not visible. Chromocenters and nucleoli can be seen (duration: seven to eight hours).

3. G2 period or gap 2 extends from the end of S phase to the beginning of mitosis (M) (duration: four to six hours). No known function is ascribed to this period.

The cell cycle composed of the above stages of the interphase nucleus terminates with mitosis proper, whose duration varies from 30 to 90 minutes. Cells entering mitosis are diploid. The chromosomes are seen only during the mitotic stages. The morphologic stages of mitosis form a continuum and are not as rigidly divided, as seen in Figure 13–2.

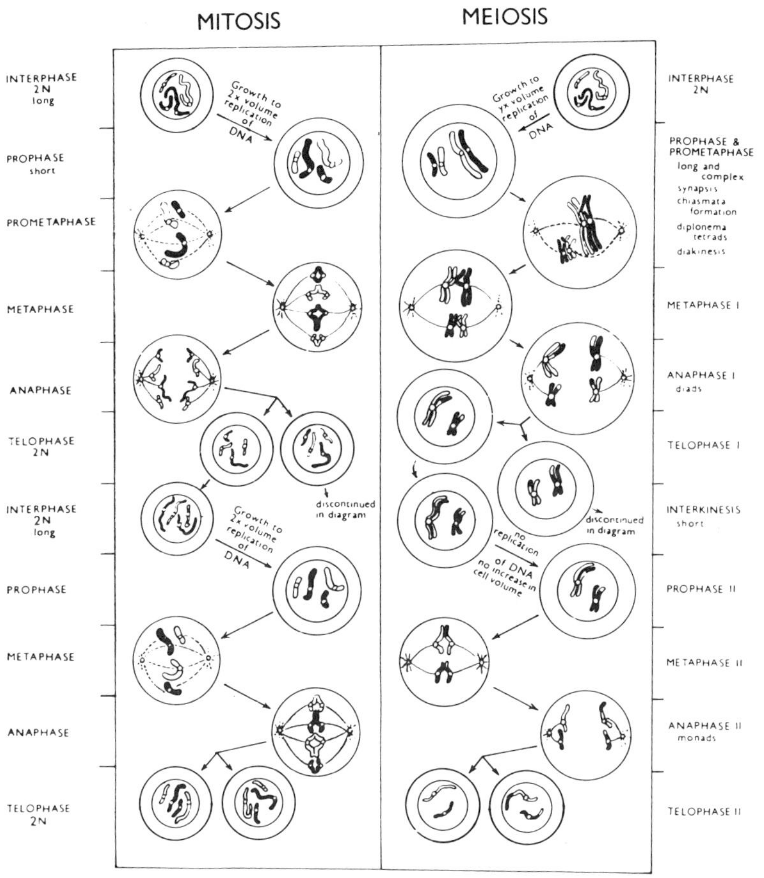

Figure 13–2. The successive stages of mitosis and meiosis. (From Hamilton, Boyd and Mossman, *Human Embryology*. London, Macmillan, London, and Basingstoke, 1972, p. 8, Fig. 2.)

1. Prophase: The elongated chromosomes start condensing, find their homologues, and join at the centromere. In the cytoplasm, the centrosome, containing two centrioles, divides, and both start migrating to their respective poles. The disappearance of the nuclear membrane heralds the start of metaphase.

2. Metaphase: Further shortening and condensation of the chromosome pairs occur. The spindle apparatus is formed. The chromosome pairs align themselves along an imaginary equatorial plate. The centromeres are the most important functional structure in the migration of the chromatids to their respective poles. Metaphase ends when the centromere splits into two units.

3. Anaphase: The homologous chromatids are repulsed from each other and migrate toward their respective poles. The centromere slides on the spindle fiber, while the chromatid arms follow passively. Fragments of chromosomes without centromeres (acentric fragments) will not be able to migrate and will be lost.

4. Telophase: The nuclear membrane reappears around every mass of chromosomes that condenses and becomes chromatin. The cytoplasm of the daughter cells reappears, and the result of mitosis is two diploid daughter cells. Thus, from one diploid cell (2n) that reduplicated its DNA (4n) developed two diploid daughter cells.

Meiosis

Meiosis is the cellular division that leads to the formation of the gametes. The main differences with mitosis are listed in Table 13–1. The general principles of cell division are similar in males and females. A diploid cell (2n) gives rise to four haploid cells (n) (Fig. 13–2). Meiosis is usually divided into two successive cell divisions:

Meiosis I has a very long prophase that, in turn, is subdivided into five different stages. This is a very important phase, which differs from mitotic prophase and is also different in mammalian males and females. DNA synthesis has been shown to occur in two phases: a major one prior to and continuing into leptotene and a minor one during pachytene. Chromosome pairing (synapsis), crossing over, and chiasma formation occur in prophase I.

Prophase of meiosis I consists of five parts:

1. Leptotene: There is an increase in nuclear volume. The chromosomes are fine, single threads with chromomeres (tightly coiled re-

Table 13–1. Differences Between Mitosis and Meiosis

	Mitosis	*Meiosis*
Cell	Somatic cells	Germ cells
Resultant	Two diploid cells (2n)	Four haploid cells (n)
DNA replication	S phase	Preleptotene and pachytene
Crossing over	–	+
Chiasma formation	–	+
Excessive repulsion of chromatids at meiosis II	–	+
Different genetic constitution of each chromatid at meiosis II	–	+

gions) separated by chromonemata. DNA replication took place just before the start of this stage. There is an increase in chromosome spiralization by the end of this stage; although chromosomes are still long, they are tightly coiled.

2. Zygotene is the beginning of synapsis of the homologous chromosomes. There is further coiling. Chromosomes become shorter and broader. Synapsis starts at any point or points along the chromosome and proceeds in a zipper fashion.

3. Pachytene ends the synapsis. It is a long stage. The number of bivalents is equal to the haploid number. There is a distinctive linear pattern of chromomeres for every bivalent. Each homologue divides into two chromatids at the end of this stage; crossing over and chiasma formation start then. The nucleolus is well seen at this stage.

4. Diplotene: There is a contraction of bivalents and repulsion of homologues except at chiasmata. The number of chiasmata is highest at the beginning of this stage; then it decreases in number by the process of terminalization. There is maximal repulsion of the homologues at the centromere.

5. Diakinesis The bivalents are further shortened. The nucleolus disappears. The bivalents migrate to the nuclear membrane, where they attach in a "bouquet" appearance.

Metaphase of meiosis I commences with disappearance of the nuclear membrane and division of the centrosome. The spindle apparatus is formed. The two centromeres of each bivalent lie on each side of the equatorial plate and attach to the spindle.

In anaphase of meiosis I, the two centromeres start moving in opposite directions. Terminalization is completed, and homologues are separated. Each homologue consists of two chromatids. Crossing over resulted in an exchange of genetic material so that, at the end of meiosis I, each chromosome has a combination of both maternal and paternal genes.

Meiosis I ends with telophase. After arrival at their respective poles, the chromosomes may uncoil. The nuclear membrane forms. More frequently, this does not occur, and the second meiotic division proceeds immediately. Interphase is very brief when it exists. There is no DNA synthesis or complete chromosome uncoiling.

Meiosis II begins with prophase, and contraction and coiling of chromosomes occur. Meiosis II differs from mitosis in two aspects: First, the chromatids show repulsion except at the centromeres, and second, their genetic constitution depends on the number of chiasmata previously present. The chromosomes resemble Maltese crosses. No DNA duplication takes place.

In metaphase and anaphase of meiosis II, the nuclear membrane disappears, and the spindle apparatus forms. The chromosomes orient themselves as in mitosis. The centromeres divide, and each chromatid moves to opposite pole. Every daughter cell has a haploid number of chromosomes at the end of telophase.

Gametogenesis

Gametogenesis is the formation of gametes through mitotic then meiotic divisions. In the female, the major portions of prophase I are accomplished in fetal life. The fetal ovary, at birth, contains 2 million oocytes, arrested in meiosis during prophase I. No oogonia are available after birth. The completion of meiosis I occurs only at ovulation; the oocyte will then be at metaphase II. The completion of meiosis II is triggered by sperm entry and terminates at fertilization. At puberty, there are approximately 300,000 ova. A woman normally ovulates 300–400 times during her reproductive years. The constant decrease of oocyte number, at birth and at puberty, is due to controlled atresia. The result of female oogenesis is one functional ovum and three nonfunctional polar bodies.

Nothing such as this happens in the testes (Table 13–2). Spermatogonia start proliferating at puberty and reproduce themselves from puberty on. The number of germ cells is unlimited. Sper-

Table 13–2. Gametogenesis in Males and Females

	Male	*Female*
Prophase I	Puberty	Fetal Life
Germ cells number	Indefinite source of spermatogonia	At birth, only primary oocytes are found (2 million)
Length of meiosis	75 hours	Years
Meiosis I terminates		At ovulation
Meiosis II terminates		At sperm entry
XY bivalent	+	–
Result of meiosis	4 spermatids	1 ovum and 3 polar bodies

matogenesis takes 60–75 hours for a completion of a cycle, resulting in four functional spermatozoa from one spermatogonium. Male meiosis starts at the primary spermatocyte stage and terminates with the production of four spermatids. Spermatogonia proliferate by mitosis to the level of primary spermatocyte, whereas the change from spermatid to spermatozoa is a complicated metamorphosis referred to as spermiogenesis.

The Human Chromosomes

The diploid human chromosomal complement is 2n = (46). It is made of 22 pairs of autosomes and 1 pair of sex chromosomes. A karyotype is a photomicrograph of the chromosomes of a cell and is divided into seven groups, depending on the size of the chromosome and the position of the centromere. Recently, several methods for chromosome identification have been developed and a new cytogenetic nomenclature standardized. In brief, the staining of chromosomes with fluorochromes such as quinacrine dihydrochloride revealed a banding pattern that is typical for each chromosome (Q bands) (Fig. 13–3). In addition, the distal arm of the Y chromosome fluoresces brighter than any of the rest of metaphase chromosomes. The identification of a male cell is made easy through its staining with quinacrine dye. The Y-body is also easily seen because of its fluorescence in the interphase nucleus (Fig. 13–4A). Female nuclei do not have this fluorescent Y-body. In the female, one of the X chromosomes is normally inactive or heterochromatic. In the interphase nucleus, it is seen as a triangular or oval structure at the

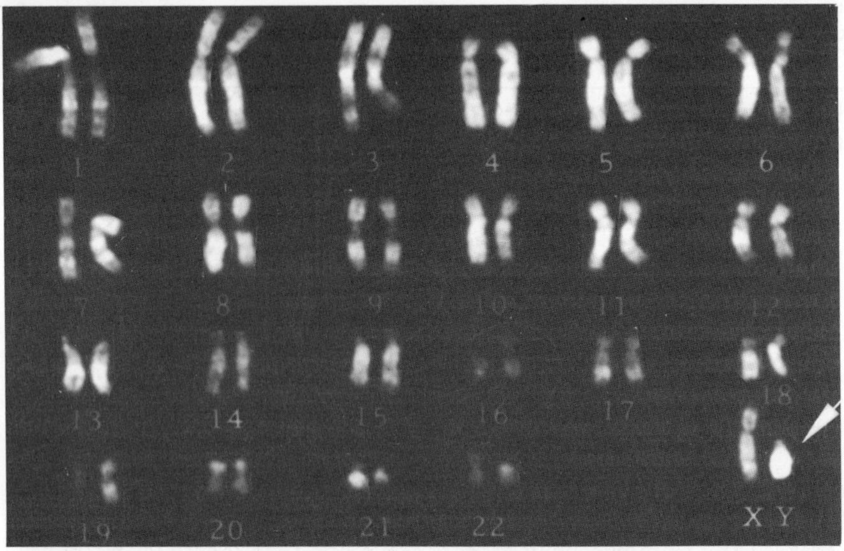

Figure 13–3. Fluorescent bands of human chromosomes (quinacrine dihydrochloride). The Y chromosome seen at arrow.

periphery of the nucleus. It is referred to as sex chromatin body or Barr body (Fig. 13–4B). The denaturation-renaturation treatment of chromosomes with heat, enzymes, or alkali and their staining with Giemsa lead to another type of banding pattern called Giemsa or G-bands (Fig. 13–5). C-bands or centromeric bands are produced by a prolonged denaturation treatment of the chromosomal heterochromatin, leading to the persistence of the centromeric heterochromatin with the rest of the chromatids appearing as faint strands (Fig. 13–6). Other banding methods are less commonly used. These methods help in identifying each chromosome and localizing accurately the exact location of a break or a translocation.

Chromosome Abnormalities

Abnormalities of meiosis and mitosis lead to structural or numerical chromosomal errors. Three types of numerical abnormalities are:

1. Aneuploidy is any deviation from normal diploid number, which is not a multiple of the haploid number and might be the result of a gain or a loss of chromosome. Examples of aneuploidy are: hypodiploidy, a number that is less than the diploid (45 or 44) and is not a multiple

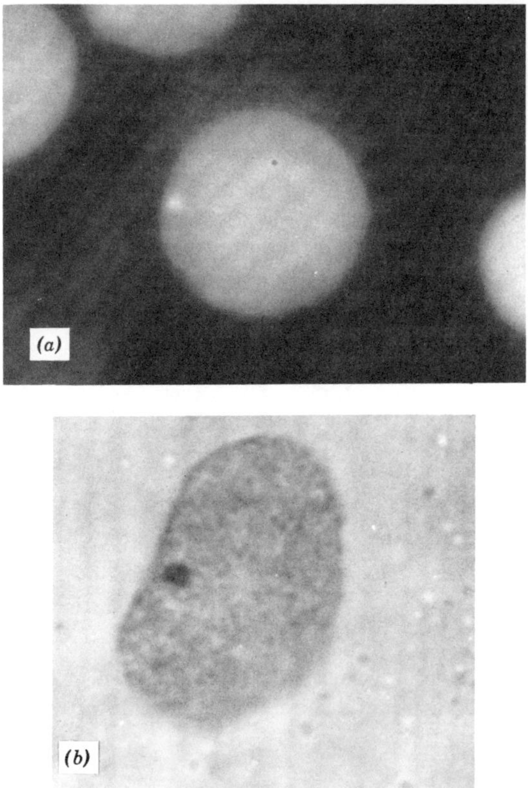

Figure 13–4. (*a*) Fluorescent spot in a lymphocyte indicating the presence of a Y chromosome. (*b*) Sex chromatin (Barr body) in the nucleus of a cell from a female.

of 23. Hyperdiploidy is a higher number than the diploid (47 or 48). A trisomy is a hyperdiploid number.

This defect arises mainly because of nondisjunction at mitosis or meiosis. The normal disjunction or separation of chromatids at anaphase might not occur, and homologous chromatids travel to only one daughter cell, increasing its chromosome number ($2n + 1$) and leaving the other cell lacking one chromosome ($2n - 1$). Anaphase lag is another mechanism whereby disjunction of the chromatids occurs but one of them travels to its respective pole; the other chromatid does not move adequately toward its own pole, lags behind the nuclear membrane formation, is left outside the enclosing nucleus, and is lost. The

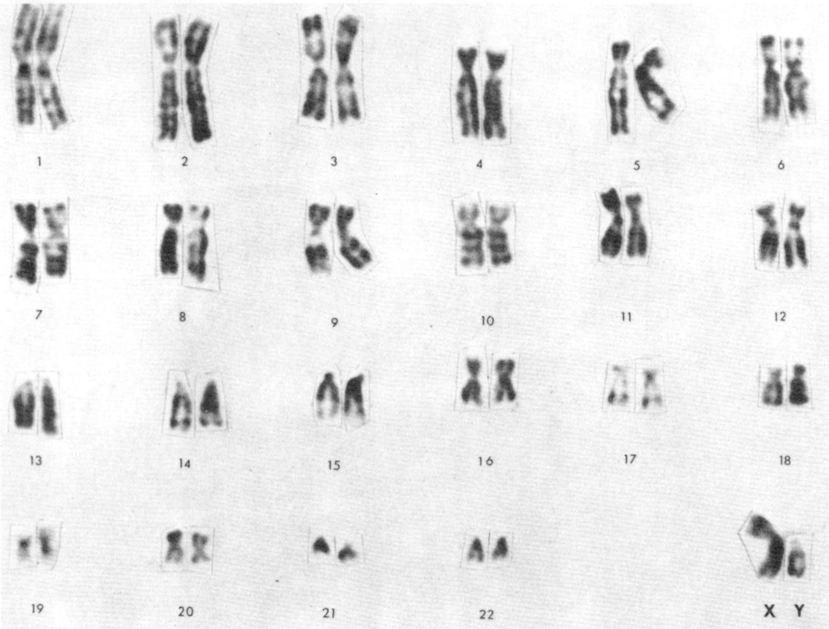

Figure 13–5. G-bands obtained after trypsinization of the human chromosomes. Dark staining areas alternate with light staining ones, giving a special pattern for each chromosome.

result is one cell with a normal diploid number (2n) and another daughter cell lacking one chromosome (2n − 1). When these errors arise at meiosis, a certain proportion of spermatozoa might be abnormal. If nondisjunction occurs during the mitotic cleavages of a zygote, it will lead to the formation of more than one cell line.

2. Polyploidy is a deviation in the number of chromosomes that is a multiple of the haploid number. The modes of origin of polyploidy are multiple. Failure of extrusion of the first or second polar body would lead to a diploid ovum that, on fertilization, will give rise to triploidy (3n-69). Tetraploidy would arise after failure of zygote cleavage or the fertilization of a diploid ovum by a diploid sperm (4n-96). Dispermy or polyspermy is another method giving rise to polyploidy.

3. Mosaicism is the presence of one or more cell lines in an individual derived from one zygote. This mixture is mainly found with sex chromosome problems such as gonadal dysgenesis. Thirty percent of such individuals are mosaic. One percent of persons with Down's syndrome

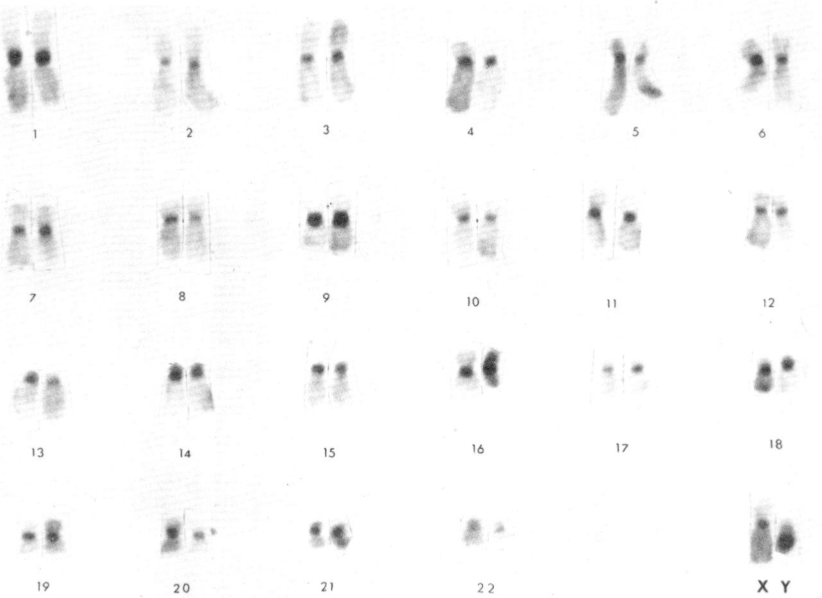

Figure 13–6. C-bands obtained after barium hydroxide treatment. The centromeres and the distal half of the long arms of the Y chromosomes are stained.

are estimated to be mosaic. In contrast, chimerism indicates the presence of one or more cell lines in an individual derived from two or more zygotes.

Structural errors in chromosomes result from breakage and reunion during G1 before chromosome reduplication. They are of two types, intrachromosomal and interchromosomal rearrangements.

Intrachromosomal rearrangements include:

1. Deletion: A loss of a part of a chromosome. It can be terminal or interstitial, involving one or two breaks. A third type of deletion would be the loss of both ends of a chromosome and the rejoining of the broken ends, forming a ring chromosome.
2. Inversion: A two-break rearrangement. The segment between the breaks becomes inverted. A pericentric inversion includes the centromere, while a paracentric inversion does not. They can be recognized at pachytene by irregularities of pairing.
3. Centromeric shift is a three-break chromosome rearrangement that involves movements of regions lying between two break points

into a position of a third break with subsequent reunion of all fragments.

Interchromosomal rearrangements involve the alteration in the position of a chromosome segment between two nonhomologues; they include:

1. Reciprocal translocation: the mutual exchange of segments between two nonhomologues, resulting in a translocation heterozygote with a fully balanced chromosome complement. Chromosome number remains unchanged.
2. Robertsonian translocation or centric fusion: the fusion of two acrocentrics with breaks on opposite sides of the centromere and rejoining of the arms. This is usually associated with a loss of a minute chromosomal segment and formation of a new chromosome with a reduction in number of the chromosomes.
3. Insertion: a rare three-break translocation that has been identified in the human only through recent cytogenetic techniques. It involves a change in position of a segment to the interstitial portion of a nonhomologue without reciprocal exchange.

Misdivision of a centromere can also produce structural abnormalities. Normally, a longitudinal division of the centromere takes place at anaphase, leading to a separation of chromatids. A transverse division of the centromere would lead to the formation of a telocentric chromosome, called *isochromosome*. This abnormal division occurs at meiosis I. This abnormal division usually affects the X chromosome in humans.

Mendelian Genetics

A pair of genes located at identical loci on homologous chromosomes are called alleles. An abnormal or mutant gene may be inherited or occur spontaneously. Except for the sex chromosomes, there is a pair of genes for each function on every pair of chromosomes. Figure 13–7 illustrates the different types of inheritance explained below.

Autosomal Recessive Inheritance

A recessive gene is a single mutant gene that does not cause any abnormality in the phenotype. The presence of a normal member of the gene pair will carry out normal function. Individuals with a single recessive gene will be heterozygotes or carriers. When both genes are

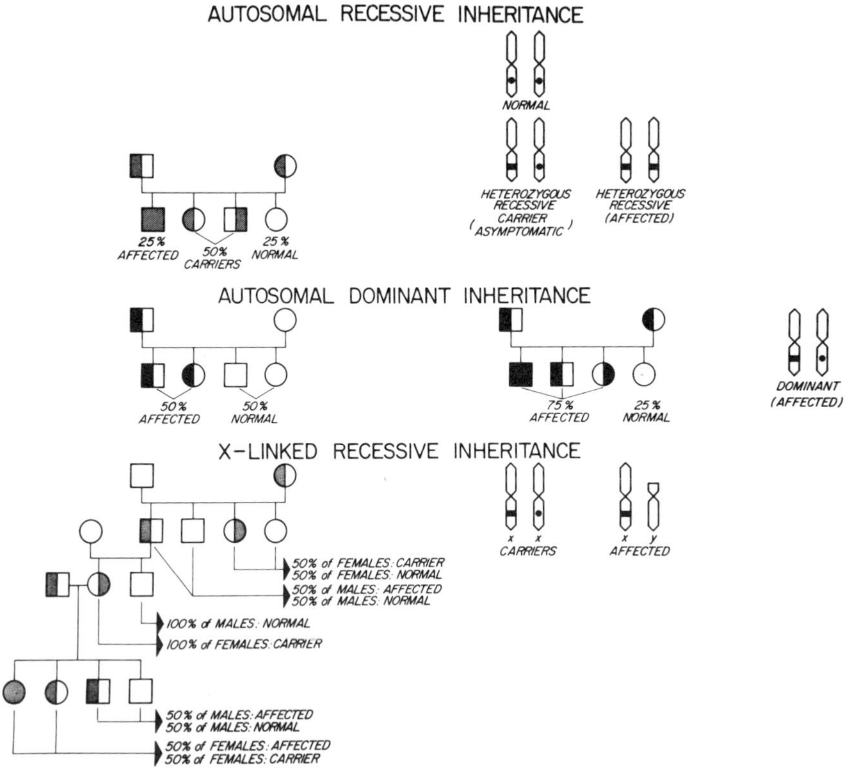

Figure 13–7. The different types of inheritance are illustrated. (See text for explanation.)

mutant genes, then their abnormal effect will show up. The individual carrying a pair of recessive mutant genes will then be homozygous. Normal heterozygous parents may produce an affected homozygous child. This would occur in 25% of their progeny. Fifty percent of their children will be heterozygous, and 25% will be normal. The risk of having an autosomal recessive disorder is increased if consanguinity exists.

Autosomal Dominant Inheritance

A single mutant gene is dominant if its presence causes an abnormality even when its partner gene is normal. If one parent has an autosomal dominant disease, 50% of the progeny will be affected and 50%

normal. If both parents are affected, 75% of the progeny will be affected and 25% normal.

X-Linked Recessive Inheritance

A single mutant gene on the X chromosome expresses itself as a "dominant gene" in males because they have only one X chromosome. It expresses itself as a "recessive gene" in females because the normal allele assumes the function of the mutant gene. If a female carrier marries a normal husband, 50% of male progeny will be affected, and 50% of female progeny will be carriers (heterozygous). If an affected male marries a normal female, all male progeny will be normal, and all female children will be carriers. The marriage between an affected male and a carrier is of rare occurrence. It leads to the birth of an affected female in the proportion seen in Figure 13–7.

X-linked dominant diseases are confused with autosomal dominant diseases. Here females are affected, and there is a more severe effect in the males. Affected males have normal sons and affected daughters.

Polygenic Inheritance

Polygenic inheritance is the common mode of inheritance in single localized defects. It is the result of multiple minor differences at many gene loci, no one of which could be held responsible for the anomaly. The incidence of recurrence of such abnormalities is 2–5%.

Chromosomal Studies in Pregnancy Wastage and Newborn

Spontaneous Abortions

A number of surveys have been carried out in order to document the role of cytogenetic errors in fetal wastage. Most studies have been in agreement and showed that the general incidence of chromosomal errors in spontaneous abortions is 25%. The earlier is the abortion, the higher the incidence of cytogenetic anomalies. The major types of chromosomal errors are listed in Figure 13–8. Trisomies seem to form the bulk of the anomalies, polyploidies, including triploidies and tetraploidies, form 25–30%, and all are lost by abortion. Monosomy X (45,X or Turner's syndrome) is the commonest single anomaly that is lost early in gestation (20–25%).

Cytogenetic errors in abortuses probably arise during spermato-

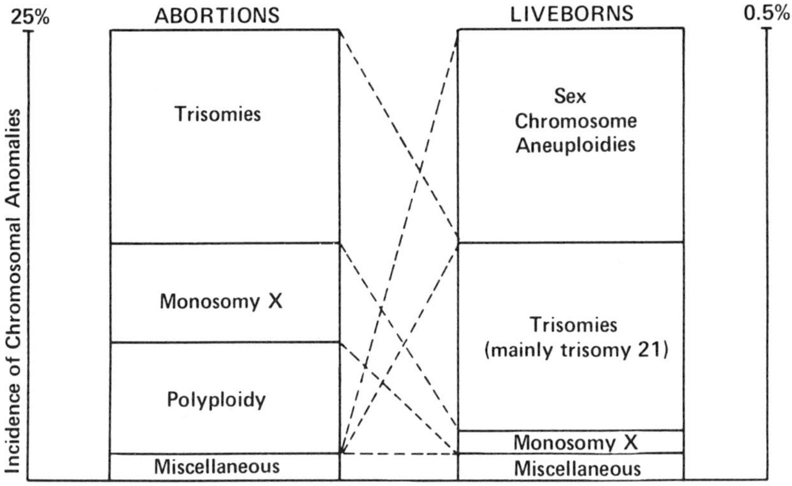

Figure 13–8. The incidence of chromosomal errors in spontaneous abortions as compared to live borns.

genesis, oogenesis, fertilization, or during the early cleavages of the zygote. Normal parents with normal karyotypes can beget chromosomally abnormal abortuses. Most of the lethal trisomies are lost early in gestation while trisomy 21 (Down's syndrome), trisomy 18 (Edward's syndrome), and trisomy D_1 (Patau's syndrome), have a higher chance to survive till neonatal life (to adulthood in the case of Down's syndrome). Practically all polyploidies are lost very early in gestation. Turner's syndrome, with its 45,X karyotype, is the commonest known monosomy in the human and is the only error of its kind that is compatible with survival. The majority of the monosomies are lost early in pregnancy, thus explaining the low incidence of Turner's syndrome among live births.

Newborns

When chromosomal anomalies are studied in liveborns at term, a different proportion of cytogenetic errors is noted. The general incidence of chromosomal disorders at term, in surveys of consecutive liveborns, is around 0.5%. No polyploidies are seen in liveborn babies; very few mosaics (diploid/triploid) with a majority of diploid cell lines have been seen in the human.The monosomy X forms a tiny minority in contrast to the large number seen in abortuses. Most trisomies have been lost in abortions, and only those that are compatible with life (trisomy 21) or with the immediate neonatal period (trisomy D_1, trisomy 18) form the majority of these disorders (40%). Nearly 50% of all chromosomal errors at this stage are sex chromosome aneuploids (XXY, XYY, XXX) that are compatible with life. They were not seen in significant proportion among abortuses (Fig. 13–8).

It is estimated that 50–60% of chromosomal errors present at birth cannot be detected by a physical examination. The incidence of cytogenetic disorders increases when babies with congenital malformations and stillbirths are included in surveys.

In parents with abnormal karyotypes, the incidence of chromosomal abnormalities in the progeny depends on the type of the rearrangement and on which parent is affected. In parents with some types of translocation, the theoretic risk of having a child with Down's syndrome is 30%, while the observed risk is 10–15% if the mother is the carrier and 2% if the father is the carrier of the translocation. On the other hand, with 21/21 translocation, the risk of producing a fetus with trisomy 21 or monosomy 21 could be 100%. All the karyotypic abnormalities in abortions and liveborns occur disregarding the genotype of the parents. They are increased when the parents have abnormal genotypes.

Sex Chromosome Anomalies

This is a brief account of the sex chromosome abnormalities that may be investigated in an obstetric, gynecologic, or infertility clinic.

Sex Chromosome Anomalies in the Male

Klinefelter's syndrome was described as a characteristic syndrome of gynecomastia and hypogonadism with hyalinized seminiferous tubules. These patients have an increase in urinary gonadotropins. They exhibit a lack of secondary sex characteristics, as evidenced by a poor androgenic response and a feminine distribution of pubic hair. Intellectual subnormality is common. Tall stature is a common sign and is contributed to by increased length of the leg rather than the trunk. The incidence is 1 per 1,000 male births, 9–20 per 1,000 mental defectives and 10–20% males attending infertility clinics. Chromosomal studies show that they are true males, but with 47 chromosomes (47,XXY). Studies of chromatin positive males have shown that 82% are 47,XXY, 11% are mosaics (46,XX/47,XXY), and 7% are 48,XXXY or 46,XX.

Sex Chromosome Anomalies in the Female

Gonadal dysgenesis is a group of disorders that affects phenotypic female individuals. The syndrome consists of primary amenorrhea, atrophic gonads, little or no pubic or axillary hair, infantile genitals, and short stature. The urinary gonadotropins are increased. The clinical features of the most severe form (Turner's syndrome) include short stature, amenorrhea, webbing of the neck, absent breast development, wide-spaced nipples with a "shield" chest, scant sexual hair, infantile genitals, and ovaries that are "streak gonads" containing no follicles.

Other anomalies associated with this syndrome are mainly congenital heart disease (coarctation of the aorta) and renal anomalies (horseshoe kidneys). Their karyotype lacks one X chromosome (45,X). The incidence varies between 1 per 2,500 and 1 per 8,000 female births.

Chromosomal studies on patients with gonadal dysgenesis have shown that it is a spectrum of karyotypes ranging from 45,X to the normal 46,XX. In between, there is a multitude of possibilities. Mosaics with different admixtures of cell lines such as 45,X/46,XX; 45,X/46,XY; 45,X/46,XX/47,XXX. Abnormal sex chromosomes may also be present. Physical abnormalities are dependent on the percentage of 45,X cell line or cell lines with abnormal chromosomes. The patient with a karyotype of 46,XX and gonadal dysgenesis is referred to as having pure gonadal dysgenesis and is expected to have fewer somatic stigmata than the patient with Turner's syndrome.

Cytogenetic Evaluation in Obstetrics and Gynecology

A chromosomal error can be ascertained through findings that are commonly or knowingly associated with a chromosomal syndrome. Table13–3 lists the incidence of chromosomal errors in certain conditions that are familiar to obstetricians. Indications for cytogenetic examination as a part of a medical evaluation are listed in Table 13–4. A sample of heparinized blood commonly serves for culture and harvesting of chromosomes. When mosaicism is suspected, other tissues such as skin, fascia, or gonad should be cultured and multiple karyotypes examined.

Prenatal diagnosis of genetic diseases has relied on amniotic fluid as the tissue of choice to study the fetus. Amniocentesis is the method by which a needle is inserted transabdominally into the amniotic cavity and 20–30 cc of fluid aspirated.

The optimal time for performing an amniocentesis is 15–16 weeks. Technically, it is easy to obtain amniotic fluid at that gestational age,

Table 13–3.

Condition	*Incidence of Chromosomal Errors (%)*
All pregnancies	6
Spontaneous abortions	25
Induced abortions	3–4
Liveborn	0.5
Stillborn	6
Congenital malformations	7–10
Growth-retarded babies	2
Habitual abortion	6
Primary amenorrhea	6–20
Azospermia	6–16

Table 13–4. Indications for Cytogenetic Evaluation

1. Amniocentesis for prenatal diagnosis of genetic diseases
2. Primary amenorrhea
3. Azospermia and severe oligospermia
4. Habitual abortion
5. Infertility of undertermined cause
6. Congenital malformations (newborns, twins)
7. Ambiguous genitalia, hermaphroditism
8. Parents of babies with chromosomal errors (Down's syndrome)
9. Unexplained stillbirths
10. Patients with germ cell tumors

and ample time would be left to get a final diagnosis and decide on the management. The incidence of problems from the procedure has been shown to be 2–3%. Ultrasound localization of the fetus and placenta should be obtained prior to amniocentesis. The procedure should then be performed under sterile technique.

Indications for amniotic fluid culture for prenatal diagnosis have been commonly chromosomal (Table 13–5). The increased risk of association of aging oocytes with nondisjunction makes the examination desirable for pregnant women above 40 years of age and advisable for women 35–39 years old. There is also an increased recurrence risk (1–2%) of trisomy 21 in women who had a previous affected infant. If either of the parents harbors a translocation, the chance of an unbalanced translocation in the progeny ranges from 2 to 100%, depending

Table 13–5. Indications for Amniocentesis in Prenatal Diagnosis

A. Chromosomal	
	1. Advanced maternal age (> 35 years)
	2. Previous trisomy
	3. Parent with chromosomal translocation
	4. Family history of trisomy
	5. Previous child with chromosomal error
	6. Miscellaneous
B. Sex-linked diseases	
C. Inborn errors of metabolism	
D. Neural tube defects	

on the sex of the parent, the type of translocation, and its behavior at meiosis (pairing, synapsis).

Sex-linked diseases are 150 in number, and the main method of diagnosis and treatment is limited to sex determination and elimination of the male progeny with the knowledge that one half of those male fetuses would be normal.

There are a large number of inborn errors of metabolism (3,000). They are rare diseases, and many of them are potentially diagnosable prenatally, but only 60–70 have been actually diagnosed (Table 13–6). Ascertainment should be carried through previously affected children or through heterozygote screening. The diagnosis should be accurately made, and the deficient enzyme well delineated in the affected individual and the heterozygotes. Accurate knowledge of the specific enzyme defect makes prenatal diagnosis of these diseases possible.

Diseases diagnosed through amniotic fluid assay include neural tube defects (NTD). The finding of elevated α-fetoprotein (AFP) in amniotic fluid during early gestation is used as an index for the detection of open NTD, such as anencephaly and spina bifida with meningomyelocele. Detection of AFP in maternal serum is another breakthrough that would help population screening. Subsequent amniocentesis would be offered those women with increased serum levels.

Advances in endoscopic procedures and in ultrasound visualization are opening new fields in the detection of anomalies. Visualization of the fetus is now possible, and fetal blood may be obtained and analyzed for diseases that cannot be assessed in amniotic fluid cells.

Table 13–6. Inborn Errors of Metabolism Diagnosed by Amniocentesis

Fabry's disease
Gaucher's disease
GM_1 Gangliosidosis (I, II)
GM_2 Gangliosidosis (I, II)
Krabbe's disease
Metachromatic leukodystrophy
Niemann-Pick disease type A
Mucopolysaccharidosis I (Hurler's)
Mucopolysaccharidosis II A (Hunter's)
Mucopolysaccharidosis III (San Filippo A)
Arginino succinic aciduria
Citrullinemia
Cystinuria
Maple syrup urine disease (infantile)
Methylmalonic aciduria
Galactosemia
Glycogen storage disease (type II, IV)
Adenosine deaminase deficiency
Congenital nephrosis
Cystinosis
Hypophosphatasia
1-cell disease
Lesch-Nyhan syndrome
Lysosomal acid phosphatase deficiency
Xeroderma pigmentosum

SUGGESTED READING

Hamerton JL: *Human Cytogenetics*, vol 1. New York, Academic Press, Inc, 1971.

Hook EB, Porter IH: *Population Cytogenetics, Studies in Humans*. New York, Academic Press, Inc, 1977.

Symposium on clinical genetics. *Clin Obstet Gynecol* 19:855, 1976.

QUESTIONS

1. The longest portion of the cell cycle
   A. is seen during interphase
   B. happens before prophase
   C. is called G1

D. all of the above
E. none of the above

2. Differences between meiosis and mitosis include:
   A. meiosis includes all divisions of oogonia and spermatogonia
   B. meiosis has a longer prophase than mitosis
   C. mitosis takes longer time than male meiosis
   D. all of the above
   E. none of the above
3. The centromere is the functional unit of the chromosome.
   A. it has two kinetochores
   B. its major function is repulsion of the chromatid
   C. acentric fragments always survive
   D. all of the above
   E. none of the above
4. Aneuploidy is:
   A. hypo and hyperdiploidy
   B. 2n + 2
   C. monosomy X, for example
   D. all of the above
   E. none of the above
5. Haploidy is found in
   A. gametes only
   B. some somatic cells
   C. 2n
   D. all of the above
   E. none of the above
6. Mosaicism is:
   A. two cell lines arising from more than one zygote
   B. two cell lines or more arising from one zygote
   C. cell lines arising in mosaic patches giving striae
   D. all of the above
   E. none of the above
7. Robertsonian translocation
   A. is acentric fusion
   B. involves acrocentric chromosomes
   C. the chromosome count is reduced
   D. all of the above
   E. none of the above
8. Klinefelter's syndrome includes:
   A. tall stature and hypergonadism
   B. short stature and hypogonadism

C. hypogonadism, hypergonadotropism, and tall stature
D. hypogonadism, hypogonadotropism, and tall stature
E. none of the above

9. Monosomy X is best represented by:
   A. tall stature and hypergonadism
   B. short stature and hypogonadism
   C. hypogonadism, hypergonadotropism, and tall stature
   D. hypogonadism, hypogonadotropism, and tall stature
   E. none of the above
10. Reciprocal translocation
    A. involves an exchange of genetic material between two chromosomes
    B. is a balanced translocation
    C. the chromosome count is unchanged
    D. all of the above
    E. none of the above

ANSWERS

1. D
2. B
3. A
4. D
5. A
6. B
7. D
8. C
9. B
10. D

chapter fourteen

fetoplacental unit and endocrine changes in pregnancy

OBJECTIVES

1. Define the normal maternal hormone levels during pregnancy, their source, and how they are produced.
2. Understand the biologic activities of the hormones of pregnancy.
3. Become familiar with the hormonal interactions of the fetoplacental unit.
4. Discuss some of the hormonal parameters that may be involved in the initiation of labor.
5. Appreciate the ontogenesis of the fetal endocrine system and its relation to normal fetal development.
6. Present some of the available data on hormones in amniotic fluid.

Following conception and implantation, a number of changes in maternal endocrine function occur that allow for maintenance of pregnancy, optimal nourishment of the fetus, delivery of the fetus at the appropriate time, and assurance of a maternal source of food for the neonate. In this chapter, the alterations in endocrine function and the cause and effect of each will be discussed.

STEROID HORMONES DURING PREGNANCY

During a normal menstrual cycle, the endometrium proliferates under the stimulatory effect of estrogens and is transformed to a secretory tissue by progesterone produced by the corpus luteum. The peak level of pituitary luteinizing hormone (LH) at ovulation stimulates formation of the corpus luteum, which, in a nonfertile cycle, has a life span of approximately 12 days. Maximum progesterone secretion is attained about seven days postovulation. In a fertile cycle, this is coincident with implantation of the blastocyst in the endometrium. After nidation, a signal is relayed to the ovary that allows for the extension of the life span of the corpus luteum. Progesterone secretion from the corpus luteum continues and increases somewhat during early pregnancy.

One factor that is known to be critical for the function of the corpus luteum of pregnancy is human chorionic gonadotropin (hCG), a glycoprotein hormone similar to pituitary LH in structure and biologic activity. This hormone is produced by trophoblastic cells and can be detected in maternal blood as early as seven days after fertilization. It is this hormone that stimulates the corpus luteum of pregnancy to produce steroid hormones.

Without adequate progesterone stimulation, the blastocyst will not implant in the endometrium or will be expelled after implantation. Progesterone suppresses uterine contractability, whereas estrogens stimulate uterine growth and blood flow. During the first few weeks of human pregnancy, these steroid hormones are predominantly produced by the corpus luteum. Removal of the corpus luteum prior to 32 days following fertilization consistently results in abortion. This can be avoided by progesterone supplementation. Luteectomy after 48 days following fertilization will not result in abortion. Thus, some abortions prior to this period may reflect inadequate progesterone secretion by the corpus luteum (corpus luteum insufficiency of pregnancy). This insufficiency may be due to an ovarian deficiency or inadequate ovarian stimulation by hCG. Causes of these inadequacies are not well understood but may be of maternal, trophoblast, or fetal origin. Those of maternal etiology may be the ones most easily managed in the future.

The change in dependency on the corpus luteum for maintenance of pregnancy is related to the continuing development of placental endocrine function and the shift of progesterone synthesis to the placenta. By seven to nine weeks' gestation (from the last menstrual period), sufficient placental steroidogenesis has been attained to maintain pregnancy without any ovarian contribution. At this same time, the fetus begins interacting with the placenta to produce estrogens. Recent studies have shown that these estrogens inhibit hCG stimulation of the corpus luteum of pregnancy, thus accounting for regression of its steroidogenic activity at this stage of pregnancy.

The placenta produces progesterone, utilizing maternal cholestrol (and, to a much lesser extent, fetal cholesterol), since it lacks the enzymes necessary for cholesterol production from acetate (Figure 14–1). Cholesterol is converted first to pregnenolone in the placenta by the activity of 20, 22 hydroxylase and a desmolase $(C_{20,22})$. Pregnenolone is transferred to the fetal and maternal circulation, and part is transformed to progesterone by placental 3-ol hydroxyl-$\Delta_{4,5}$

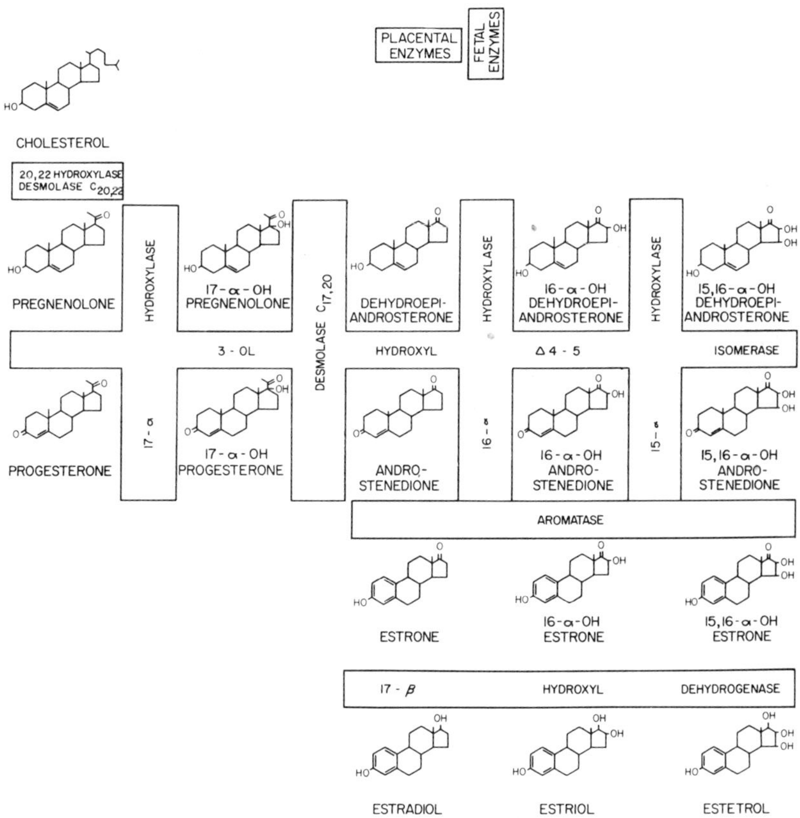

Figure 14–1. Enzymes of the steroidogenic pathways in the fetal-placental unit. (See text for discussion.)

isomerase. Thus, progesterone synthesis is dependent on maternal cholesterol and the placental enzymes named above.

Circulating levels of progesterone in maternal blood are well defined and increase up to 175 ng/ml in late pregnancy. In some studies, a slight dip at seven to nine weeks' gestation has been observed, most likely due to the decreasing ovarian contribution of progesterone while placenta production is still ascending. It is at this stage and thereafter that if placenta steroidogenic function is not adequate, abortion may result.

Estrogens are also required for the continuing progress of pregnancy. In early pregnancy, the corpus luteum can produce some estrone (E_1) and estradiol (E_2). However, the major site of estrogen production during pregnancy is from the fetoplacental unit, beginning at about six weeks fetal age. The placenta alone cannot synthesize any estrogen from cholesterol because it lacks some of the required enzymes. To supplement the placenta, the fetal liver and fetal zone of the adrenal (not the cortex) provide the missing enzymes. However, the fetus does not contain in these two tissues all of the enzymes necessary for estrogen synthesis but complements those missing in the placenta. Thus, the fetus and placenta form a critical unit for the production of estrogens in pregnancy.

After the placenta has formed pregnenolone or progesterone from maternal cholesterol, they are passed to the fetus (primarily pregnenolone; progesterone is predominantly sent to the mother). Immediately, these C-21 steroids are sulfated by the fetal liver and adrenal. This is an important step since conjugated steroids are not biologically active, and thus the fetus can distinguish between the steroids of the fetoplacental unit (nonbiologically active in the fetus) and those of the fetus's own adrenal cortex and/or gonad, which are biologically active and play an important role in the development and differentiation of the fetus's endocrine axis and sex differientiation.

The sulfated pregnenolone or progesterone is 17α-hydroxylated by the fetal liver and adrenal (Fig. 14–1). The ability to metabolize sulfated steroids exists with greatest efficiency in fetal tissues and is gradually lost in the neonatal and prepubertal period. The 17α-OH-C_{21} sulfated-steroids are converted to C_{19} steroids by the fetal adrenal zone, a specialized fetal tissue rich in the $C_{17,20}$ desmolase for sulfated steroids. Thus, in the fetal adrenal zone, 17α-OH-pregnenolone-SO_4 is converted to dihydroepiandrosterone-sulfate (DHEA-S) and the 17α-OH-progesterone-SO_4 to androstenedione. These steroids may be further hydroxylated at the 16 and/or 15 position in the fetal liver to form 16α-OH, or 15α, 16α-OH form of these steroids.

The vast majority of these steroids do not undergo further metabolism in the fetus, but on recirculation in the placenta, the sulfate is cleaved by the very active placenta sulfatase to produce the unconjugated steroids. The placenta can then convert the unconjugated dehydroepiandrosterone or the hydroxylated forms of dehydroepiandrosterone to androstenedione and its hydroxylated forms by activity of the 3-ol-hydroxyl$\Delta_{4\text{-}5}$ isomerase enzyme, similar to that described for the conversion of pregnenolone to progesterone. This enzyme is effective only after the sulfate has been cleaved and is present in the placenta with high activity. The resulting androstenedione and hydroxylated androstenediones can be further metabolized by a very active placenta aromatase enzyme yielding estrone, 16α-OH or 15α, 16α-OH estrone. Further hydroxylation of these steroids by the 17β-OH hydroxylase will produce estradiol, estriol (16α-OH estradiol = E_3) or estetrol (15α, 16α-OH estradiol = E_4).

Maternal tissues have little ability to 15α or 16α-hydroxylate steroids as compared to the fetal liver. Thus, without a functioning fetal liver, the precursors for estriol and estetrol formation are greatly reduced. The estrogens produced by the placenta, like progesterone, are transferred predominantly to the mother. For this reason, these two estrogens, estriol or estetrol, in maternal blood or urine are considered indices of fetal well-being and are used in conjunction with other clinical parameters to monitor complicated pregnancies.

The functions of these high circulating maternal steroid levels during pregnancy are multiple. As mentioned previously, progesterone is critical for suppressing uterine contractions and thus acts to maintain pregnancy. It also acts to further develop breast tissue, while it inhibits the enzyme required for lactose synthesis. In this manner, it prepares the breast for postpartum lactation, while it suppresses lactation during pregnancy even in the presence of greatly elevated prolactin (PRL) levels. Progesterone also plays a role in inducing insulin production and the peripheral insulin resistance observed in the last two trimesters of pregnancy (discussed in more detail later in this chapter). An action of progesterone on the kidney to alter water retention is discussed in detail in Chapter 16.

Estrogens play a primary role in inducing uterine growth. It is of significance that estrogen production in pregnancy is related to fetal function, thus, as the fetus develops, estrogen production increases, and so does uterine growth. Another biologic activity of estrogen is to induce uterine contractions. The most potent of the three major estrogens in pregnancy is estradiol. The dominant estrogen, estriol, has only $\frac{1}{100}$ of the activity of estradiol. The activity of progesterone counteracts the uterine contraction activity of estrogens through most of pregnancy. Estrogens are also involved in inducing breast development during pregnancy by a direct mammary action and in stimulating maternal PRL production, while concomitantly inhibiting (at these high levels) lactation. Thus, breast preparation for lactation postpartum is initiated during pregnancy partly due to the action of estrogens. Other actions of estrogens are on the pancreas (discussed in more detail later in this chapter) and on the kidney.

PROTEIN HORMONES DURING PREGNANCY

The protein hormones of pregnancy are produced by maternal endocrine glands and the placenta. In this section, the levels and functions of each hormone during pregnancy will be discussed.

The maternal pituitary has both increased and decreased functions during pregnancy. A weight gain from 500 mg to approximately 800 mg at term reflects predominantly the increase in lactotrophs and PRL production in pregnancy. Circulating levels of PRL increase from 20 to 200 ng/ml during gestation. This increase is most likely estrogen stimulated, and the high PRL acts to stimulate breast development for postpartum lactation. Other possible actions of PRL on glucose homeostasis or water metabolism are still under investigation.

A decreased pituitary function observed during pregnancy is that of the gonadotrophs. LH and follicle stimulating hormone (FSH) productions are very low, and stimulation with luteinizing hormone releasing factor (LRF) is ineffective. The high levels of circulating estrogens and progesterone are the most likely cause for this pituitary inhibition. Pituitary growth hormone (GH) is not elevated during pregnancy, and its release to stimuli progressively declines. It is noteworthy that for both LH and GH, the placenta produces two protein hormones that have similar biologic activity and thus take over their needed activities during pregnancy.

A slight increase in thyroid stimulating hormone (TSH) is observed in pregnancy. However, this is most probably due to the detection of a cross-reacting protein, human chorionic thyrotropin (hCT) produced by the placenta. Thyroid function in pregnancy is increased. However, the total increase is much greater than that of only the free active hormone because of the concomitant increase in thyroid binding globulin induced by the high circulating estrogens during pregnancy. Thus, it is important when assessing thyroid function in pregnancy to compare the results to that for normal pregnancy and not to interpret the findings in relation to the nonpregnant adult.

Cortisol is also elevated in pregnancy, but whether this is due to increased ACTH production by the maternal pituitary and/or stimulation by a placental ACTH-like molecule is not known. Once again, as for thyroid hormone, the total plasma cortisol is more increased than the free plasma cortisol because of the elevated corticosteroid binding globulin (CBG) induced by high circulating steroids during pregnancy.

During pregnancy, and clearly apparent in the second and third trimester of pregnancy, maternal insulin production increases; however, glucose tolerance decreases. This is primarily related to the peripheral antagonist effects of a placental protein hormone, human chorionic somatomammotropin (hCS), which is similar to GH and PRL. Also, these effects are induced by the high steroid levels during pregnancy acting on the pancreas and peripheral tissue to increase insulin and decrease peripheral response to it.

The renin-angiotensin-aldosterone system in the pregnant state is stimulated; however, the relative degree of responsiveness is similar to that observed in the nonpregnant state. (Chapter 16 contains an extensive discussion on this subject.)

As already alluded to, the placenta is not only very active in steroidogenesis but also produces a number of protein hormones that are critical for normal pregnancy. Human chorionic gonadotropin (hCG) is a glycoprotein, consisting of two subunits, α and β. The α subunit is identical to that of LH, FSH, and TSH, and the β subunit is very similar to, although immunologically distinguishable from, the β subunit of LH. The biologic activities of LH and hCG are essentially the same. Thus, separate measurement of these two hormones must be on a physiochemical or immunologic basis.

Human chorionic gonadotropin is produced by the synciotrophoblast and is secreted predominantly into the maternal circulation. Very high levels of this potent gonadotropin circulate in the mother throughout pregnancy. Early in gestation, hCG is critical in stimulating ovarian steroidogenesis. Coincident with the ovarian-placental shift, peak levels of hCG in the maternal blood are attained. Thereafter, hCG declines rapidly with a smaller second increase toward term. This hormone is thought to control placental steroidogenesis; however, the mechanism of stimulation and feedback regulation is not understood. Recently, some factors, LRF and cyclic-AMP, have been shown to stimulate hCG production in vitro. It is of possible significance that at least one of the mechanisms of LRF action on the pituitary gonadotrophs is via cyclic-AMP stimulation. In addition, huge quantities of

LRF have been found in the human placenta. Thus, a possible role for this hormone in the control of hCG production in vivo seems a possibility.

A second protein hormone of the placenta is hCS, also known as placental lactogen (hPL). Like hCG, hCS is produced by the syntiocytotrophoblast. By radioimmunoassay, it is first detectable in maternal blood about three weeks following implantation. It is predominantly secreted into the maternal circulation and increases with increasing gestational age. This hormone is biochemically and biologically similar to GH and PRL, although it is not immunologically or biochemically identical to either of them. In pregnancy, its most important role appears to be its effect on glucose homeostasis. It is primarily through this hormone that the normal glucose utilization is reduced in the mother by its contrainsulin effects on peripheral tissue, thus providing more glucose for the fetus. HCS also enhances lipolysis, elevates free fatty acids, and thus provides the mother with fats for metabolic utilization, while it conserves nonlipid nutrients, like glucose and gluconeogenic precursors, for fetal growth. HCS also plays an important role in breast development during pregnancy. It shows biologic activities similar to PRL in this aspect.

Another glycoprotein hormone of the placenta that has been incompletely characterized is hCT. It is thought to stimulate maternal thyroid function during pregnancy. Immunologically, it is indistinguishable from bovine TSH and cross-reacts partially with human TSH. Further purification and characterization is necessary for more complete definition.

Another peptide hormone, thyrotropin releasing factor (TRF), has also been found in the human placenta. However, so far, no reports as to an activity in vivo or in vitro have appeared. In adults, it affects the release of TSH and PRL. Other peptide hormones of the placenta have been indicated, that is, ACTH, MSH; however, only very preliminary data exist, and many more studies are needed.

PARTURITION AND POSTPARTUM LACTATION

A number of hormonal factors have been indicated to play a role in parturition. They are related to placental and fetal functions. In many species, including the human, a decline in progesterone has been observed just prior to the onset of labor. In addition, an increase in cortisol and decrease in estriol (the least potent in inducing uterine contractions of three dominant estrogens) occurs. Also, fetal oxytocin has been reported to increase during labor. Of the former hormonal changes, progesterone is a placental factor. The hypothesis that a predetermined gestational life span is coded in the placenta is consistent with the observation that in a number of species, in the events of fetal death or removal of the fetus, the placenta will deliver itself at term. The other hormonal changes occurring at parturition are related to fetal function or fetoplacental function and suggest that labor ensues when fetal signals indicate. Data from anencephalic fetuses or fetuses with pituitary ablation support this concept. In these cases, where fetal endocrine function is abnormal, pregnancy is prolonged. Thus, it appears that both placental and fetal factors may indicate the appropriate time for parturition. Which factor or factors are of primary significance or how they interrelate remains unknown.

Following delivery of the fetus and placenta, maternal hormonal levels undergo very rapid changes. The high levels of estrogens and progesterone decline during the first few days postpartum; hCG and hCS begin falling immediately after delivery of the placenta, with hCS reaching undetectable levels within a few days postpartum; hCG takes at least one week due to its longer half-life. The elevated PRL levels also decline from 200 to 20 ng/ml in the first few days postpartum in the nonlactating mother. If lactation is initiated, PRL will only decline to 60 ng/ml. During suckling, PRL levels are greatly increased, and the degree and duration of elevated basal PRL may well be related to the frequency of suckling. By three months' postpartum, basal PRL is about 30 ng/ml but is still greatly stimulated by suckling. By six months postpartum, when the frequency of suckling is usually decreased, basal levels of PRL return to nonpregnant levels. In the lactating mother, during the earlier postpartum period when PRL is chronically elevated, fertility is inhibited. During this period, exogenous gonadotropins do not induce normal ovarian function. However, as PRL levels decline, reproductive ovarian function returns, that is, menses accompanied by ovulation and adequate luteinization occur. Therefore, in later phases of postpartum lactation, when PRL is not chronically elevated, fertility may exist.

FETAL ENDOCRINE FUNCTION

Fetal endocrine function is established early in gestation. Pituitary morphologic differentiation is well defined by 8 weeks fetal age. Endocrine competent cells in the anlage of the anterior pituitary gland, Rathke's pouch, are present as early as 5 weeks. Pituitary hormones in fetal serum have been detected at 9 weeks, the earliest gestational age studied. Recently, hypothalamic LRF, TRF, and somatostatin have been measured from 10 weeks fetal age (earliest age studied). The pattern or hormonal content in each tissue may vary with the sex and gestational age of the fetus, indicating development of the endocrine feedback systems very early in fetal life. The gonads produce biologically active steroids (not sulfated) during gestation. The testes have a peak in steroidogenesis at 11–17 weeks that may be stimulated by a small amount of hCG transferred to the fetus as well as by fetal pituitary LH. Even as the hCG in the fetal circulation declines (at 10 weeks and thereafter), fetal LH increases. Normal gonadal function in the second and third trimester is dependent on fetal pituitary gonadotropins. Ovarian estradiol production in early gestation is not as apparent, however. By the second trimester, the fetal ovary is a functionally developed tissue.

Establishment of the hypothalamic-pituitary-gonadal axis system in the fetus is indicated in the second trimester. This is totally independent of maternal or placental hormones since protein hormones of the mother and placenta are not transferred to the fetus in significant quantities and placental steroids are sulfated and, thus, biologically inert. Actually, this compartmental system provides a safeguard for the normal programming and differentiation of male and female offspring.

Other fetal endocrine tissues develop, and their function is critical in utero, such as the thyroid, which is stimulated and dependent on fetal pituitary TSH. Fetal thyroid aplasia may result in cretinism. Even though thyroid hormones pass from the mother to the fetus, they may not be adequate to compensate for fetal thyroid failure. The converse may also apply. Maternal hypothyroidism may deplete the fetus of thyroid hormone, which can pass from the fetus to the mother, and result in retarded fetal development. Similarly, fetal adrenal and pancreatic functions and feedback systems are established in utero.

Fetal endocrine systems not only develop and function in utero, but also their normal functions are critical for the direction of normal fetal development. The fetal endocrine system is autonomous and is influenced by maternal or placental hormones or nutrients only to the extent that biologically active substances may be passed to the fetus. In the case of gonadal hormones, this is normally insignificant. However, as for thyroid hormones or glucose that do cross the placenta in active form, the fetal endocrine system may be affected by the mother.

HORMONES IN AMNIOTIC FLUID

A number of hormones, both protein and steroid, are present in amniotic fluid. The sources of these hormones are multiple—placental, fetal, and possibly from embryonic membranes. The physiologic significance of these hormones for normal pregnancy is a current question. The very high levels (up to 10 μg/ml) of PRL, which may be produced predominantly by the chorion, and some by the fetus might play a role in the regulation of amniotic fluid osmolarity and volume. Other protein hormones, glucagon and GH, are present in amniotic fluid. Their most likely source is fetal. Both hCG and hCS can reach the amniotic fluid through the chorion and amnion.

Similarly, steroid hormones, both in free and conjugated forms, are present in the amniotic fluid. Clinical significance of these various hormones is being investigated at this time. Since in many cases the steroid hormones are predominantly of fetal origin, they may be used as an index of fetal well-being. Unconjugated testosterone in amniotic fluid is higher in male than in female fetuses. Future studies of hormones in amniotic fluid promise to provide a more complete understanding of the physiology of pregnancy along with new and possibly better parameters for determining fetal well-being.

An understanding of the endocrinology of pregnancy requires the study of a number of endocrine compartments—maternal, placental, fetal, and amniotic fluids. Further complicating the study is an understanding of the interactions of these compartments and how they influence each other. In general, the endocrine functions of pregnancy provide for the regulation of a system capable of maintaining, nourishing, and producing normal offspring.

SUGGESTED READING

Chard T, Silman RE, Rees LH: The fetal hypothalamus and pituitary in the initiation of labour, in Landon JL (ed): *The Fetus and Birth* (Ciba Foundation Symposium #47). New York, Elsevier, 1976, pp 359–378.

Dawood MY: Hormones in amniotic fluid. *Am J Obstet Gynecol* 128:576, 1977.

Josimovich JB: Placental protein hormones in pregnancy. *Clin Obstet Gynecol* 16:46, 1973.

Ryan KJ: Endocrine control of gestational length—a time to be born. *Am J Obstet Gynecol* 109:299, 1971.

Ryan KJ, Hopper BR: Placental biosynthesis and metabolism of steroid hormones in primates. *Contrib Primatol* 3:258, 1974.

Yen SSC: Endocrine regulation of metabolic homeostasis during pregnancy. *Clin Obstet Gynecol* 16:130, 1973.

QUESTIONS

1. Is postpartum lactation always associated with infertility? If not, when may it not be?
2. When do peak maternal plasma hCG levels occur?
3. What are the steroids of the fetoplacental unit and which are dependent on placental and/or fetal enzymes?
4. Which hormone or hormones are useful as an index of fetal well-being and why?
5. What are the biologic functions of hCS in pregnancy?
6. Is maternal thyroid function altered in pregnancy? If so, how?
7. What is the function of the corpus luteum of pregnancy?
8. How is the fetal endocrine system protected from the activity of the high amount of placental steroids sent to the fetus?
9. What hormonal changes occur at or during parturition, and how might they affect the normal progress of labor?
10. How is full lactation inhibited during pregnancy while breast development is induced?

ANSWERS

1. No. When basal PRL levels drop to near normal, pituitary gonadotrophins rise and can stimulate ovulation
2. Coincident with the shift in steroidogenesis from the corpus luteum to the placenta
3. Progesterone and estrogens
4. Estriol and estetrol because maternal tissues generally lack enzymes to 15 and 16 hydroxylate estrogens
5. Preparation of the breasts for lactation and provide glucose for the fetus by antagonizing the action of maternal insulin. It also provides fatty acids for use as fuel by the mother

6. Yes. Increased thyroid binding globulin and some production of placental TSH
7. Maintenance of early pregnancy through production of estrogen and progesterone
8. Placental steroids are rapidly sulfated and rendered inactive in the fetus
9. Slight decrease in progesterone and estriol; slight rise in cortisol—fetal oxytocin increases
10. By a local blocking action of estrogen on breast tissue

chapter fifteen

the placenta

OBJECTIVES

1. Understand the anatomy of the placenta and the circulation on the maternal and fetal sides.
2. Describe mechanisms of transfer from mother to fetus and vice versa.
3. Know representative pathologic lesions that affect the well-being of the fetus.
4. Recognize that the fetus and placenta form one unit, but lesions that affect the placenta may or may not be reflected in the fetus.

The human placenta is derived from trophoblast of the embryonic pole of the blastocyst. It is an organ that links the mother to her fetus. It functions as a lung by providing oxygen in exchange for carbon dioxide; as a kidney, by taking care of wastes; and as an endocrine gland, by synthesizing peptide and steroid hormones for the maintenance of pregnancy. It is a unique organ possessing all these functions and regulating the transfer of nutrients and drugs through multiple transfer systems.

MORPHOLOGY OF THE PLACENTA

The term placenta is a discoid structure that weighs 450 gm and measures 18 cm in diameter and 1.5–2 cm in thickness. The fetal surface has a bluish metallic sheen with prominent fetal vessels dispersed throughout the area and converging toward the insertion of the umbilical cord. The fetal surface is covered with the fetal membranes amnion and chorion, which are inserted at its periphery. Both membranes form a sac that lines the uterine cavity. The fetal surface is a part of the amniotic sac that contains amniotic fluid and the fetus attached by its umbilical cord to the placenta. The liquor amnii varies in amount during pregnancy and reaches 700–1,000 ml at term. The amnion is an avascular membrane that lines the amniotic sac and the fetal aspect of the placenta. The chorion is a vascular membrane that is apposed to the outer aspect of the amnion and attaches to the decidua.

The maternal surface of the placenta is meaty and red brown in color. It is divided into lobes composed of cotyledons. Maternal spiral arterioles open at the base of the cotyledons. The number of these arterioles varies between 25 and 100 or even more. Each spiral arteriole supplies one or a group of fetal cotyledons. A fetal cotyledon resembles a tree, its trunk being the primary stem villus with large fetal vessels and its branches the ramifications of the villi. A tertiary villus is a terminal villus that is surrounded by a double layer of trophoblastic cells. The outer syncytiotrophoblast (large multinucleated cells) is the site of protein and steroid hormone production. The inner cytotrophoblast (unicellular) is highly mitotic and thought to produce the syncytiotrophoblast layer. Villous stroma contains mesoderm, macrophages (Hofbauer cells), and fetal vessels (Fig. 15–1).

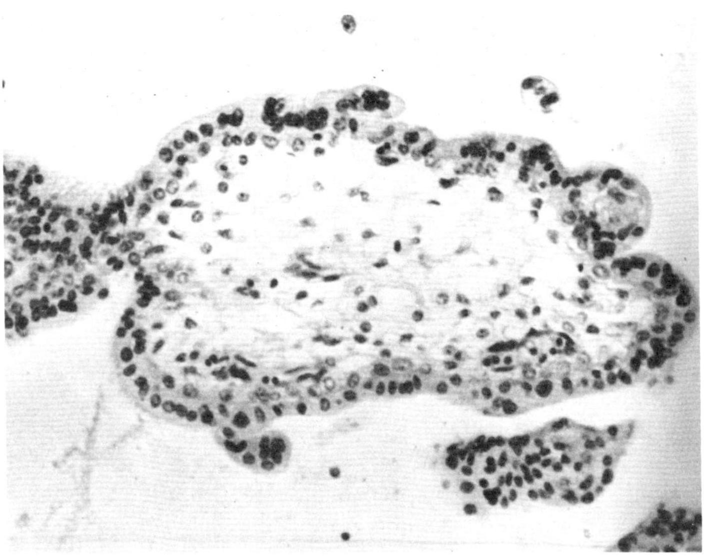

Figure 15–1. An early placental villus, displaying its two cell layers of trophoblast. (H & E, magnification × 400)

The villi are immersed in maternal blood coming from the spiral arterioles into the intervillous space.

The umbilical cord varies in length from 50–75 cm and is commonly inserted on the fetal surface of the placenta in eccentric fashion, though central and marginal insertions are also seen in normal conditions. The cord contains two umbilical arteries, which arise from the hypogastric arteries of the fetus, and an umbilical vein that enters the liver through the ductus venosus.

Fetal vessels run in a spiral fashion in the cord and are embedded in a soft connective tissue referred to as Wharton's jelly. Vestigial structures that can be seen occassionally on histologic sections are the allantois and the omphalomesentric duct and vessels. The yolk sac may be seen as a small 0.2–0.5 cm flat whitish calcified nodule between the amnion and chorion, at the periphery of the placenta or in the wall of the amniotic sac.

PHYSIOLOGY OF THE PLACENTA

The average volume of a placenta is 500 ml. The intervillous space capacity is 150 ml. Maternal blood feeds the intervillous space at a pressure of 70–80 mm Hg. Intervillous space pressure is estimated to be 12 mm Hg. Maternal vein pressure at the exit from the placenta is 8 mm Hg. The variation in pressure is well delineated in the fetal lamb where there is a stepwise decrease in pressure from the aorta to the umbilical veins and to the inferior vena cava.

The total area of chorionic villi exposed to maternal blood is 12.6 m^2 at term. The diffusion distance for an oxygen molecule, from a maternal erythrocyte in the intervillous space to a fetal erythrocyte in fetal capillary, is about 11 μ. The placental barrier or diffusion membrane has a mean thickness of 3.5 μ.

One or more fetal cotyledons may form a unit, surrounded by placental septae and supplied at the base by one or more spiral arterioles. Maternal blood is ejected from the arterioles and showers the villi. The opening pressure of the arterioles is high enough to push blood toward the chorionic plate. Maternal blood drains through openings in the basal plate that lead to maternal veins. The intervillous space is fed by the spiral arterioles at the rate of 600 ml/min at term.

The fetal and maternal circulations are separated by one to six cell layers, depending on the species concerned. Placentas may be classified as:

- Hemochorial: maternal blood in contact with trophoblast (man, monkey)
- Endotheliochorial: maternal endothelium in contact with trophoblast (cat, dog)
- Hemoendothelial: maternal blood in contact with placental blood vessels (rabbit, rat)
- Epitheliochorial: maternal epithelium in contact with trophoblast (horse, cattle)

The human placenta is a hemochorial placenta. This classification is thought to relate to the rate of transfer of substances from mother to fetus.

Multiple transfer systems are at play in the placenta:

Simple Diffusion: The rate of transfer is proportional to the concentration of the substance across the membrane and does not involve energy consumption. Oxygen and carbon dioxide are two essential gasses that are transferred by this mechanism. All drugs with a molecular weight of less than 1,000 are transferred by simple diffusion. Other factors that affect diffusion are lipid solubility, degree of ionization, and plasma binding.

Facilitated Diffusion: Involves a carrier substance that will increase the transfer rate above the predicted by simple diffusion. There is no requirement for energy utilization. The direction of the transfer is determined by the concentration gradient. Glucose and similar substances are transferred through this mechanism.

Active Transport: Requires the expenditures of energy for transfer against a concentration gradient. Vitamins, amino acids, sodium, and potassium utilize this mechanism.

Pinocytosis: Is a mechanism where a substance is engulfed by the cell membrane and transfer by vacuole formation occurs. Globulins are transferred by this method.

Breaks: Disruption of cells would produce leaks in the villous tissue, where erythrocytes are transferred to the maternal circulation. This is one way that Rhesus isoimmunization is induced.

PATHOLOGY OF THE PLACENTA

Several lesions of academic and clinical importance have been described extensively. We shall stress lesions that affect the placenta and its appendages. The lesions can be classified as infectious, hemorrhagic, vascular, and tumorous.

Abnormal forms of many kinds are seen in human placentas: horseshoe, bilobed, trilobed, and circummarginate. The circumvallate placenta (Fig. 15–2) is most interesting due to numerous theories pertaining to its formation. The fetal surface of this placenta appears narrowed by a grayish thick plica formed by the reflected membranes. The plica of the circumvallation contains dead villi and decidual tissue embedded in fibrin. This abnormal placentation has been associated with repeated episodes of antepartum bleeding and has also been described as a common cause of second trimester abortion. It is associated with an increased incidence of malformations in the newborn.

Large placentas are usually associated with large babies. The placental/fetal ratio is around 1:6 or 1:7. If this ratio increases to 1:3 or more, the presence of a hydropic fetus and placenta is suspected.

Hydrops fetalis and placentalis has been most often associated with

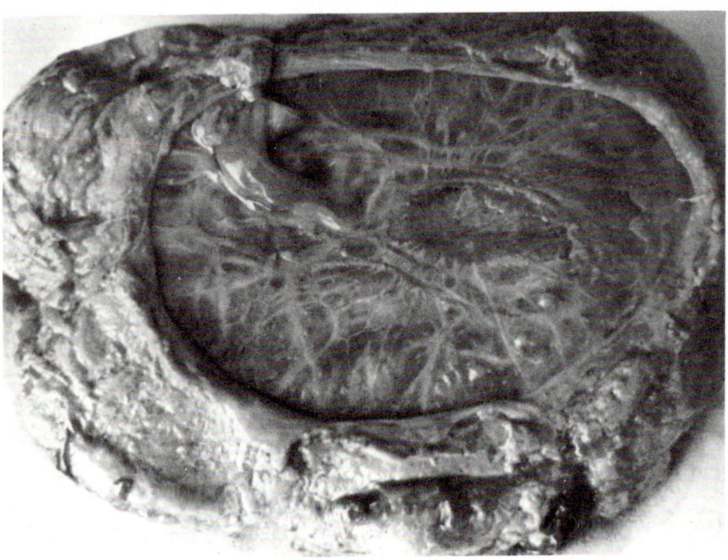

Figure 15–2. Circumvallate placenta with a fresh fibrinous ring that was associated with a premature stillborn and severe antepartum hemorrhage.

Rhesus isoimmunization. The placenta is unusually large and pale. The villi are friable, and the umbilical cord and membranes are edematous and stained a yellowish color by bilirubin pigments. Histologically, enlarged edematous villi with occasional foci of extramedullary hematopoiesis are typical of this disease. Large placentas associated with diabetes do not have any characteristic histologic features. Small placentas are associated with vascular disease, prematurity, or intrauterine growth retardation.

Vascular lesions are common findings in normal placentas and usually consist of infarcts or intervillous thrombosis. The pathology of the placenta is quantitative rather than qualitative. For example, the normal senescence of the term placenta produces some vascular lesions, an increased amount of fibrin deposits, and calcifications that are not harmful to the fetus. These lesions that exist in the normal term placenta may, under other circumstances, increase in number to an extent that they will eventually affect the well-being of the fetus.

Placental infarcts are seen in conjunction with maternal toxemia or vascular and renal disease. They arise secondarily to narrowing and thrombosis of the spiral arterioles, which leads to death of a segment of villous tissue (Fig. 15–3). Large areas, amounting to 50% and more of the placenta, can die and affect the well-being of the baby. Intervillous thrombosis and fetal vessel thrombosis are common features of the placenta of the stillborn infant.

Bleeding from the placenta occurs from three main causes: placenta previa, abruptio placentae, and placenta accreta.

Abnormalities of implantation lead to low insertion of the placenta over the lower uterine segment. The placenta may completely or partially cover the internal os of the cervix (central or partial placenta previa). Dilatation of the cervix induces separation of placental tissue leading to hemorrhage.

Abruptio placentae is a condition where the placenta separates from the decidual surface before delivery of the infant. Large amounts of blood may collect behind the placenta (concealed hemorrhage) or may dissect between the decidua and the membranes and be recognized as overt vaginal bleeding. Consumption coagulopathy forms a part of the clinical picture. Abruption is associated with vascular disease, toxemia, trauma, and decompression of the uterus. When the uterine muscle becomes infiltrated by trapped blood, a bluish mottled appearance of the uterus (Couvelaire uterus) results. There may be associated uterine hypocontractility. Both placenta previa and abruptio placentae are associated with an increased maternal and perinatal morbidity and mortality (Fig. 15–4).

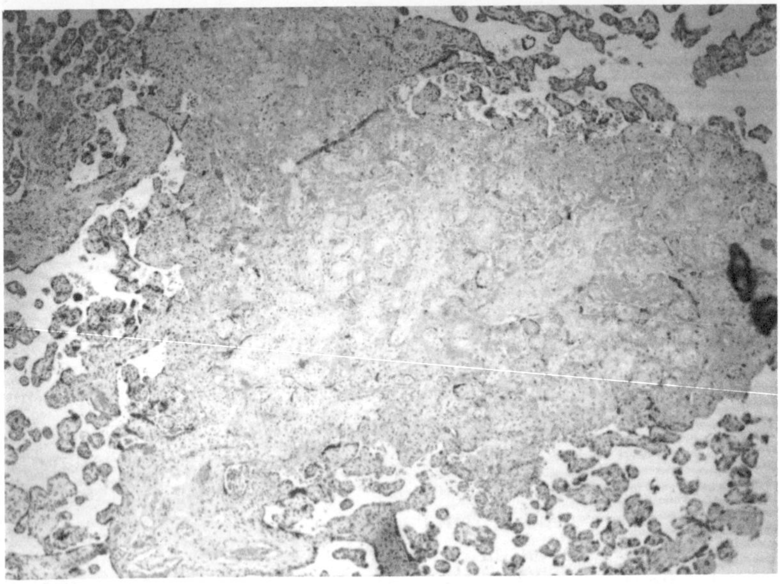

Figure 15–3. Microscopic section of a placental infarct shows dead villi embedded in fibrin. The infarct is surrounded by normal term villi. (H & E, magnification × 25)

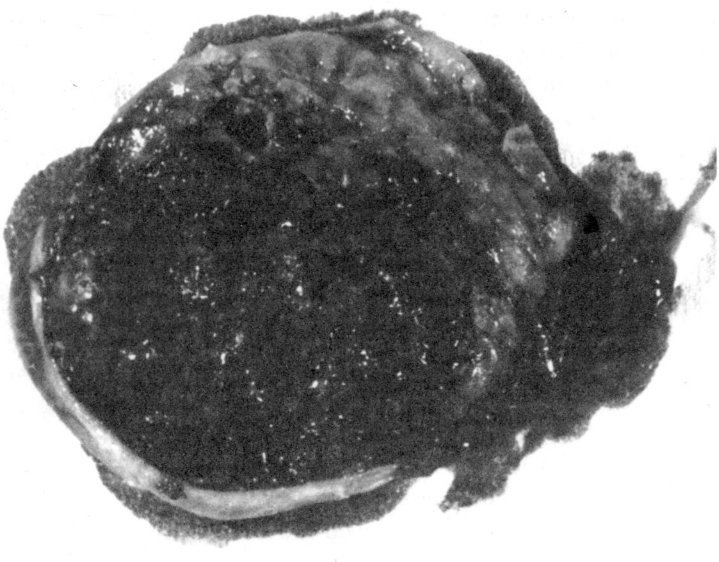

Figure 15–4. Abruptio placentae that was associated with an intrauterine fetal death. Note that 80–90% of the surface area is covered by the blood clot that impinges markedly on the maternal surface.

Placenta accreta is a disorder of placental adherence in which the anchoring villi invade the uterine muscle (increta) and may reach the uterine serosa (percreta). This defect of attachment may be due to a lack of decidual tissue. The condition is recognized after the delivery of the baby, when the placenta fails to separate. No plane of cleavage can be found between the placenta and the uterine wall. Incomplete separation leads to vaginal bleeding that may necessitate hysterectomy.

Infections of the placenta may be bacterial, viral, fungal, or protozoal. When an infection ascends from the cervix or vagina, it usually affects the amniotic fluid, membranes, fetus, umbilical cord, and the chorionic plate of the placenta. The villi are usually intact. If the infection is severe, the placental villi will be infiltrated, and abscesses may form. The usual picture of ascending bacterial or fungal infections would then be chorioamnionitis with infiltration of the amnion and chorion by leukocytes (Fig. 15–5). Maternal transplacental infection primarily affects placental villi and the maternal-fetal interface of the placenta (tuberculosis, viruses). Generally, viruses produce nonspecific focal lesions in the villi referred to as chronic villitis, which destroys small groups of villi in different areas of the placenta (Fig. 15–6).

Macroscopically, severely infected placentas are yellow-green in color and foul smelling. Purulent discharge may cover the membranes

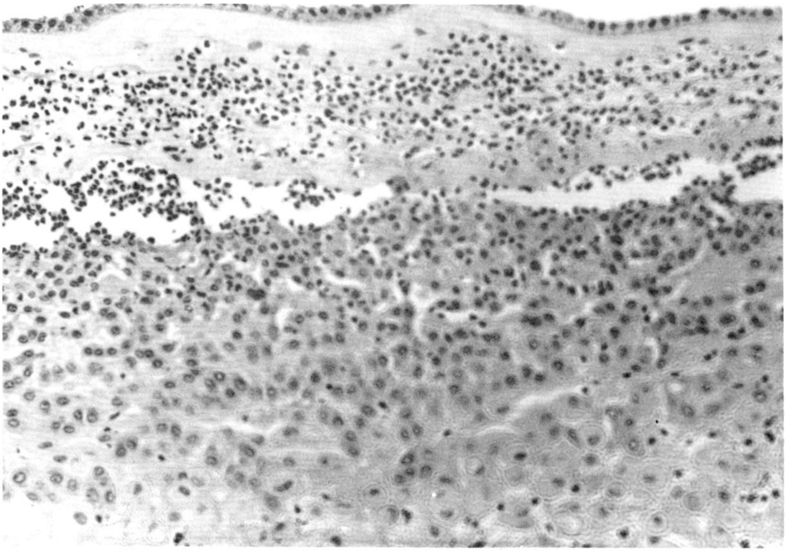

Figure 15–5. Diffuse leukocytic infiltrates in the amnion and chorion. (H & E, magnification × 100)

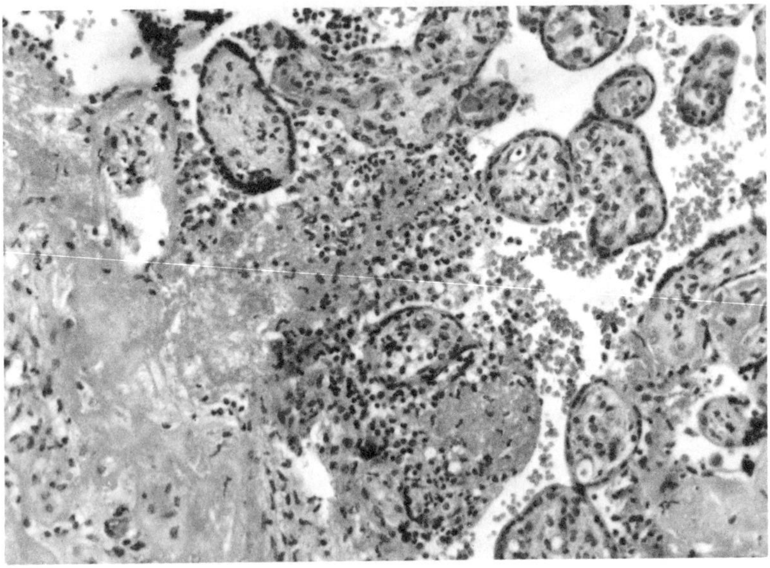

Figure 15–6. Chorionic focal villitis. Note the round cell infiltrate inducing a focal destruction of a group of villi. (H & E, magnification × 400)

and the fetal surface. The placental substance may contain microabscesses, depending on the nature of the invading agents. These changes are found in association with bacterial infections. Viral infections do not induce such macroscopic change.

The membranes can be affected by infections and meconium, hemosiderin, or bilirubin pigments. One lesion should be noted because of its association with congenital malformations—amnion nodosum (Fig. 15–7). These are tiny nodules that are made up of compressed squamous cells, vernix, and lanugo hair. They are found over the fetal surface and the amniotic membrane in association with oligohydramnios (reduction in amniotic fluid volume) and renal anomalies in the newborn (mainly renal aplasia).

Hydramnios or polyhydramnios is a condition in which the amniotic fluid volume is greater than 2,000 ml. There is an increased association with defects of swallowing that may be of central origin (anencephaly) or of local origin in the upper gastrointestinal tract (esophageal atresia). It may also accompany multiple pregnancy, diabetes mellitus, or erythroblastosis fetalis. Oligohydramnios is related to anomalies involving the genitourinary system (renal aplasia, urethral stenosis), mainly obstructive anomalies.

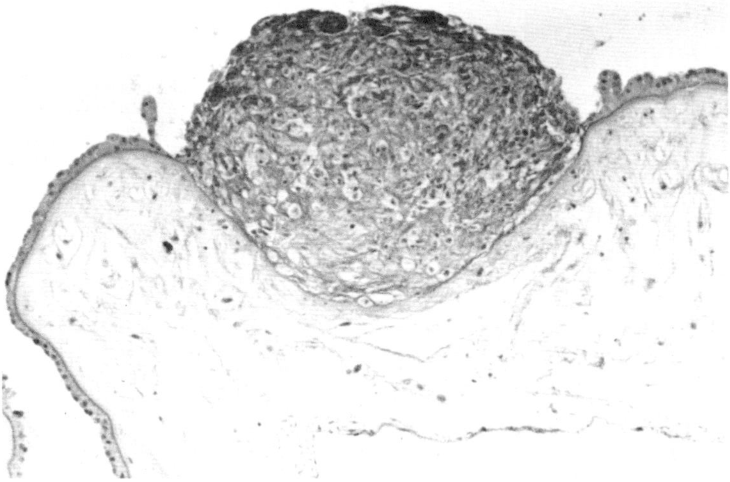

Figure 15–7. Amnion nodosum. (H & E, magnification × 400)

Velamentous insertion of the umbilical cord occurs when the cord vessels insert via the fetal membranes instead of into placental disc. The fetal vessels run through the membranes, devoid of their Wharton's jelly covering, into the placental disc. The lack of protection of the fetal vessels makes them prone to injury, leading to fetal death. When the vessels pass over the cervical os, the condition is known as vasa previa. Rupture of these vessels during amniotomy leads to fetal death by exsanguination.

Single umbilical artery (SUA) is an entity that occurs in 1% of deliveries. The incidence is increased in abortions and in multiple pregnancies. Approximately 40% of babies with SUA have an associated congenital malformation of genitourinary, cardiovascular, or musculoskeletal system.

The placenta has few benign tumors. The commonest is chorangioma. This is a hemangiomatous tumor that does not affect the fetus if it is small but may act as an arteriovenous fistula if the tumor is large or multiple. Tumors in the mother may, on rare occasions, transfer to the fetus and placenta. The metastatic tumors to the placenta have been mainly melanomas.

Trophoblastic tumors are uncommon in the United States but are frequent in the Far East. There are three different forms of these tumors. Hydatidiform mole is a benign lesion formed of multiple vesi-

Figure 15–8. Hydatidiform mole. Note its grapelike appearance with different sizes of vesicles.

cles resembling a "bunch of grapes" that totally replace the placenta (Fig. 15–8). Usually, a fetus does not develop in association with the tumor. Histologically, the vesicles are swollen villi containing edema fluid and no fetal vessels. The trophoblastic cells proliferate extensively.

The incidence is 1 in 2,000 pregnancies in the United States and 1 in 200 in Asia. The main clinical signs and symptoms are first and second trimester bleeding. No fetal parts are visualized. The uterine size is often larger than the size expected from the dates. Laboratory data reveals high chorionic gonadotrophin (hCG) values. After evacuation of the tumor, the patient should be followed with hormonal measurements for a year.

Chorioadenoma destruens or invasive mole follows expulsion of a hydatidiform mole in about 2% of moles. Usually, the lesion is located deep in the myometrium. Histologically, it is formed of invasive sheets of trophoblast (both cyto- and syncytiotrophoblast) and villi. Symptoms include continuous vaginal bleeding after molar evacuation. The persistence of an increased hCG titer may indicate the presence of an invasive mole. The treatment is mainly chemotherapy.

Choriocarcinoma is a rare and highly malignant tumor that may follow the evacuation of a mole. The signs and symptoms may range from vaginal bleeding or uterine rupture to symptoms related to the site of metastasis (vagina, lungs, liver, brain). If untreated, death ensues within few months. Histologically, the tumor is formed by trophoblastic elements. Typically, no villi are present. Gonadotropin titers are increased. The management is mainly chemotherapeutic. Hysterectomy is considered depending on clinical circumstances. (Trophoblastic disease is also discussed in Chapter 6.)

SUGGESTED READING

Benirschke K: *Pathology of the Human Placenta*. New York, Springer-Verlag, 1967.

Gruenwald P: *The Placenta*. Baltimore, University Park Press, 1975.

Woodling BA, Kroener JM, Puffer HW, et al: Gross examination of the placenta. *Clin Obstet Gynecol* 19:21, 1976.

QUESTIONS

Choose one or more answers.

1. The human placenta is formed of:
   A. maternal cells
   B. trophoblast
   C. septae with x-cells
   D. decidual tissue
   E. endothelial cells
2. The umbilical cord
   A. contains two arteries and one vein
   B. is the site of fetomaternal transfer
   C. transfers maternal blood to the fetal side of the placenta
   D. has a large amount of Wharton's jelly
3. The membranes
   A. line the fetal surface of the placenta
   B. are formed by an inner avascular chorion and an outer amnion

C. adhere to the decidual tissue
D. are formed by an inner vascular amnion

4. The fetal cotyledon is the important unit of the placenta.
   A. it is the major area of placental transfer
   B. it is formed by villi of all sizes
   C. it is bathed by maternal blood
   D. it contains fetal blood
5. IgG passes the placental barrier
   A. by facilitated diffusion
   B. in cases of Rh isoimmunization
   C. through placental breaks
   D. by pinocytosis, probably
6. Maternal blood
   A. diffuses through the fetal circulation
   B. surrounds and bathes the fetal villi
   C. is forcefully ejected from the maternal circulation
   D. none of the above
7. Polyhydramnios is associated with
   A. defects of swallowing
   B. multiple pregnancy
   C. obstructive urinary anomaly
   D. all of the above
8. Oligohydramnios
   A. is normal at 20 weeks
   B. indicates gestational diabetes
   C. may be associated with urethral stenosis
   D. reflects a decreased swallowing mechanism
   E. all of the above
9. Congenital malformations may be associated with
   A. normal placentas
   B. circumvallate placentas
   C. single umbilical arteries
10. Hydatidiform mole
    A. is usually benign
    B. causes malformations in the newborn
    C. is formed of trophoblast but no villi
    D. hysterectomy is the treatment of choice

ANSWERS

1. A, B, C, D, E
2. A, D
3. A, C
4. A, B, C, D
5. B, D
6. B, C
7. A, B
8. C
9. A, B, C
10. A

chapter sixteen

physiologic adjustments in pregnancy

OBJECTIVES

1. Describe changes in the vasculature of the ovary, uterus, and vagina.
2. Describe changes in esophageal, gastric, liver, and gall bladder function.
3. Describe alterations in pulmonary anatomy and function.
4. Describe anatomic changes in kidney, ureter, and bladder and changes in renal function.
5. Describe changes in salt and water metabolism during pregnancy.
6. Describe the renin-angiotensin-aldosterone system and its changes with pregnancy.
7. Identify the magnitude of change in maternal blood volume and identify when in pregnancy the changes in plasma volume and red cell volume occur.
8. Identify benefits of changes in blood volume.
9. State the pregnancy requirement for iron.
10. Extrapolate from what you will be taught about the iron stores in nonpregnant women the reason that supplemental iron is given to pregnant women.
11. State when in pregnancy the iron requirements are greatest.
12. State the relationship between maternal anemia and the fetal hematologic status.

13. Identify various hematologic factors that may be increased or decreased in normal pregnancy.
14. Identify changes that occur in cardiac output and peripheral resistance during pregnancy.
15. Explain the "supine hypotension syndrome."
16. Describe changes in adrenal, thyroid, and pancreatic function during pregnancy.

The maternal organism must adjust to allow normal growth and development of a fetus. Understanding normal physiologic changes in pregnant women allows one to assess pathophysiologic changes that occur in disease states during pregnancy.

Reproductive System

Vagina

During pregnancy the vaginal mucosa becomes edematous, hypertrophied, and hyperemic; increased secretion of a white, thick discharge occurs. The pH is acidic due to the growth of *Lactobacillus acidophilus,* which thrives on the increased glycogen content of vaginal epithelial cells. The lowered pH is bacteriostatic for many pathogenic bacteria in the vaginal tract. The cytologic picture of the vaginal epithelial cells is similar to that seen in the luteal phase.

Uterine Vasculature and Blood Flow

Uterine blood flow (UBF) increases markedly from the time of implantation to the end of the first trimester. From that time on, UBF per unit weight of tissue remains constant. At term, the UBF is 500–750 ml/min with 80% going to the placental site. The UBF behaves as though it is a maximally dilated system responding only and directly to perfusion pressure (cardiac output). The uterine vascular bed, however, responds to vasoconstrictive stimuli (endogenous catecholamines or drugs). During labor, UBF (particularly placental blood flow) varies inversely with the intensity and duration of myometrial contractions. The spiral arteries that are initially invaded and "tapped" by the cytotrophoblast during placental formation are vastly altered. At the decidual and myometrial levels, most of the fibroelastic material of the vessel wall is replaced by fibrous and fibrinoid material allowing wide dilatation, thus effecting maximum blood flow to the placental intervillous space. These changes are brought about by replacement of endothelial cells by trophoblast cells and their further invasion into vessel walls.

Ovary

The alteration in cyclic function brought about by pregnancy is the maintenance of the corpus luteum (CL) function beyond the 10th–12th postovulatory day. Human chorionic gonadotropin (hCG) secreted by trophoblast supports continued CL function and is detectable within three days after nidation. The time when CL function wanes and placental function takes over the endocrine support of human pregnancy has been the subject of debate. Csapo has indicated that the luteal-placental shift occurs between the fifth and seventh postovulatory weeks or the seventh to ninth week from the start of the last menstrual period. Following this shift, removal of the CL will not disrupt gestation. A small decrease in plasma progesterone occurs at the time of this shift to the placenta for steroid support of the pregnancy. Up to 20% of normal pregnant women notice slight to moderate amounts of vaginal bleeding coincident with the shift.

Endocrine System

Hypothalamus

The hypothalamus, which has been suppressed by estradiol and progesterone produced by the CL, remains suppressed after nidation because of the luteotrophic effect of hCG and continued estrogen and progesterone synthesis.

Pituitary

In the absence of hypothalamic stimulation, there is no rise in either FSH or LH production by the pituitary. As a result of gonadotropin inhibition, prolactin inhibition is released. Prolactin levels become elevated above normal by eight weeks' gestation and rise progressively toward term. At term, serum prolactin levels are approximately 10 times normal nonpregnant levels. In spite of high prolactin levels, lactation is not initiated during pregnancy. Recent evidence suggests that a local blocking effect of estrogen on the breast tissue prevents lactation during pregnancy. The pituitary gland enlarges during pregnancy.

Thyroid

During normal pregnancy, increased urinary excretion of iodine may result in iodine deficiency with goiter formation if dietary iodine is low. Estrogen causes an increase of thyroid binding globulin. Levels of bound T_4 increase, while levels of free T_4 remain the same, presumably because of decreased degradation of free T_4. Levels of bound T_3 rise slowly during pregnancy, peaking in the second trimester. Free T_3 levels are below normal for the nonpregnant state. There appears to be no increase in thyroid hormone synthesis during normal pregnancy. Further dynamics of thyroid function during pregnancy will be discussed under the topic of thyroid disease in Chapter 20.

Adrenal

Changes in adrenal function in pregnancy are similar in mechanism to those noted for the thyroid; there is no increase in cortisol synthesis, rather an alteration in binding and degradation. Placental estrogens cause an increase in cortisol binding globulins so that total serum cortisol levels are increased. The fraction of free as well as bound cortisol is higher in the pregnant than in the nonpregnant state presumably because the half-life is longer in pregnancy. There is no evidence for increased ACTH secretion in pregnancy, and the response to exogenous ACTH administration appears to be blunted.

Aldosterone secretion is increased markedly in pregnancy. This is due to elevated levels of the components of the renin-angiotensin system in pregnancy. The increased aldosterone levels function to increase sodium reabsorption from the renal tubules to facilitate the needs of pregnancy.

Pancreas

Pregnancy has been referred to as a diabetogenic state. Pancreatic function and normal relationships of carbohydrate metabolism undergo marked changes from the nonpregnant state. Many of the changes appear to be related to the effects of human chorionic somatomammotrophin (hCS; human placental lactogen), which has anti-insulin properties. Energy metabolism of the pregnant woman utilizes fatty acids rather than carbohydrates. Fatty acids appear to interfere with peripheral insulin effects to promote carbohydrate utilization. Carbohydrates then are spared for use by the fetus. Further dynamics of carbohydrate metabolism will be discussed under the topic of diabetes in Chapter 20.

Alimentary System

The major change in the gastrointestinal system is reduced tone and motility. The slower transit time of food through the gut allows more efficient nutrient absorption but contributes to the common complaint of constipation due to excessive water absorption. Esophageal reflux of gastric contents due to increased intragastric pressure and diminished esophageal sphincter tone produces discomforting symptoms of heartburn. Gastric acid production decreases mainly in the midtrimester, and patients with peptic ulcers usually improve during pregnancy. However, patients with peptic ulcers can have an exacerbation of their symptoms during the postpartum period.

Liver function is slightly altered. Bromsulphophthalein (BSP) retention is increased by 12% at 45 minutes. Bilirubin conjugation and excretion are slightly impaired. Alkaline phosphatase levels are increased about four times in pregnancy. This is mainly due to placental production of *heat-stable* alkaline phosphatase.

Two pregnancy changes contribute to an increased incidence of gallstones in parous women: hypercholesterolemia due to hCS results in increased cholesterol secretion into bile, and decreased peristalsis of the bile duct and slower emptying time of the gall bladder promote stasis of bile.

Pulmonary System

Lung Volumes

Functional residual capacity (FRC) progressively decreases from the sixth month to term. At term, the FRC has decreased 20%. This is due to a decrease of both expiratory reserve volume (ERV), maximum volume exhaled from the resting end-expiratory position, and residual volume (RV), volume remaining in lung at end of maximal expiration. The decrease in ERV is primarily due to decreased recoil of the chest wall, which is related to increased intra-abdominal pressure on the diaphragm. This is evidenced by anatomic changes of an increased substernal angle and increased anteroposterior diameter of the chest. There is no change in the vital capacity (VC), so that the tidal volume (TV) increases at the expense of decreased ERV. The larger TV drawn into a reduced RV allows more efficient gas exchange and improves alveolar ventilation by 65%. Since the respiratory rate (RR) is increased (15%), the minute ventilation (RR × TV) is greatly enhanced (55%) (Fig. 16–1).

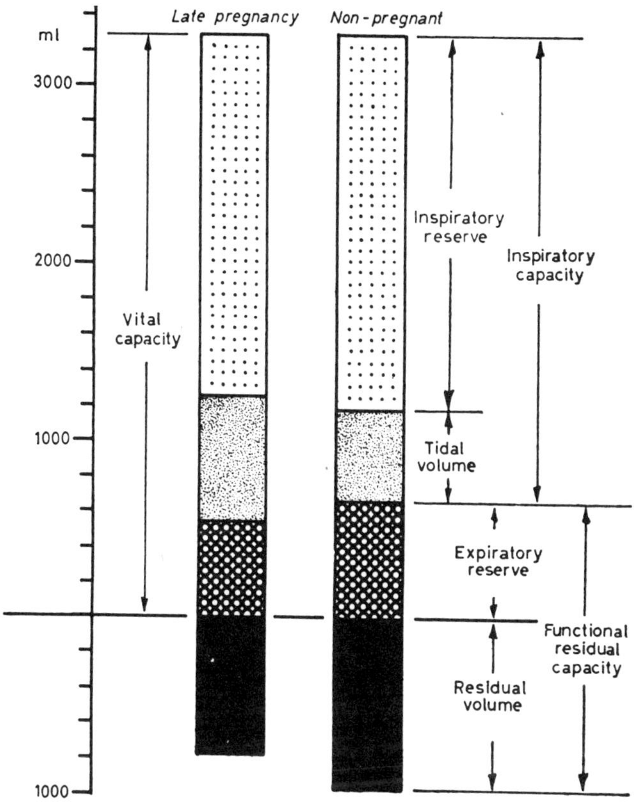

Figure 16–1. The components of lung volume in late pregnancy compared with those in the nonpregnant woman. (From Hytten FE, Leitch I: *The Physiology of Human Pregnancy,* ed, 2. Oxford, Blackwell Scientific Publications, 1971.)

"Hyperventilation of Pregnancy"

Increased minute ventilation is one of the most notable and consistent pulmonary changes of pregnancy. It begins as early as the ninth week and may initially be related to the increased RR. It is out of proportion to the increased metabolic demand of pregnancy. The resting O_2 uptake in late pregnancy increases 20%. Hyperventilation results in the lowering of alveolar partial pressure of carbon dioxide (pCO_2). Although arterial pCO_2 is decreased, reflecting decreased alveolar pCO_2, there is no change in arterial pH. This is due to compensatory renal excretion of bicarbonate, lowering the total base by 5 mEq/liter. The medullary chemoreceptors are reset to discharge at a lower arterial pCO_2. The respiratory center has increased sensitivity to pCO_2, resulting in an increased RR. These changes are probably related to increased progesterone levels.

Lowering of the arterial pCO_2 and bicarbonates allows an optimum environment for fetal respiratory exchange. This is important for normal fetal development since it has been shown that increasing pCO_2 or acidosis can be deleterious to the fetus.

In the first two trimesters, 60–70% of pregnant women complain of breathlessness. They behave as though they are receiving a high altitude or an alveolar-capillary block stimulation of ventilation, yet no pathologic alterations have been found. No explanation can be offered.

Patients with pulmonary impairment who are not dyspneic at rest usually tolerate pregnancy quite well. Vital capacity measurements indicate the degree of pulmonary restriction. Patients with a VC of 1 liter or more can be expected to tolerate pregnancy successfully. The compensatory changes to maintain normal blood gas tensions are achieved by increased pulmonary and cardiac work.

Increased pulmonary ventilation together with decreased functional residual capacity permits more rapid changes in the concentration of gases in the lungs. Both induction of and recovery from inhalation anesthesia are more rapid, and arterial blood levels of pO_2 and pCO_2 undergo greater changes during the breathing cycle. Induced hyperventilation during anesthesia may rapidly result in an arterial pCO_2 as low as 10–15 mm Hg.

Renal System

Pyelograms and post-mortem examination of pregnant women have demonstrated dilatation of the ureters above the pelvic brim up to and including the renal pelvis. This dilatation begins as early as the 10th week of gestation and does not return to normal until 12 weeks postpartum. There is always more dilatation on the right side of the renal system. The ureter is frequently thrown into curves of varying sizes due to elongation. These changes are thought to be caused by progesterone and mechanical factors. High levels of progesterone during pregnancy result in generalized ureteral atony, which is enhanced when the uterus enlarges and compresses the ureters at the pelvic brim, more on the right ureter due to uterine dextrorotation. Studies of ureteral tone in the supine position have shown normal ureteral tone in the lower part of the ureter, whereas the ureteral tone above the pelvic brim was higher than normal during pregnancy. This increased tone fell when the women were turned on their side and returned to normal levels immediately after delivery during cesarean section. Neither frequency of ureteral contraction nor mean contractile pressures were affected by pregnancy. Mechanical theories do not explain the ureteral dilatation seen early in gestation before the uterus rises above the pelvic brim. A venographic study, however, has demonstrated that compression of the ureters (particularly on the right side) is due to dilatation of the ovarian venous plexus that surrounds them in the region of the pelvic brim. This was demonstrated prior to significant enlargement of the uterus.

Bladder tone is unaffected by pregnancy, although the trigone of the bladder becomes thickened and is elevated due to the enlarging uterus. This may result in frequency of micturition and vesicoureteral reflux. The volume of the collecting system is doubled by the above changes, and the resulting stasis and reflux contribute significantly to the susceptibility of pregnant women to urinary tract infection.

The mechanism of alteration of renal function in pregnancy is unknown, although changes in renal blood flow, total blood volume, and in the concentration of several hormones, including hCS, estrogens, aldosterone, and adrenal cortical compounds may be contributing factors. Renal blood flow (RBF) is elevated in early pregnancy and is related to lowered vascular resistance within the kidneys. The increase of renal plasma flow (RPF) is about 45% (from 500 to 725 ml/min). This occurs as early as the ninth week of gestation and continues through the second trimester, to decline slightly near term. The mean increase of glomerular filtration rate (GFR) is about 60% (from 90 to 145 ml/min near term). Since GFR increases more than renal plasma flow, the filtration fraction (GFR/RPF) increases during pregnancy. This increase in filtration fraction increases the amount of plasma constituents presented to the renal tubules and effects some alterations in their blood levels. The advantages of improved clearance rate are also related to fetal excretory needs handled by the mother after transplacental passage of fetal metabolites to her circulation. Creatinine is filtered by the glomerulus but is not absorbed by the renal tubules. Plasma levels (0.4 mg/dl) in pregnancy are lower than nonpregnant levels (0.8 mg/dl). Although there are slight differences between the creatinine clearance and inulin clearance, the endogenous creatinine clearance can be used as a reliable index of renal functional status in pregnancy. Creatinine clearance is determined by an accurate 24-hour urine collection with blood and urine creatinine determinations during the same time period. The following formula is used for calculations:

$$\text{Creatinine clearance} = \frac{\text{Total mg of urinary creatinine in 24 hours}}{\text{Plasma creatinine mg/ml} \times 1{,}440 \text{ (minutes in 24 hours)}} = \text{ml/min}$$

Normal pregnancy values range from 90 to 180 ml/min. Urea is filtered by the glomerulus, but some is passively reabsorbed by the tubules. The amount of urea excreted is greater during pregnancy, which accounts for a blood level below the nonpregnant blood levels. The upper limit of normal blood urea nitrogen (BUN) in pregnancy is 10 mg/dl, compared to the nonpregnant levels of 15 mg/dl. Uric acid is also lower in pregnancy (3 mg/dl as opposed to 6 mg/dl). Although nitrogen retention and storage are predominant trends during pregnancy, there is increase in the urinary excretion of amino acids. The maximal tubular reabsorption of glucose is unaltered in pregnancy, but as the filtered load of glucose is increased, the tubular maximum is liable to be exceeded at a relatively low plasma glucose level. Thus, about 20% of normal pregnant women demonstrate glycosuria. In contrast, the maximum tubular reabsorption for sodium increases to match the elevated filtration of sodium, thereby maintaining sodium balance.

Renal function studies performed with pregnant women have demonstrated that there is a significant decrease in renal blood flow in the supine position, thereby altering the outcome of the tests. This impairment of renal blood flow in the supine position in pregnancy should be kept in mind during the performance of any critical renal function studies.

Salt and Water Metabolism in Pregnancy

During pregnancy, there is an accumulation of salt and water without significant change in their absolute or relative concentrations in maternal tissues or serum. The mean total body water accumulation is 8.5 liters, nearly 6 liters of which is present in the fetus, placenta, amniotic fluid, uterus, and maternal plasma. The remainder is contained in the maternal extracellular spaces. To maintain sodium balance, 700–800 mEq of total exchangeable sodium is accumulated during pregnancy. This is accomplished as a direct consequence of renal hemodynamics and hormonal changes. The GFR determines the sodium load presented to the renal tubules and maintains gross control over sodium output. Changes in tubular reabsorption of sodium exert a more sensitive control over sodium excretion. The level of tubular sodium reabsorption is largely influenced by aldosterone and progesterone. Aldosterone enhances sodium absorption and potassium loss, whereas progesterone counteracts the aldosterone effect and has an overall influence of potassium sparing. As gestation advances, both aldosterone and progesterone levels progressively increase.

Posture is an important determinant of sodium and water balance in pregnancy. When a pregnant woman turns from her side to her back, sodium and water excretion falls by more than 50%. This has been related to the reduced cardiac output caused by obstruction to the inferior vena cava by the gravid uterus, and perhaps other causes. Recent work has shown that even in supine subjects in whom the GFR does not fall, sodium excretion is diminished.

The upright position has an antidiuretic effect in that the diurnal water excretion pattern is reversed in pregnant women. Maximum diuresis occurs at night when the woman is recumbent. This pattern is most likely due to sequestration of water in the tissues of the lower extremities while upright with mobilization of this water while recumbent.

Free water clearance is greatly enhanced in early pregnancy, only to decline as pregnancy progresses. Ingestion of 1 liter of water in early pregnancy will lead to increased urine output and a decrease in the serum osmolality. Later in gestation, there is no response of increased urine output and decreased serum osmolality in response to ingestion of 1 liter of water. This suggests that ingested water rapidly leaves the blood without triggering a diuretic response.

Renin-Angiotensin System

Activity of the renin-angiotensin system increases in pregnant women. Plasma renin, angiotensin, angiotensinogen, and angiotensinases are all elevated when compared to nonpregnant levels. It is not clearly understood whether the initiating factors are related to hormone stimulation (progesterone and estrogen), the relatively expanded intravascular bed, the increased sodium requirements of pregnancy, or a combination of these or other causes. The effective end result of this increased activity of the renin-angiotensin system is high levels of circulating aldosterone.

Angiotensin II acts directly on the zona glomerulosa of the adrenal cortex to increase aldosterone production. Aldosterone, in turn, acts on the renal distal tubules to promote sodium reabsorption. Increased sodium expands the blood volume, which acts to decrease the renin output from the juxtaglomerular apparatus cells. Extrarenal sources of renin have been described. The pregnant uterus is a source of high levels of reninlike material that may be contributory to increased activity of the system. Angiotensin II is a powerful vasopressor substance. Pressor response to infusions of angiotensin II has been shown to be blunted in pregnant patients. No hypertensive effect has ever been demonstrated in normal pregnancy secondary to increased activity of the renin-angiotensin system. We assume its major effect is to increase aldosterone secretion.

Cardiovascular System

Blood Volume and Blood Components

Pregnancy induces a hypervolemic state. Blood volume begins to increase in the first trimester, increases rapidly in the second trimester, and then slightly increases in the third trimester. The magnitude of increase varies from woman to woman, with some almost doubling their blood volume. Mean values from a carefully done comparative study of blood volumes in term pregnant and nonpregnant women are shown in Table 16–1.

This dramatic increase in blood volume has several beneficial effects for the pregnant woman. It helps to meet the increasing metabolic demands of the conceptus and the large vascular uterus, it helps protect the mother and fetus from impaired venous return in late pregnancy, and it helps safeguard the mother against adverse effects of blood loss at delivery. Many theories have been advanced to explain the reason for pregnancy-induced hypervolemia, but none fully explain it. Aldosteronism is associated with an increased blood volume. High doses of estrogen can increase blood volume. Pregnancy is associated with both of these hormonal changes. Arteriovenous fistulas are asso-

Table 16–1. Blood Changes in Late Pregnancy

	Nonpregnant	*Pregnant*	*Change (%)*
Blood volume	3,250 ml	4,820 ml	↑ 48
Red cell volume	1,355 ml	1,790 ml	↑ 32
Hematocrit	41.7%	37.0%	↓ 11

From Pritchard JA: *Anesthesiology* 26:393, 1965.

ciated with an increased blood volume, and the placenta is essentially a low-resistance arteriolar-venous circuit. Whatever the cause, the presence of a fetus is not required since hypervolemia occurs in molar pregnancies.

The increase in blood volume results from an increase in plasma volume and red cell volume, with the plasma volume increasing earlier and to a greater degree than the red cell volume. This disparity of increase in plasma volume over red cell volume is most pronounced in the second trimester and results in a dilutional lowering of the hemoglobin and hematocrit that has been termed the "physiologic anemia" of pregnancy. During the second half of pregnancy, approximately 450 ml of erythrocytes are added to the circulation. This increase is due to increased marrow production stimulated by increased erythropoietin levels in maternal plasma. Human placental lactogen may also play a role in the accelerated erythropoiesis of late pregnancy.

Since there is 1.1 mg of iron in each milliliter of red cells or 3.4 mg of iron in each gram of hemoglobin, the addition of 450 ml of red cells to the circulation of pregnant women requires 500 mg of iron. In addition, the fetus and placenta require approximately 300 mg of iron, making the total requirement for pregnancy around 800 mg. Most of this requirement is needed in the third trimester.

All adults lose 0.5–1 mg of iron per day from the skin and gastrointestinal mucosa, but women lose an average of 40 mg of iron per month in menstrual flow (approximately 500 mg per year). In an elegant study, Pritchard compared the iron stores in a population of college women who had no history of bleeding, surgery, pregnancy, blood donations, or exogenous iron therapy with those of college men and found that in women the iron stores averaged only 347 mg, whereas in men the iron stores averaged nearly 1 gm. Pritchard also compared the amount of stainable iron in bone marrow aspirates from young healthy nonpregnant women to standard laboratory tests (Table 16–2).

These data indicate that apparently healthy women do not have enough iron in storage to meet the demands of pregnancy, nor will simple lab tests identify women at risk to develop anemia during pregnancy. Thus, the 800 mg of iron demanded by pregnancy must be provided by a dietary intake of 6–7 mg of iron per day during the last half of pregnancy. Since diet rarely contains enough iron to meet this demand, 30 mg of iron in the form of a simple iron salt such as ferrous sulfate, gluconate, or fumarate ingested daily during the latter half of pregnancy should meet the demands of pregnancy and protect any preexisting iron stores. Twenty percent of ferrous sulfate is elemental iron, so 200 mg of ferrous sulfate daily should provide preventive maintenance during pregnancy.

Table 16–2. Comparison of Standard Hematologic Lab Tests According to the Amount of Bone Marrow Iron in Young Healthy Women

Amount of bone marrow iron	2 + *to* 4 + (39 *pts.*)	1 + (39 *pts.*)	*No Iron* (27 *pts.*)
Hemoglobin/Hct	13.7/42	13.6/42	13.0/40.5
Serum iron	91 μg/100 ml	86	63
Iron binding capacity	305 μg/100 ml	327	377
MCV	89	90	89
MCHC	33	32	32
MCH	29	29	29

Data from Scott DE, Pritchard JA: *JAMA* 199:897, 1967.

Anemia in the nonpregnant state is defined as a hemoglobin concentration of less than 12 gm%, whereas, in pregnancy, a hemoglobin concentration of less than 10 gm% is diagnostic of anemia due to the physiologic dilution that occurs. Iron deficiency anemia is the most common type of anemia in pregnant women. Therapy requires oral iron in the dose range of 200 mg/day. This can be supplied by 900 mg ferrous sulfate a day. Therapy should be continued for at least three months after hemoglobin concentrations are normal in order to build up adequate iron stores.

The normal folic acid requirement in pregnancy is unknown but is thought to be higher than the 50–100 μg daily requirement of nonpregnant women. Folic acid is found in green leafy vegetables, and a deficiency caused by dietary inadequacy results in an uncommon complication of pregnancy in the United States, megaloblastic anemia. Folic acid deficiency results in impairment of erythrogenesis and is evidenced by megaloblasts in the bone marrow and macrocytic erythrocytes in the peripheral blood. Anemia may be profound with hemoglobin concentrations of 4–8 gm%, often accompanied by thrombocytopenia, neutropenia, and nucleated erythrocytes. The neutrophils exhibit nuclear hypersegmentation (usually $>$ 3 lobes). The diagnosis is supported by low serum folate levels and bone marrow aspiration (Fig. 16–2).

The treatment of megaloblastic anemia in pregnancy should include oral folic acid, a well-balanced diet, and oral iron. Folic acid given in a 1 mg daily dose results in a dramatic hematologic response within three to five days. This is evidenced by a striking reticulocytosis. The serum iron levels are usually normal with megaloblastic anemia due to

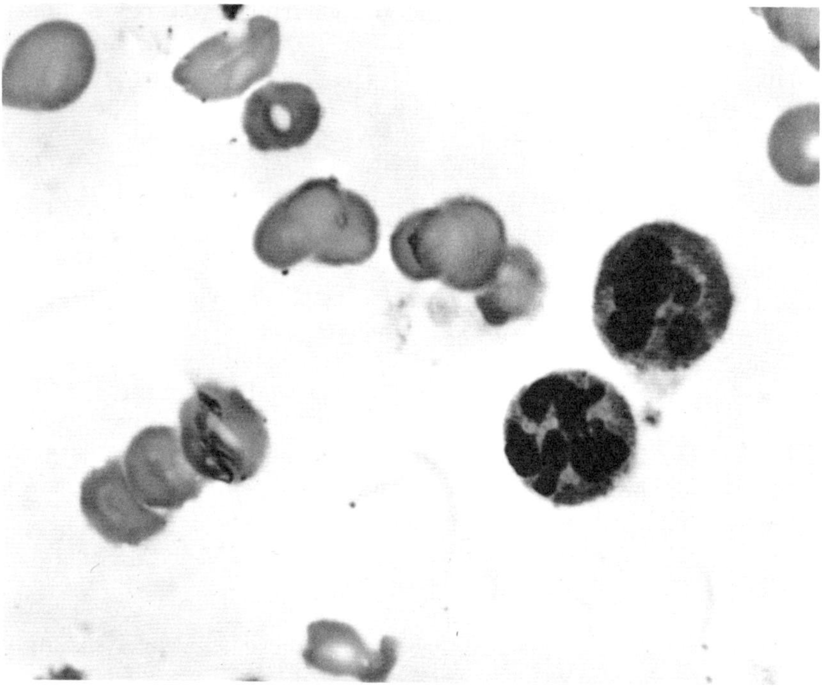

Figure 16–2. Blood from a pregnant patient with folate deficiency anemia demonstrating hypersegmentation of the neutrophils.

decreased utilization. However, iron therapy is required to support the increased hematopoiesis during folate replacement.

Folic acid deficiency in pregnancy has been implicated as an etiology for various disorders, including abortions, hypertension, abruptio placentae, and congenital anomalies; however, these have never been satisfactorily proven. A case could be made for the supplemental administration of folic acid to pregnant women, but at present it is not routinely found in prenatal vitamins. The other cause of megaloblastic anemia, vitamin B12 deficiency, is exceedingly rare in the reproductive age group.

The fetus of a mother with iron deficiency or megaloblastic anemia is rarely, if ever, anemic at birth. Iron and folic acid preferentially cross the placenta to support fetal hematopoiesis despite maternal deficiencies. Fetal problems are few unless the maternal hemoglobin is less than 4 gm%, and these may relate more to general dietary indiscretions by the mother.

The leukocyte count varies considerably during pregnancy, but most often it is elevated to between 12 and 15,000 per cubic millimeter during labor and postpartum periods. The practical implication is that infection may be harder to detect by relying on the leukocyte count.

Some clotting factors increase in pregnancy. Fibrinogen increases from a prepregnancy level of 300 mg% to 450 mg%, an increase of 50%. Factor VIII, the antihemophilia globulin, is also elevated, often lessening the severity of von Willebrand's disease during pregnancy. Factors VII, IX, and X are also increased appreciably. The fibrinolytic activity of plasma is prolonged during pregnancy. The platelet count and function usually remain unchanged. These changes suggest that the blood of pregnant women is hypercoagulable; yet this is not apparent since the whole blood clotting time, prothrombin time, and partial thromboplastin time are unchanged. It is noteworthy that after delivery, when these clotting factors are returning to normal, the patient is more likely to develop venous thrombosis and thromboembolism than at any other time in her reproductive career.

Cardiac Output

In pregnancy, the cardiac output begins to increase as early as eight weeks' gestation and reaches a maximum at 24–28 weeks' gestation. The increase may vary from 27 to 64%, but in most women the increase is 40% above their nonpregnant levels. After 30 weeks of pregnancy, the cardiac output remains stable at this elevated level until just after delivery when it temporarily decreases. The cardiac output increases again during the early postpartum period when extravascular tissue fluid is mobilized into the vascular system to initiate postpartum diuresis.

Cardiac output increases by increase in heart rate or stroke volume or both. In pregnancy, the heart rate increases modestly (10%); therefore, most of the increase in cardiac output is due to a significant increase in stroke volume. A concomitant decrease in peripheral resistance accompanies the increased cardiac output to maintain normal peripheral blood pressure during pregnancy.

The reasons for this increased cardiac output are poorly understood. Several explanations have been offered. The metabolic demands of the fetus in late pregnancy increase oxygen consumption by 20%. This demand could easily have been met by less than a 40% increase in cardiac output; also, as the demand increases, the cardiac output remains relatively unchanged. The placenta behaves as a low resistance shunt, and such shunts are associated with an increase in cardiac output. It seems unlikely, however, that the placenta at eight weeks' gestation could provide enough of a shunt to result in a rise of cardiac output. The hypervolemia of pregnancy may play a role in increasing the stroke volume. Studies have shown that isovolemic contraction time decreases in response to increased myocardial fiber length, which results from an increase in blood volume. Increased levels of steroid hormones may be involved since estrogen given to nonpregnant sheep increases cardiac output by decreasing peripheral resistance and increases vascular flow to the uterus. Women taking oral contraceptive steroid hormones also have an increased cardiac output due to an increase in stroke volume.

When the pregnant woman in her third trimester is placed in the supine position, the weight and shape of the gravid uterus compresses the inferior vena cava against the vertebral column. This can result in a significant decrease in venous return and decreased cardiac output. In some women, this decrease is sufficient to cause symptoms of faintness, anxiety, and hypotension. This sequence of events is termed "supine hypotension syndrome" and is corrected by positioning the patient on her side or by manually displacing the uterus off the vena cava. The majority of women are able to compensate for this decrease in venous return via the vena cava by enlarging paravertebral venous collateral channels. Also, peripheral resistance increases with no increase in heart rate to maintain a normal blood pressure during the supine positioning. Only about 11% of pregnant women experience the supine hypotension syndrome. What is unique about these women is unknown.

The profound physiologic adjustments that the maternal organism must undergo to maintain a healthy conceptus have been reviewed in this chapter. These changes must be kept in mind in the management of normal prenatal care as well as understanding the pathophysiologic mechanisms of disease states in pregnancy.

SUGGESTED READING

Bellina JH, Dougherty CM, Michal A: Pyeloureteral dilatation and pregnancy. *Am J Obstet Gynecol* 108:356, 1970.

Brosens I, Robertson WB, Dixon HG: The physiological response of vessels of the placental bed to normal pregnancy. *J Path Bact* 95:569, 1967.

Burwell CS: The placenta as a modified arteriovenous fistula considered in relation to the circulating adjustments to pregnancy. *Am J Med Sci* 195:1, 1938.

Chesley LC, Sloan DM: The effect of posture on renal function in late pregnancy. *Am J Obstet Gynecol* 89:754, 1964.

Greiss FC, Jr: A clinical concept of uterine blood flow during pregnancy. *Obstet Gynecol* 30:595, 1967.

Howard BK, Goodson JH, Mengert W: Supine hypotensive syndrome in late pregnancy. *Obstet Gynecol* 1:371, 1953.

Kitay DZ: Folic acid in pregnancy. *Am J Obstet Gynecol* 104:1067, 1969.

Lindheimer MD, Weston PV: Effects on hypotonic expansion on sodium, water and urea excretion in late pregnancy: the influence of posture on these results. *J Clin Invest* 48:947, 1969.

Lund, CJ, Donovan JC: Blood volume during pregnancy. Significance of plasma and red cell volume. *Am J Obstet Gynecol* 98:393, 1967.

Metcalfe J, Ueland K: Maternal cardiovascular adjustments to pregnancy. *Prog Cardiovasc Dis* 16:363, 1974.

Pritchard JA: Changes in blood volume during pregnancy and delivery. *Anesthesiology* 26:393, 1965.

Prowse CM, Gaensler EA: Respiratory and acid-base changes during pregnancy. *Anesthesiology* 26:381, 1965.

Rubi RA, Sala NL: Ureteral function in pregnant women. III. Effects of different positions and of fetal delivery upon ureteral tonus. *Am J Obstet Gynecol* 101:230, 1968.

Scott DE, Pritchard JA: Iron deficiency in healthy young college women. *JAMA* 199:897, 1967.

Sims EA, Krantz KE: Serial studies of renal function during pregnancy and the puerperium in normal women. *J Clin Invest* 37:1764, 1958.

Walters WAW, Lim YL: Cardiovascular dynamics in women receiving oral contraceptive therapy. *Lancet* 2:879, 1969.

Watanake M, Meeks CI, Gray MJ, et al: Secretion rate of aldosterone in normal pregnancy. *J Clin Invest* 42:1619, 1963.

Zarowitz H, Newhouse S: Renal glycosuria in normoglycemia glycosuric pregnancy: a quantitative study. *Metabolism* 22:755, 1973.

QUESTIONS

1. True statements about the vaginal pH of a pregnant woman include:
   A. a result of increased glycogen content of epithelial cells
   B. should be acidic

C. may change to alkaline as a result of the presence of ruptured membranes
D. allows the growth of many bacteria
E. allows the evolvement of acidophilic staining vaginal epithelial cells (follicular phase of menstrual cycle)

2. Uterine blood flow:
A. 500–750 ml/min at term, with 80% going to placenta
B. is a submaximally dilated system
C. is resistant to vasopressor drugs
D. varies inversely with uterine contractions
E. is the cause of dilated spiral arteries

3. The major changes in the alimentary system during pregnancy include:
A. ↑ gastric motility and ↓ gastric acid
B. ↑ esophageal reflux
C. ↓ tone and ↓ motility
D. ↑ pepsin in gastric juices and ↑ gastric ulcers
E. ↑ heartburn due to ↑ gastric acidity

4. All of the following changes result in ↑ alveolar ventilation *EXCEPT:*
A. ↓ functional residual capacity
B. ↑ tidal volume
C. ↑ affinity of hemoglobin for oxygen
D. widening of the transverse diameter of the chest
E. almost no change in vital capacity

5. The following changes occur in pregnancy:
A. complaints of being short of breath
B. ↑ oxygen consumption
C. timed vital capacity decreases
D. ↑ of basal metabolic rate
E. ↑ anesthetic gas uptake

6. The volume in the urinary collecting system increases with pregnancy because:
A. increased GFR and RPF
B. decreased filtration fraction
C. progesterone and mechanical obstruction of growing tissue
D. ↑ thickness of bladder trigone and reflux
E. bladder infections

7. Glucosuria occurs in pregnancy because of all of the following *EXCEPT:*
A. diabetes

B. ↑ TmG
C. ↑ GFR/RPF
D. ↑ creatinine clearance
E. ↑ cardiac output

8. Na balance in pregnancy results from:
   A. a fixed TmNa
   B. progesterone effect
   C. increased Na in diet
   D. aldosterone
   E. an increase in extracellular space
9. Renin increases in pregnancy because:
   A. reason unknown for the most part
   B. of increased effective circulating blood volume
   C. not enough Na reabsorbed to shut off its release
   D. all of the above
   E. none of the above
10. Angiotensin II does all the following *EXCEPT:*
   A. ↑ aldosterone output 20 times in pregnant women
   B. has a blunted pressor effect in pregnant women
   C. is broken down by angiotensinase
   D. countereffects the effect on the tubular function of progesterone directly
   E. a powerful vasopressor
11. To maximize renal function and naturesis a patient should

    ____
12. The most notable change in pulmonary function in pregnancy is

    ____
13. The average increase in blood volume in normal pregnancy is ________%.
14. Which increases earlier in pregnancy, plasma volume or red cell volume? ______________________________
15. Blood volume expands most rapidly during which trimester of pregnancy? ______________________________
16. Increase in red cell volume is caused by:
   A. increased life span of red cells in pregnancy
   B. accelerated production of red cells in pregnancy
17. The total iron requirement of normal pregnancy is ________mg.
    The fetus and placenta require ________________________mg.
    The increased red cell volume requires ________________mg.

18. The total iron content of normal adult women is around____gm.
19. A normal hemoglobin and hematocrit means a woman has normal iron stores. TRUE FALSE
20. A normal serum iron and iron binding capacity means a woman has normal iron stores. TRUE FALSE
21. "Prophylactic" administration of iron to all pregnant women is justified. TRUE FALSE
22. Most of the iron required for a pregnancy is needed in the last half of pregnancy. TRUE FALSE
23. If a patient has iron or folate deficiency anemia, her child will most likely be born anemic. TRUE FALSE
24. Indicate which of the following are often increased or decreased during pregnancy:

	Increased	*Decreased*
A. Blood volume	________	________
B. Hematocrit	________	________
C. Leukocytes	________	________
D. Fibrinogen	________	________
E. Factor VIII	________	________

25. What are three protein hormones and four steroid hormones synthesized by the placenta?____

____

____

26. How early in pregnancy can hCG be detected in the serum?

____

27. What appears to be the general role of hCS in pregnancy?

____

____

28. Which of the three estrogenic hormones is the best index of fetal well-being and why? ____

____

29. Name three basic changes that occur in the hypothalamic-pituitary ovarian axis during pregnancy.____

____

____

30. What phenomenon accounts for increased plasma levels of

thyroxine and cortisol during pregnancy? ______________________________

____

____

31. Which placental hormone has the greatest effect on glucose metabolism in pregnancy? ______________________________

ANSWERS

1. A, B, C
2. A, D
3. B, C
4. C
5. A, B, D, E
6. C
7. B
8. D
9. D
10. D
11. be placed in lateral recumbency
12. ↑ minute volume—"hyperventilation of pregnancy"
13. 50
14. plasma volume
15. second
16. B
17. 800, 300, 500
18. 2–2½
19. FALSE
20. FALSE
21. TRUE
22. TRUE
23. FALSE
24. A. ↑
    B. ↓
    C. ↑
    D. ↑
    E. ↑

25. hCG
    hCS
    hCT
    Progesterone
    Estrone
    Estradiol
    Estriol
26. Approximately 1–3 days postimplantation or 8–11 days after the LH peak.
27. Alteration in maternal metabolic processes to favor fetal nutrition
28. Estriol; it is synthesized from a precursor that is made in the fetal adrenal.
29. FSH and LH secretion are suppressed
    Prolactin secretion is increased
    Ovarian function is suppressed except for androstenedione secretion
30. Estrogen produces an increase in the levels of binding globulin for thyroxine and cortisol
31. hCS

chapter seventeen
normal pregnancy

OBJECTIVES

1. Describe the maternal physiologic and anatomic changes that occur in the first 12 weeks of gestation that are useful in making the presumptive diagnosis of pregnancy.
2. List and state the time of appearance of the major manifestations of the fetus that are useful in diagnosing pregnancy.
3. List the nutritional requirements of a normal pregnant woman in terms of calories, protein, and iron.
4. List the sources of weight acquired by a normal pregnant woman and state the amounts contributed by the conceptus and the maternal adaptations.
5. State three purposes of prenatal care. List the major conclusions drawn from the initial prenatal evaluation.
6. Describe the content of prenatal patient education programs.
7. State common complaints encountered in normal pregnancy.
8. Discuss psychological adaptations to pregnancy that are commonly encountered during each trimester of pregnancy and postpartum.

Most pregnancies are normal and terminate with the delivery of a healthy infant by a healthy mother. This chapter will focus on the diagnosis and duration of pregnancy, maternal nutrition and weight gain, the scope of prenatal care, common problems during pregnancy, and maternal psychologic adaptations to pregnancy.

DIAGNOSIS AND DURATION OF PREGNANCY

The diagnosis of pregnancy ultimately rests on the detection of a fetus; the presumptive diagnosis of pregnancy can be made by detecting maternal changes induced by pregnancy or by the presence of trophoblast.

Maternal Changes

Failure to menstruate is the first evidence of pregnancy noticed by most women. Below the age of 50, secondary amenorrhea is presumed to be due to pregnancy until proven otherwise. True cyclic bleeding during pregnancy is extremely rare. Occasionally, light bleeding may be noted at the time of implantation, and bleeding may also be detected at the time of the luteal-placental shift in progesterone production (seven to nine weeks after the last menstrual period). Generally, bleeding during pregnancy must be considered abnormal and the cause of bleeding investigated.

Amenorrhea associated with pregnancy will not respond to a progestin challenge test (Chapter 4) because the endometrium is under the control of estrogen and progesterone produced by the corpus luteum or placenta. This failure to bleed after progestin administration has been used as a pregnancy test but is now contraindicated because of the possibility of producing limb abnormalities in the fetus.

Nausea and vomiting occur in most pregnant women, usually associated with an empty stomach—"morning sickness." This usually begins in the 5th or 6th week after the last menstrual period (LMP) and abates by the 12th week. The cause of morning sickness is unknown but may be related to the high levels of chorionic gonadotrophin (hCG).

Breast enlargement and tenderness and increased areolar pigmentation are noticed between the 8th and 12th week after the LMP. Sebaceous glands in the areola (Montgomery's glands) also enlarge at this time.

Physical examination of the woman in early pregnancy may reveal several other changes suggestive of pregnancy. A cyanotic appearance of the upper third of the vagina (Chadwick's sign) and softening and cyanosis of the cervix have been described as early signs of pregnancy. The changes are not reliable. Loss of the ferning pattern on microscopic examination of dried cervical mucus implies a progesterone influence compatible with pregnancy.

Alterations in uterine size, shape, and consistency also provide early suggestive evidence of pregnancy. About six weeks after the LMP, the isthmus of the uterus begins to soften (Hegar's sign). The corpus of the uterus may also begin to soften at this time, but definite uterine enlargement may not be detected until seven or eight weeks. The shape of the corpus is usually globular by eight weeks after the LMP, and uterine softening is apparent. Errors in clinical assessment of these changes are usually associated with an inadequate examination due to obesity, uterine retroversion, or discomfort. Occasionally, other causes of uterine enlargement, such as myomas or adenomyosis, may be misleading. Comparison with a recent previous examination or a repeat examination in 10–14 days may be helpful in evaluating uterine changes if the examiner is uncertain.

Trophoblast

The detection of hCG produced by trophoblast forms the basis for most pregnancy tests. Biologic tests done by injecting the woman's urine into a test animal have now been replaced by more sensitive and reliable tests.

Several test kits are available that contain latex particles or red cells that are agglutinated by hCG antibodies. The kits are designed so that the presence of hCG in a woman's urine inhibits the agglutination process. Failure of agglutination is a positive test for hCG. The tests are sensitive to small amounts of hCG (1,000–4,000 IU/liter of urine) and they become positive 10–14 days after the first missed menstrual period. False negative tests may occur if the urine is dilute, so the first urine of the morning should be selected for testing. False positive tests are more difficult to explain but may occur. Reliability, rapidity, convenience, and low cost have made these immunologic tests very satisfactory.

Recently, it has been possible to separate the α and β subunits of hCG and to determine their structure. The α subunit of hCG is similar to the α subunit of luteinizing hormone (LH), follicle stimulating hormone (FSH), and thyroid stimulating hormone (TSH), but the β subunit is different for each of the hormones. Antisera to the β subunit of hCG have been produced that make possible the detection of minute amounts of hCG. With this new immunologic assay, it has been possible to detect hCG in the pregnant woman within a week of conception. The expense of this test has limited its availability.

Fetus

The definitive diagnosis of pregnancy is based on detection of a fetus. Maternal diseases can mimic early maternal changes, and trophoblastic tumors can produce hCG.

The fetal heart can usually be detected by a head stethoscope around the 20th week of gestation. Doppler techniques may detect the fetal heart beat as early as the 8th week after the LMP and can usually detect it by the 10th week. Palpitation of fetal parts or fetal movement by the examiner is a positive sign of pregnancy but is usually not present until after 20 weeks' gestation. The fetal skeleton may be visible in roentgenograms after the 16th week of pregnancy, but x-rays are not advocated to diagnose pregnancy. Ultrasonography can detect a gestational sac within the uterus by six weeks after the LMP. The gestational sac is lost to sonography by 10–12 weeks, and the sonographic diagnosis of pregnancy is somewhat unreliable at that time. From 13 weeks onward, the fetal head can be seen within the uterus by sonography and confirms the presence of a pregnancy.

Once pregnancy has been presumptively or definitely diagnosed, it is customary to calculate the expected date of delivery (EDD). The average pregnancy terminates 40 weeks after the first day of the LMP. In different large series, this has been a consistent finding, with a two-week span on either side of the EDD including almost all pregnancies. Naegele's rule is often employed to obtain the EDD by adding one week to the first day of the LMP and then subtracting three months. For example, if the woman's LMP began on June 15, the expected date of delivery would be March 22. Clinicians date a pregnancy from the start of the LMP, but the student may encounter confusion in his reading because embryologists date pregnancies from the time of fertilization, typically two weeks after the LMP. Clinicians often divide the pregnancy into more or less equal trimesters. The first trimester contains the period of embryogenesis and is the time during which the uterus is still a pelvic organ. The second trimester is the time in which the uterus becomes palpable abdominally and fetal movement is perceptible to the mother. Generally, fetuses born in the second trimester of pregnancy have very little chance for survival. The third trimester begins at approximately 27 weeks' gestation with a fetus weighing 1,000 gm and having a fair chance of survival outside the uterus. Most fetal weight gain is in the third trimester, and this is the time of the greatest metabolic and physical drain of pregnancy on the mother.

WEIGHT GAIN AND NUTRITION

Pregnancy is a state of accretion of mass that results from acquisition of the conceptus (fetal, placenta, amniotic fluid) and from maternal physiologic adaptations (blood volume, uterus, breasts). However, the pregnant adult, like the nonpregnant adult, represents an open-ended energy system, meaning that if she eats more calories than she needs, she will accumulate fat. If she eats less than necessary, she will lose weight.

Most pregnant women in the United States gain about 24 lb during a pregnancy, but deviations from this value are frequent.

The major sources of this weight gain are:

Fetus	3,300 gm
Placenta	650 gm
Amniotic fluid	800 gm
Uterus	900 gm
Breasts	400 gm
Maternal blood	1,250 gm
Interstitial water	1,500 gm
Total	8,800 gm = 19.4 lb

The remainder of the 24 lb is deposited as fat. Usually 3 lb are gained in the first trimester and 10 and 11 lb in each of the later trimesters. Many women retain fluid in their legs because of increased femoral venous pressure. The fluid may account for 3–5 lb of weight before becoming evident as dependent edema.

There is no evidence that toxemia of pregnancy is caused by excessive weight gain (although it might be manifested by sudden weight gain due to fluid retention, especially in the last trimester). As long as sufficient protein, vitamins, and minerals are supplied, normal women should be allowed considerable freedom. For example, obese women may restrict their gain if they wish, and underweight women may gain more than the "normal" 24 lb. There is no value to increasing the anxieties of a pregnant woman by permitting her to imagine that each mouthful of food is critical in shaping the physical, emotional, and mental destinies of her child. To do so only enhances the guilt reaction in a few women whose pregnancies are unsuccessful in producing living, healthy babies. Emphasis should be placed on nutritional quality of food rather than arbitrary limitations on weight.

Excessive gain, failure to gain, or actual weight loss may be noted during pregnancy. Weight is an important measure of health and disease in adults, particularly change in weight. In pregnant patients with chronic or acute disease, it is important to recognize that the situation is more complex than in nonpregnant adults. The fetus serves as a large energy "sink" with a positive balance for amino acids and utilizes carbohydrate as its energy source. Although glucose traverses the placenta readily, and the concentration of glucose is almost identical on the two sides of the maternal-placental interface, the net direction of flux is, of course, from mother to fetus. Free fatty acids are in extremely low concentration in the fetus and do not begin to rise until after birth. In diabetic pregnancy when the maternal glucose concentration is poorly controlled, the concentration of glucose in the fetus is similarly elevated. It is suggested that this overnutrition of the fetus of the poorly controlled diabetic is the cause of fetal macrosomia so common in these newborns. The hypothesis has not been proven, but it will do until more facts come along. Thus, the changes in total body weight during pregnancy are the result of three phenoma: energy balance, the synthesis of a conceptus, and the necessary maternal adaptations and water retention.

When caring for a pregnant patient who gains much more or much less than the average, one must ask: Is there an imbalance in energy metabolism or water metabolism, or possibly both? For example, in a mother with severe untreated hyperthyroidism, weight gain can be nominal because of her increased basal metabolic requirement. On the other hand, if a pregnancy occurs in a woman with chronic renal failure and azotemia or a woman with cyanotic heart disease (hematocrit over 60%), the patient always gains little weight (unless water retention occurs). The uterus grows very slowly, the fetus suffers intrauterine growth retardation, and labor may begin prematurely. A normally functioning and growing conceptus is a stimulus to the maternal adaptations of pregnancy—the increase in blood volume and mass of the uterus and breasts. If growth of the placenta and fetus is inadequate, the maternal adaptations will not occur to the degree that they usually do; thus, the patient will gain little weight.

The woman who "eats for two" obviously will have an excess of caloric intake over energy utilization and will store fat, resulting in greater than average weight gain during pregnancy. When the patient gains excessively and has little evidence of significant water retention, suggesting accumulation of fat, then one must consider her prepregnant height and weight. If her weight was subnormal for her height, this excess weight gain is harmless; in fact, it may be useful. If she was obese to start, then she will be even fatter as the result of the pregnancy experience.

Thus, when a woman gains weight too quickly or too slowly, inquire about her dietary habits, examine her for evidence of water retention or of hypermetabolism or inadequate uterine growth, and reach a diagnosis so that appropriate treatment can be recommended.

The nutritional requirements of normal pregnancy have been the subject of great interest and continuing debate. The Food and Nutrition Board of the National Academy of Sciences-National Research Council periodically recommends daily dietary allowances for pregnant and lactating women of average body size (163 cm or 65 in tall and weighing 58 kg or 128 lb).

Current recommendations include increasing caloric intake from 2,100 to 2,400 kcal during pregnancy as well as increasing protein intake from 48 to 78 gm a day. Vitamin A intake should be increased from 4,000 to 5,000 IU daily; other vitamin requirements increase modestly. Mineral intake of calcium and phosphorous does not change, but iodine, magnesium, and zinc intake should be increased slightly. The iron requirements of normal pregnancy are so great that they often cannot be achieved by dietary intake, and iron supplementation should be provided as 30 mg of iron daily in a preparation such as ferrous sulfate.

If the woman is less than 18 years old, she should have 2,500 kcal a day and twice as much vitamin D as the nonpregnant woman—800 IU a day.

The problems concerned with evaluating nutrition in human pregnancy are many. Most animal studies have dealt with restrictive diets that do not generally occur in humans, and human experimentation by dietary restriction would be unethical.

Dietary supplementation given to poor women in Central America whose average diet consisted of 1,500 kcal with 40 gm of protein daily was able to reduce the incidence of low birth weight infants and increase the mean birth weight by almost 300 gm. Caloric supplementation alone was as effective as caloric supplementation and protein supplementation at increasing birth weight.

There is some evidence from epidemiologic studies that reproductive performance in women relates better to their father's socioeconomic status than to their husbands'. This suggests that some factors affecting reproductive performance operate in utero and during early childhood. It may seem farfetched and somewhat mystical to assume that events in utero may determine the female fetus's reproductive performance 20 years later, but when we consider that active cell replication in many organs of the body terminates at or before birth and that much growth after birth is hypertrophy, it is not so foolish to believe that if the hyperplastic process is interfered with, it cannot be made up after birth. For example, 40% of children weighing less than 2,500 gm at birth are smaller than their classmates when they reach first grade, and they probably never do catch up.

The effects of maternal nutritional deprivation on the newborn have been studied as they related to Holland during World War II. For a well-defined period of time, the food supply to urban Holland was severely curtailed, and generalized undernutrition resulted. In Western Holland, pregnant women received as little as 1,145 kcal with 34 gm of protein a day. Associated with these severe restrictions, the birth weight of infants decreased between 8 and 9% or a median of 240 gm. The greatest decrease in birth weight was associated with famine during the last trimester of pregnancy. Once food was restored, birth weights promptly returned to normal. Despite the famine conditions, premature birth increased only slightly (4.98–6.3%), stillbirths decreased slightly, and the neonatal death rate did not change. Examination of 19-year-old males born during the famine revealed no effects on nervous system function.

This study was not presented to advocate starvation of pregnant women. In general, birth weight parallels maternal weight gain, and the best infant outcome is associated with a maternal weight gain of 20–24 pounds. Vitamin supplements are often given but rarely needed by pregnant women receiving an adequate diet; iron supplementation is necessary. Pregnant women should eat what they want seasoned to taste. Protein intake should approximate 1 gm/kg of body weight a day, and the woman should be weighed at each visit to be sure weight gain approximates normal.

PRENATAL CARE

Prenatal care is medical assistance given to mother and unborn child prior to the onset of labor. To be judged adequate, it should begin prior to the 20th week of pregnancy and should be rendered at regular intervals thereafter. Prenatal care has three purposes.

First is the prevention and/or early detection and treatment of obstetric complications. Many conditions that threaten the health of mother and child become evident during pregnancy. Conditions such as preeclampsia, Rh isoimmunization, and multiple pregnancy are detectable, and the outcome may be modified by appropriate treatment. A way to detect and prevent impending premature labor is greatly needed.

Second is the early detection and treatment of medical and surgical problems. Some conditions may be present when the pregnancy occurs, and the patient may or may not be aware of their existence. The effect of pregnancy on the disease process as well as the effect of the disease on pregnancy should be analyzed and understood as early as possible. Such conditions as heart disease, diabetes, and chronic renal disease are commonly seen. Cancer of the cervix or breast is occasionally found early in pregnancy. Of equal importance to the final pregnancy outcome are those conditions that arise or become manifest during gestation. Urinary tract infections, anemia, and diabetes may not become overt yet still have a serious impact.

Third is the preparation of the family for birth and the postnatal period. Most parents-to-be have informational and emotional needs that should be met by the prenatal care system. The aim of the preparation is to secure the woman's informed cooperation in a program of health care and to decrease the impact of common fears and anxieties by developing confidence in her own strengths and abilities as well as confidence in the concern and skills of those caring for her.

The purpose of the initial prenatal evaluation is to determine the state of physical and mental health that the patient brings to her reproductive experience. This data will be useful in correcting current deficits, anticipating and preventing future problems, and as a data base against which future phenomena can be compared. The prognostic accuracy of this initial evaluation is very high. In addition, the encouragement and confidence gained by the woman as a result of a thorough examination early in her pregnancy are most important benefits.

During the initial history, the status of the woman since her last menstrual period is reviewed in detail. Symptoms, problems, and attitudes are explored. A review of systems emphasizing gastrointestinal, genitourinary, respiratory, and cardiovascular systems is recorded. The woman's past history, with particular emphasis on past obstetric experience, is obtained. Previous unsatisfactory reproductive experience, such as abortions and fetal losses, are dealt with in some detail. If necessary, records from other practitioners or institutions are requested. The frequency with which prematurity, fetal death, and neonatal death recur in subsequent pregnancies requires careful attention to this item. A family history of diabetes, epilepsy, hypertension, renal disease, and bleeding disorders may be important information.

Initially, a complete physical is performed, including optic fundi, external auditory meati, and tympanic membranes. A thorough chest examination, including auscultation, is performed. The state of the breasts and nipples is noted. The heart is examined and the presence and character of any murmurs described. The abdomen is carefully examined, and the presence or absence of uterine fundus is noted. If the fundus is identified, it is measured with a tape measure from the pubic symphysis vertically up the midline to the top of the uterus. Asymmetry of the uterus or presence of other abdominal masses are noted. Pelvic examination includes a gonococcal culture and cytologic smear from the cervix. A careful evaluation of uterine size and the adnexal structures is made. A rectal examination is done. The extremities are evaluated for lesions, deep tendon reflexes, and edema. Varicosities and other vascular changes are also noted.

The initial laboratory evaluation should include serologic test for syphilis, blood grouping and Rh typing, determination of hemoglobin and/or hematocrit, white blood cell count, rubella antibody titer, urinalysis, and the before-mentioned cytologic smear. Chest film is obtained from high-risk patients or patients with a significant history. Cervical and anal cultures for the gonococcus are often added. Additional laboratory work is ordered to identify any problems that might have been detected by history or physical examination.

From data obtained on history and physical examination, one should determine the presence or absence of a pregnancy and the duration of pregnancy based on the date of the last menstrual period as well as on the size of the uterus. Any discrepancy between gestational dates and pregnancy size is noted, but the LMP is almost always used as the basis for calculation of the expected date of delivery. All departures from normal health status, both physical and mental, are listed and appropriate evaluation planned. Previous health-care records are requested, consultations obtained, and an estimate made of maternal and fetal risk. Most patients receive a nutritional supplement containing iron, vitamins, and minerals. The value of this supplement is questionable except for iron. Because of the iron loss of menstruation and with previous pregnancies and because of fetal requirements, supplementation of elemental iron orally is considered useful. Patients are given general dietary advice emphasizing protein intake.

Women are seen for subsequent prenatal examinations at increasingly shorter intervals during the remainder of the pregnancy. For the first 28–30 weeks, normal patients are seen once each month and thereafter at two- to three-week intervals until the final month. At this time, weekly visits are made. At each visit, the woman is weighed, and blood pressure is taken in a sitting position. A careful interview establishes the status of her health during the interval since her last examination and pertinent data are recorded. Specific questions—how do you feel, are you sleeping, how are your bowels, do you have any pain, is there anything you wish to ask me?—will detect physical or emotional problems. A specimen of urine is obtained and tested for sugar and albumin. In most clinics, the woman's abdomen is exposed and the uterine fundus palpated and compared to what is usual for that period of gestation. At 18 weeks, the uterus should be near the umbilicus, and at this time the fetal heart is listened for. Commonly, the fetal heart is first heard around 20 weeks and helps confirm the previously recorded EDD. Fetal movement is usually detected by the woman between the 17th and 20th week. The woman is asked to remember and report when she first perceives fetal movement. At about 28 weeks, the fetus can be outlined and its relationship to the mother determined. The relationship of fluid to fetus also is noted, as is the consistency of the uterus and areas of tenderness. Under normal circumstances, the reflexes are not tested, but observations are made of the face, hands, legs, and feet for edema. This kind of information may be recorded in many forms. Some clinics use a flow sheet; others record in a narrative form.

In addition to the chemical urinalysis performed on each visit, the hemoglobin/hematocrit is determined at 27 or 28 weeks and again at 34–35 weeks. If the woman is Rh negative, antibody titers are obtained as required. Some clinics now perform a repeat serologic test for syphilis at 35–36 weeks, so that if syphilis has been acquired in the interval since the first visit, treatment can be instituted prior to delivery. Toward the end of the pregnancy, a copy of the prenatal record is sent to the hospital where the patient has chosen to have her baby in order that this information will be available when the woman is admitted in labor. In addition to this, some clinics give their patients a data sheet on which certain information is recorded each visit. The patients then give this to the attendants when they report to the hospital.

During each visit, the woman's attendants attempt to ascertain and deal with her questions, fears, and concerns. Also, they keep her informed as to the progress of her pregnancy and give her whatever instructions are necessary as clearly and as carefully as possible. Every attempt is made to maintain the rapport established between patient and her attendants. Failure to follow instructions should not be the cause for a disturbed relationship between patient and physician. Every effort must be made to maintain the relationship and ensure whatever prenatal care the woman is able to accept.

Whenever possible, a planned program of education for both parents-to-be is instituted. The educational program usually focuses on the physical and emotional phenomena of pregnancy. Instruction in basic reproductive biology, including anatomy, conception, and fertilization, fetal growth and development, maternal changes of pregnancy, psychologic adaptations and stresses encountered by both mother and father, problems of other children in the home, coital activity during pregnancy, and other hygienic measures are discussed. When presented to groups in a formal fashion, this kind of information takes two or three one-hour sessions plus time for questions and answers. Various means of family planning and the health benefits of child spacing should also be discussed with both parents.

The hospital experience, including labor and delivery, are explained in detail. The means of pain relief, the various examinations and procedures done to and for women in labor, the probable method of delivery with the possible choices, the roles of house staff, nursing staff, and the relationship between the obstetrician and his patient are all carefully outlined. If possible, the parents-to-be are given a tour of the hospital facilities to reduce their anxieties. The duration of stay in the hospital, the puerperal experience, and what can be expected in return of activity and function are clearly explained. Hospital expenses are frequently discussed during this time. Care of the newborn, including bathing, methods of feeding, and danger signals, is fully discussed. Breast-feeding should be encouraged.

COMMON PROBLEMS OF NORMAL PREGNANCY

Many minor ailments and disabilities occur in normal pregnancy but are not serious enough to constitute a complication. These will be considered in the organ systems involved.

Gastrointestinal

Nausea and vomiting occur in early pregnancy very commonly. It is somewhat related to an empty stomach, and frequent small feedings are often helpful in reducing the intensity of the symptom. It usually disappears by the 14th week. Helpful drugs are available, but women rarely demonstrate weight loss or ketonuria. When this occurs, rehydration by intravenous fluids is advisable.

Constipation is a very common problem often related to insufficient fluid intake or the deferring of bowel movements. It is nearly always possible to reeducate women in good bowel habits, proper fluid intake, and the use of fruits and cereals in maintaining good bowel function. Laxatives should be avoided if at all possible.

Heartburn usually occurs in late pregnancy and is probably related to terminal esophagitis. Unless warned, some pregnant women may ingest large amounts of sodium bicarbonate in an attempt to treat this symptom. Antacids not containing sodium, such as aluminum hydroxide or magnesium tricilicate, may be used.

Genitourinary

Urinary frequency is experienced by nearly all pregnant women during the entire period of gestation. Because urinary tract infection is so common and may be relatively asymptomatic, careful inquiries of additional symptoms, such as dysuria and hematuria, are often necessary. Women should not restrict their fluids because of urinary frequency.

The loss of urine during coughing, sneezing, crying, climbing stairs, or straining is a common problem associated with the latter months of pregnancy in women who have previously borne children. Women may be encouraged that this symptom will usually go away when the pregnancy is ended unless the symptom was present prior to the onset of the pregnancy. Treatment is not warranted.

Vaginal discharge is a normal component of pregnancy because there is an increased amount of cervical mucus that appears at the vaginal introitus. As long as this produces no other symptoms, treatment is not warranted. When itching, burning, or discomfort occur, examination is necessary in order to determine the cause. Appropriate therapy should be instituted because of the discomfort associated with these symptoms.

Musculoskeletal

Backache associated with the increased lumbar lordosis is noted by most pregnant women. Instruction in proper posture and shoes and in techniques of lifting and simple exercises may be helpful. When severe, it may be necessary for the patient to wear a maternity girdle in order to lift the protuberant abdomen and therefore compensate somewhat for the lumbar lordosis. If the backache is early morning in nature, a firmer mattress may be needed. At times, nerve root distribution of pain similar to that seen with disc disease is found. When severe, this may require hospitalization. Pain in the upper back (T 1-5) is frequently experienced as a result of stresses compensatory to the lumbar lordosis. Here again, training in posture, analgesics, and heat should be helpful.

Varicose veins are frequently seen in women who have had children and are most common in the saphenous distribution along the inner aspect of the thigh, the popliteal fossa, and over the calf. Although they usually do not produce discomfort, elastic stockings may be worn for cosmetic purposes. When varicosities are sufficiently high on the upper leg or on the vulva, it may be necessary to wear a leotard garment to keep them compressed. Surgical treatment is rarely indicated in pregnancy, and rupture of a varix rarely occurs.

PSYCHOLOGIC ADAPTATIONS TO PREGNANCY

Pregnancy is a crisis situation in which the woman must accomplish both physiologic and psychologic adaptations if she is to be successful in producing and rearing a child. This is not her first such "crisis," for puberty and heterosexual maturation have preceded the pregnancy. The first pregnancy tends to revive problems and unsettle the resolution of psychologic conflicts of earlier years. Thus, pregnancy, particularly the first, represents a new disequilibrium state requiring a solution through maturational reorganization. This unsettling of the woman is not "bad." It is necessary if she is to mature from a wife role to a wife and mother role. Furthermore, this disequilibrium state makes for ready acceptance of a new transference figure—MD, RN, MSW—provided the transference figure is understanding and sympathetic. While the woman may exhibit pathologicic symptomatology at times, the disequilibrium state usually responds to brief superficial therapy as the adult woman evolves from self-dependence (unmarried) followed by interdependence (with husband) to her new responsibility of motherhood.

The advent of pregnancy, like puberty and the menopause, is another testing ground of psychologic health. The woman, with a modicum of professional support, handles this well if she is healthy. However, if she is emotionally unhealthy, and the environment is unfavorable, she may proceed to neurotic or psychotic solutions. All women experience this crisis, but the intensity of the crisis and the degree of success in its resolution are determined by personal resources and social and cultural forces. In summary, preventive psychologic management requires recognition of these facts and incorporating them into the prenatal care program, for the degree of resolution of this crisis determines the set of the mother-child relationship. This is particularly true for the woman pregnant for the first time.

The basic relationships of the family are expressed in the triad of husband, wife, and child, but the start of it all begins in marriage. In pregnancy, a significant representation of the love object (the conceptus) becomes a part of the woman herself. The physical changes of early pregnancy, nausea and vomiting, amenorrhea, breast and abdominal enlargement, enforce her narcissistic libido. She tends to concentrate on herself and considers the conceptus as an integral part of her own body, almost as a new organ. At the same time, it is a threat to the woman's body image. No woman finds joy in swollen belly and breast or morning nausea. As you can see, there is much opportunity for conflict here between husband and wife and within the wife herself as she turns inward. Early pregnancy is rarely accomplished without some evidences of conflict in these respects.

Most women report that after feeling the baby move (about 18 weeks), they usually feel very well, emotionally and physically, from that time until they become quite ponderous. While the patients do not relate these events as cause and effect, they do exist in that relationship. With the onset of fetal movement, the biologically fanciful concept of the conceptus as a maternal organ ends, and a new maternal-fetal relationship begins where she sees the fetus as separate (and separate means unequal). She now knows that it *is* a fetus; it *will* be a newborn; it *will* leave her. This is a new level of fusion of narcissistic and libidinal strivings. The shift is from herself to a new love object soon to be outside herself. As a patient once said, "I won't be Jeannie much longer, but I'll be a mother forever and ever." How successfully a woman adapts to pregnancy can be predicted by her relationship to her husband (specifically) and men (generally); her modes of receiving, retaining, or releasing and her love-object relationships.

The preceding paragraphs described the fundamental psychologic processes of adaptation in pregnancy and the puerperium. However, the patient does not exist in a vacuum but in a society in which her subculture and her educational, social, and economic sets function as significant modifiers of her response to impregnation. Thus, we see the tremendous ambivalence of women who are pregnant out of wedlock, with the concept of a child as an expression of their femaleness and the existence of their child as destructive to almost every one of their life goals. Almost any medical student-husband can set his wife to vomiting in early pregnancy by making it clear that: he is displeased with the pregnancy; it is her "fault"; this event puts great economic pressure on him because she won't be able to work; he will probably never graduate; she has already begun to change physically; her features are coarser and swollen; she is quite unattractive; and the secretaries at the school look better to him all the time. Similarly, the 45-year-old woman with three children whose youngest is 12 years old and who has become pregnant in spite of contraception has little that is positive, for she has been satisfied with her previous obstetric experiences. She can find no positive expression in this unwanted event, only negative responses. Thus, in all early pregnancies, the woman's innate personal responses will be considerably modified by the economic and social realities of the event and its impact on interpersonal relationships, particularly with her husband.

From a clinical point of view, the woman in early pregnancy does have real symptoms and signs: missed periods, enlarging breasts, enlarging abdomen, nausea, hunger, urinary frequency and nocturia, constipation, and 20% of the patients have some spotting as well. All these real events reinforce the concern with herself in the absence of any real evidence of pregnancy. All these signs and symptoms, including the physician's reassurances, enhance the probability that she is pregnant, but none of them are absolute signs to her. Thus, her responses to all these symptoms and signs and their severity will be determined by the external and her internal hostility to the pregnancy.

However, once fetal movement occurs, with attendant recognition of the "separateness of the fetus," in the emotionally healthy woman whose conception is consistent with her personal goals and supported by her family environment, a period of peace and good feelings evolves. The woman who has multiple symptoms without disease after the appearance of fetal life has not made an adequate psychologic adaptation to pregnancy or has an undetected physical ailment. The patient who has made a good adaptation becomes concerned about "normalcy" (Am I? Is the baby?) and has a growing interest in the birth process. Unpleasant dreams are common but usually relate to an abnormal fetus or injury to herself at delivery. Physicians and nurses can be of help to her by supplying information about pregnancy, the birth process, and newborn care. Often patients will ask questions of nurses they are loath to ask of physicians. ("I didn't want to waste his time.") The woman wants to be in an environment that will accept "stupid" questions and offer valid answers. She is often curious about analgesia and anesthesia. Both doctor and nurse can be informational resources to the patient. The normal well-adjusted woman after the 20th week of pregnancy is tuned to "receive mode," and it is a pity that a great learning opportunity in preventive medicine for our children is lost so often.

The woman has no identification and no pride with her big belly and usually has a progressively decreasing tolerance to discomfort as term approaches. Patients will often verbalize some coital anxiety if they are in a professional atmosphere in which they feel free to speak and often need some reassurance. More than 30 years ago, it was very easy to advise patients about coitus in late pregnancy. The advice was "Do not have coitus during the last six (eight–ten–twelve) weeks of pregnancy." When someone finally asked, "Why?" and received no answer, women were then advised that coitus is quite permissible at all times during late pregnancy. Recently, some studies have been published to suggest that female orgasm may be associated with active uterine contractility and may induce various sensations of discomfort in the patient. It has been suggested that almost 50% of women in late pregnancy decrease the frequency of orgasm simply to avoid these symptoms. We are concerned that orgasm may be a factor in the initiation of premature labor. Obviously, the last word has not been written on this subject, but at the moment, it is a little difficult to accept this suggestion, for the women with the highest incidence of orgasm and the highest incidence of multiple orgasms (the upper middle class) have the lowest incidence of premature labor. Certainly coitus should be avoided in the last trimester by women with past histories of premature labor or current history of vaginal bleeding or ruptured membranes.

If adapted emotionally, as the woman reaches term, she becomes ever more ready to turn the fetus loose. She becomes impatient with waiting. The reactions may be interpreted by the professional as anxiety about the birth experience, but this is generally not true. The patient simply is ready for the baby to leave her body; she wishes to hold it and examine it. Because 50% of women will not be delivered by EDD, it is helpful to remind the woman frequently throughout pregnancy that her chances of delivering two weeks after the EDD are almost as good as her chances of delivering two weeks before it.

Once delivered, the normal woman's rate of recovery for the experience is more determined by the anesthetic than by the specific experience of labor and delivery. Once she has seen and examined her child, she is usually elated, pleased with her accomplishment, pleased with her child, and relieved of her anxieties about her and her child's health. It is curious to note that almost every woman who receives her child in her own bed postpartum for the first time will undress the infant and examine it totally. However, once the party is over, about 48–72 hours postpartum, many women get the "blues," manifested by unexplained depression and accompanied by crying without apparent cause. Again, in the normal woman, this phase is mild and short-lived, and soon she gets on with her work of being a mother. There are probably many reasons for this postpartum episode of the blues, but the explanations are basically speculative: steroid hormone withdrawal, the face-to-face reality of the child with all the attendant limits and the responsibilities this fetus imposes, guilt reaction to sexual gratification in nursing, being forced to share with her child the attention of husband and family, guilt reactions for ambivalent feelings about the pregnancy and the child, etc. Whatever the reason, this phase is brief in normal women, usually lasting no more than a day. However, if the woman's depression is deep, continues, and is associated with insomnia, the physician should suspect a developing postpartum psychosis. Puerperal psychoses appear at a frequency of about 1 per 400–1,000 pregnancies. It is written that 3–10% of all psychotic females admitted to neuropsychiatric hospitals have had a recent childbirth.

A final note on unsuccessful pregnancy: spontaneous abortion, stillbirth, malformation, and neonatal death. All of these experiences are interpreted by the woman as a failure in her fundamental female role. We know the woman's behavior and thoughts do not create these losses. She is guiltless; however, because of the ambivalence every woman has toward pregnancy, she feels guilty and has a sense of failure and a sense of loss as well. She and her husband need help from a professional. She needs education from her physician to help relieve her guilt and reassurances for future successful childbearing or adoption, whichever is appropriate. Her husband also needs education both about the facts of the problem and about the very significant supportive role he can play in her recovery. If uninstructed, a "wall of silence" often develops between husband and wife on this subject. This failure of communication aggravates the problem. However, the investment of a brief amount of time by the physician to help the couple understand what has happened to them and why, to accept their blamelessness, have some understanding of their future childbearing potential, and the like, requires only a short period of time and is a very effective preventive to psychopathology and marital discord.

SUGGESTED READING

Lechtig A, Habicht JP, Delgado H, et al: Effect of food supplementation during pregnancy on birthweight. *Pediatrics* 56:508, 1975.

Pritchard JA, MacDonald PC: *Williams Obstetrics*, ed. 15. New York, Appleton-Century-Crofts, 1976.

Smith CA: Effects of maternal undernutrition upon the newborn infant in Holland (1944–1945). *J Pediat* 30:229, 1947.

QUESTIONS

1. List three maternal physiologic and/or anatomic changes that are useful in making the presumptive diagnosis of pregnancy in the first 12 weeks of gestation: ______________________________

____

2. List three fetal and/or trophoblastic changes and their time of appearance that are useful in making the diagnosis of pregnancy.

____

____

____

3. Give the following daily nutritional requirements of a normal pregnant woman: calories________; protein________; iron ________
4. List seven major sources for the weight gain associated with normal term pregnancy. ________, ________, ________, ________, ________, ________, ________.
5. List three purposes of prenatal medical supervision:

____

____

____

6. List the major conclusions drawn from the initial prenatal evaluation. ______________________________

____

____

7. List four major items usually included in prenatal educational efforts:

____

____

____

____

ANSWERS

1. Amenorrhea, uterine softening, uterine enlargement
2. a. hCG production—14 days after first missed menstrual period by commonly used tests.
   b. fetal heart beat—8–10 weeks by Doppler technique—20 weeks by stethoscope
   c. fetal parts—5–6 weeks by sonar, 16 weeks by x-ray, over 20 weeks by palpation
3. a. calories 2,400
   b. protein 75 gm
   c. iron 30 mg supplement to diet
4. fetus, placenta, amniotic fluid, uterus, breasts, blood, water, fat
5. a. prevention or early detection of obstetric problems
   b. detection and treatment of medical and surgical complications
   c. preparation of the family for birth and postnatal life
6. a. confirmation of pregnancy
   b. duration of the pregnancy based on uterine size and LMP
   c. estimated date of delivery
   d. general health status of the woman
7. a. physical and emotional phenomena of pregnancy
   b. hospital experience
   c. care of the newborn
   d. family planning

chapter eighteen
high-risk pregnancy

OBJECTIVES

1. Define "high-risk pregnancy."
2. Define maternal mortality, fetal death, neonatal death, and perinatal mortality.
3. Define abortion, immature birth, premature birth, term birth, and postdate birth.
4. List factors that have stimulated development of the concept of high-risk pregnancy as it applies to the delivery of reproductive health services.
5. List the major causes of maternal mortality and identify factors that are known to be associated with increased maternal risk.
6. List the major causes of perinatal mortality and identify maternal factors known to be associated with increased perinatal risk.
7. Appreciate how a health-care system can be developed to meet and serve a pregnant woman at her level of need.

High-risk pregnancy may be defined as any pregnancy that has an increased probability of maternal or perinatal death or damage. Over half of pregnancies that will have this poor outcome can be identified prior to pregnancy or early in gestation.

Before proceeding in the discussion of high-risk pregnancy, it is necessary to define some terms that will be used in the chapter:

- *Maternal mortality rate*. The number of maternal deaths that occur as a result of pregnancy or in the early puerperium. The rate is calculated per 100,000 live births.

- *Neonatal death rate*. The death of a newborn infant within the first 28 days of life. The rate is calculated per 1,000 live births. (*Note*: Some authors will vary this to include newborns of a certain weight or will include only deaths in the first 7, 10, or 14 days of life.)
- *Fetal death rate*. The death in utero of a fetus over 20 weeks in gestational age and/or over 500 gm in weight. The rate is expressed per 1,000 births.
- *Perinatal mortality rate*. The sum of neonatal and fetal death rates.
- *Abortion*. A pregnancy terminated before 20 weeks of gestational age. (Some states say a fetal weight less than 500 gm.)
- *Immature birth*. A pregnancy terminated between 20 weeks and the end of the 27th week and/or the delivery of an infant weighing between 500 and 999 gm.
- *Premature birth*. Delivery between the 28th week and the end of the 36th week.
- *Term birth*. Delivery between the 37th week and the end of the 41st week.
- *Postdate birth*. Delivery after the 42nd week of pregnancy.
- *Low-birth-weight infant*. This is a general term for infants born weighing less than 2,500 gm. About two-thirds of these infants are premature by dates, and the other third represent term births of growth retarded infants.

The concept of high-risk pregnancy has been stimulated from several areas. In recent years, the maternal and perinatal death rates have not dropped as sharply as in preceding decades. Studies have indicated that two-thirds of maternal and one-third to one-half of perinatal deaths might have been prevented if modern diagnostic and therapeutic measures had been fully brought to bear. In addition, it is a basic principle of perinatology that any condition that produces death also increases the incidence of permanent damage among survivors. For example, the highest incidence of brain-damage babies occurs among survivors of premature birth.

It is neither possible nor necessary to provide every woman and newborn the depth of care required to prevent all losses. For example, there are certain obstacles to care, such as cultural and social customs, that make modern methods embarrassing or frightening to some women. Economic and geographic barriers to care often exist. Some tests are expensive, time-consuming, and uncomfortable. Even if these factors were not operative, large numbers of pregnant patients have to be seen and evaluated in order to identify the 30% of women who produce 60% of the problem pregnancies. It is then necessary to give to these selected patients the increased services that could make a difference.

The concept of high-risk pregnancy was developed in order to provide maximum utilization of all resources to those who would benefit most, with a view to reducing preventable death and damage. The identification of these high-risk patients enables us to provide them with increased care while not expending expensive and limited resources on patients of lower risk. To be effective, these measures must begin prior to conception and continue throughout pregnancy, labor, and delivery and on into the first weeks of life.

MATERNAL HIGH RISK

Table 18–1 illustrates the dramatic decrease in maternal mortality that has been achieved in the United States within the last 60 years. This has been due to many factors, including the practice of hospital delivery, blood transfusions, antibiotics, local and state review of maternal deaths, and an increase in the number and quality of training programs in obstetrics.

Table 18–1. Maternal Mortality Rates in the United States for Whites and All Others (Figured Per 100,000 Live Births)

Year	*Total*	*Whites*	*All Others*
1925–1929	668.6	615.0	1163.7
1935–1939	493.9	439.9	875.5
1943	245.2	210.5	509.5
1948	116.6	89.4	301.0
1953	61.1	44.1	166.1
1958	37.6	26.3	101.8
1963	35.8	24.0	96.9
1968	24.5	16.6	63.6
1973	15.2	10.7	34.6

From *Vital Statistics of the United States 1973*, vol. II, Mortality, Part A, Tables 1–16.

Table 18–2. Causes of Maternal Mortality in 1973 and the Maternal Mortality Rate for Each Cause

Cause	*Number of Deaths*	*Rate per 100,000 Live Births*
Toxemia	111	3.5
Sepsis	98	3.1
Hemorrhage	60	1.9
Ectopic pregnancy	46	1.5
Abortion	36	1.1
All other causes	124	4.0
Unknown	2	0.1
Total	477	15.2

From *Vital Statistics of the United States 1973,* vol. II, Mortality, Part A, Tables 1–15.

Despite this decrease in the number of maternal deaths, the three main causes of maternal death have remained the same: toxemia, sepsis, and hemorrhage. Table 18–2 presents the causes of maternal mortality in 1973. Causes of maternal mortality may vary in selected areas, such as university hospitals, where heart disease and anesthetic deaths may assume a relatively greater role.

Characteristics of women at higher risk of dying as a result of pregnancy include race, age, parity, and past history as well as factors detected by a physical examination or laboratory evaluation.

Nonwhite populations in the United States have higher maternal mortality rates than the white population (Table 18–1); presently, it is three times as high; in the recent past, it has been four times higher. The difference in rates is usually explained by social and economic factors operating to deny optimum reproductive performance and care.

There is an increasing risk to pregnancy as the age of the mother increases. The maternal mortality rate almost triples for women aged 35–39, is five times higher for women aged 40–44, and is 10 times higher for women pregnant at age 45 or more. Age is related to maternal death primarily through toxemia, heart disease, placenta previa, and, perhaps, ruptured uterus. Although preeclampsia and eclampsia occur more frequently in the very young pregnant woman, the superimposition of toxemia on preexisting chronic hypertension is a more serious threat to a woman's life. Superimposition is more common in the older woman because chronic cardiovascular disease is more common.

Women of high parity who are in their sixth or subsequent pregnancy are also at greater risk. Ruptured uterus and postpartum hemorrhage both occur more frequently in women of high parity. Multiple pregnancy, infants of excessive size, and abnormalities of fetal presentation in labor are also common and can produce maternal complications.

Past obstetric history of cesarean section, puerperal hemorrhage requiring transfusion, and toxemia in two or more pregnancies all indicate an increased likelihood of serious complications. Likewise, a history of chronic hypertension, chronic renal disease, heart disease, pulmonary disease, or cerebrovascular disease is often encountered in patients who subsequently die in pregnancy.

Hypertensive changes in the optic fundi indicate increased risk. An overdistended uterus (twins, polyhydramnios, or large fetus) indicates an increased chance of postpartum hemorrhage. Large infants, abnormal fetal presentations, and small bony pelvis indicate a possible abnormality of labor that could result in an operative delivery with an increased chance for hemorrhage, infection, or anesthetic complications.

Laboratory findings of severe anemia or albuminuria indicate trouble, as do chest roentgenograms confirming cardiac or pulmonary disease.

PERINATAL HIGH RISK

The neonatal mortality rate, fetal death rate, and resulting perinatal mortality rate have all exhibited a downward trend in recent years, as shown in Table 18–3. As in the case with maternal mortality, the etiology of perinatal deaths will vary with the population served. Table 18–4 shows the causes of perinatal deaths over 1,000 gm at the Robert B. Green Hospital in San Antonio, Texas, for the five years 1971–1975. During this time, there were 23,915 births, the perinatal mortality rate dropped from 26.4 per 1,000 to 17.3 per 1,000, and the percentage of low-birth-weight infants born decreased from 10.9 to 8.8%. The cause of death in about one-fourth of cases cannot be determined by examination of the mother or infant. However, in the remaining three-fourths, it is apparent that the sequellae of prematurity constitute a large portion of the total. Babies dying of respiratory distress, intracranial hemorrhage, or sepsis are usually prematures. Maternal

Table 18–3. Neonatal Mortality Rate, Fetal Death Rate and Perinatal Mortality Rate (Figured Per 1000 Live Births)

Year	*Neonatal Mortality Rate*	*Fetal Death Rate*	*Perinatal Mortality Rate*
1915–1919	43.4	—	—
1935–1939	31.0	—	—
1943	24.7	24.5	49.2
1948	22.1	20.6	42.7
1953	19.6	17.8	37.4
1958	19.5	16.5	36.0
1963	18.2	15.8	34.0
1968	16.1	15.8	31.9
1973	13.0	12.2	25.2

From *Vital Statistics of the United States 1973*, vol. II, Mortality, Part A, Tables 2–1 and 3–2.

Table 18–4. Causes of Perinatal Mortality (> 1000 gm), Robert B. Green Memorial Hospital 1971–1975, Total Deaths: 383

No abnormal state in the infant			146	38%
Without maternal disease	90	(23%)		
With maternal toxemia	32	(8%)		
With maternal diabetes mellitus	13	(3%)		
With other maternal diseases	11	(3%)		
Hypoxia			71	19%
Placental disease	32	(8%)		
Complications of labor	23	(6%)		
Cord compression	11	(3%)		
Other	5	(1%)		
Respiratory distress syndrome[a]			54	14%
Malformation			63	16%
Infection[a]			32	8%
Erythroblastosis fetalis			3	1%
Birth injury			5	1%
Meconium aspiration			9	2%
				99%

[a]Related to prematurity.

illnesses, such as toxemia and diabetes mellitus, constitute a large identifiable cause of perinatal mortality.

Characteristics of pregnancies destined to result in perinatal death can be identified by race, age, parity, past obstetric history, past medical history, family history, and physical examination, and laboratory evaluation.

As in the case with maternal mortality, perinatal mortality in the United States is much higher in the nonwhite population, almost twice as high in 1973. Again, this is usually explained by economic and social factors that impede access of nonwhite parents into an effective health-care system.

The reproductively very young or very old woman is prone to develop toxemia of pregnancy that may kill by intrauterine demise of complications of a premature delivery. Older patients have a higher incidence of placenta previa, which may result in premature delivery or maternal shock, which are harmful to the fetus. The chance for chromosomally abnormal infants also increases with increasing maternal age.

The babies of primagravidas are at increased risk because of toxemia and dystocia. The multipara has a higher incidence of multiple gestations with its 50% prematurity rate. The maternal hazards of grand multiparity previously mentioned (ruptured uterus, mal presentation) have a direct effect on perinatal mortality.

A past obstetric history of prematurity or fetal death is significant. When a patient has had two or more pregnancy losses past 20 weeks, the risk is increased and grows rapidly thereafter. Careful documentation of clinical and autopsy records can be most helpful in assessing the degree of risk. Previous cesarean sections or repeated difficult deliveries are also important clues. Specific neonatal diseases, such as respiratory distress syndrome and erythroblastosis, are valuable pieces of information.

Past medical history of chronic or recurrent renal disease, diabetes, and chronic cardiopulmonary disabilities has significant impact on perinatal prognosis and is encountered rather commonly.

A family history of diabetes mellitus may indicate subclinical diabetes in the patient, and tests for diabetes are usually carried out. Multiple gestation is somewhat familial. A history of repetitive familial disorders, such as bleeding tendencies, malformations, or anemia, may warrant investigation.

Evidence of cardiovascular or renal disease on physical examination is serious. Failure of the uterus and its contents to grow as the pregnancy progresses might indicate intrauterine growth retardation from either maternal or fetal causes. Conversely, a uterus larger than expected may contain more than one fetus or excessive amounts of fluid. Appropriate diagnostic and therapeutic measures could prove useful in preventing premature labor in such circumstances. The development of hypertension in the latter weeks of pregnancy indicates the probability of preeclampsia or, perhaps, previously unrecognized chronic hypertension, both of which have a deleterious effect on the infant.

Presence of proteinuria or glycosuria indicates renal disease or diabetes. Gestational diabetes may be detected by a glucose tolerance test. Presence of Rh antibodies in the blood of an Rh negative woman indicates a previous sensitizing exposure to Rh positive blood. Likewise, the presence of a positive serology indicating untreated syphilis is significant.

In practice, an assessment and prognosis for the pregnancy are made at the initial prenatal visit. At this time, the combination of history, physical, and laboratory examinations should identify most of the 30% of women who will produce 60% of the perinatal and maternal losses. These patients should then be sequestered and arrangements made to give them intensive prenatal care. Where large groups of patients have to be cared for by few attendants, it is possible for the selected normal patients to be seen only occasionally by physicians. A large part of their prenatal supervision may safely be conducted by nurses or nurse-midwives.

From this low-risk population, an additional group of patients at increased risk to themselves or their infants will be identified as pregnancy proceeds. Hypertension, bleeding in the third trimester, intrauterine growth retardation, or pyelonephritis may occur. Anemia, proteinuria, glycosuria, or Rh isoimmunization may be recognized. These patients then join the previously selected high-risk group for increased observation for the remainder of the pregnancy.

Appropriate treatment is offered throughout pregnancy. Most often, this is given to the mother but often results in a useful effect on the fetus. Eradication of maternal infections, such as syphilis or pyelonephritis, will directly benefit the unborn infant. There is a direct relationship between control of endocrinologic disorders (i.e., diabetes and thyroid dysfunction) and fetal health. Surgical repair of a faulty cervix or medication to stop premature labor may be useful. Prolonged bed rest for patients carrying twins or triplets is probably helpful. Medical and surgical illnesses detected during the pregnancy are treated as indicated, therefore maintaining as optimal a state of health as possible.

One of the more exciting recent advances in obstetrics has been the direct treatment of the fetus. This began with the sampling of amniotic fluid for diagnosis of fetal well-being. Soon therafter, donor red blood cells were deposited in the abdominal cavity of the erythroblastotic fetus without producing premature labor, and intrauterine transfusion was born. At present, this is the only direct fetal therapy in clinical practice. It is likely, however, that other fetal illnesses will be treated similarly in the near future.

As the end of gestation approaches, new questions arise. Should the pregnancy be terminated prior to term? Preterm delivery in the interest of the infant is often required in pregnancies complicated by diabetes. Should the pregnancy continue to term but labor be avoided? Patients with two or more previous cesarean sections or other uterine surgery are usually scheduled for abdominal delivery at or near term. If labor is to be awaited, what circumstances require special or different management at the onset? For example, patients known to have a small pelvis might be managed differently if the presentation on admission was other than cephalic. Patients with a past history of postpartum hemorrhage or with anemia or an overdistended uterus would be prepared for transfusion.

Once labor is established, would electronic monitoring of fetal health be useful? Should labor be managed differently regarding analgesia, duration, or oxytocin augmentation? What method of delivery and what anesthetic would be best? Is a depressed infant requiring resuscitation expected? Will prompt transfusion of the newborn be required? Careful attention to such questions will almost always allow the team to anticipate and prepare to take care of problems.

The care of the high-risk pregnancy does not end in the delivery room. Intensive care nurseries, equipped and staffed for the complete care of infants born at increased risk, are rapidly replacing the "premature" nursery. To be sure, most premature infants need special care, but so do babies from diabetic mothers, undergrown term infants, babies shocked by maternal hemorrhage, and many other infants.

Implementation of the high-risk -pregnancy concept of prenatal care pays great dividends in decreasing maternal and perinatal mortality.

The opening of the University of Texas Medical School in San Antonio provided an opportunity for the study of the impact of a medical school on medical care in a county hospital. On July 1, 1968, the faculty of the clinical departments of the medical school became the attending staff of the hospitals. Although training programs had existed in the county hospitals prior to that date, the first group of interns and residents to be selected by the medical school started their training on July 1, 1969. Systematic collection of information on maternal and perinatal mortality and morbidity at the Robert B. Green

Memorial Hospital began on January 1, 1970. The county hospital provides the only perinatal service for indigent patients in the county.

The county hospital's obstetric and neonatal services offer an organized system of perinatal care. Multiple portals of entry to care are provided via city-county clinics, obstetric intake services, emergency room of the county hospital, and the labor and delivery suite. First contact and first judgment regarding the quality of care required for each woman is usually made by nurses trained to identify patients at increased risk for maternal or perinatal loss. These nurses have the authority and opportunity to obtain prompt physician consultation in ambulatory or inpatient environments, as needed. Special diagnostic tools, such as ultrasonography, amniotic fluid analysis, and hormone assays are available to inpatients and outpatients. Medical care appropriate to need is provided by a medical student/house officer/faculty team.

Women judged to have high-risk pregnancies are seen in a "complicated obstetrics" clinic by senior house officers under faculty supervision. Women with low-risk pregnancies are seen in neighborhood clinics or by junior house officers.

This same structuring of health care to meet the needs of the pregnant woman is carried out on the inpatient service, the labor and delivery area, and in the nursery.

For the six-year period 1970–1975, the number of births at the county hospital increased from 4,100 in 1970 to 5,258 in 1975. The percentage of births at the county hospital increased from 21 to 29% of the total births in the county. Despite this increase in services to a high-risk group of patients, the perinatal mortality rate at the Robert B. Green Memorial Hospital fell from 34.2 per 1,000 in 1970 to 17.3 per 1,000 in 1975—almost 50%. By 1975, the county hospital had a lower perinatal mortality rate than the private hospitals or the military hospitals in the county. This rate was also considerably below the 1973 perinatal mortality rate for the United States. Thus, the concepts of identification, sequestration, and intensified perinatal care for women at high risk have paid concrete dividends in reduced perinatal mortality.

SUGGESTED READING

Gibbs CE, Locke WE: Maternal deaths in Texas, 1969 to 1973. *Am J Obstet Gynecol* 126:687, 1976.

Harper RG, Sokal MM, Sokal S, et al: The high-risk perinatal registry. *Obstet Gynecol* 50:264, 1977.

Pearson JW: The management of high-risk pregnancy. *JAMA* 229:1439, 1974.

Ryan GM: Toward improving the outcome of pregnancy. *Obstet Gynecol* 46:375, 1975.

QUESTIONS

1. List three factors that have stimulated the development of the concept of high-risk pregnancy as it applies to the delivery of reproductive health services:

   ____

   ____

   ____

2. List the three major causes of maternal death:

   ____

   ____

   ____

3. List four factors known to be associated with increased maternal risk:

   ____

   ____

   ____

   ____

4. List five major causes of perinatal mortality:

   ____

   ____

   ____

   ____

   ____

5. List five maternal factors known to be associated with increased perinatal risk:

____

____

____

____

____

ANSWERS

1. a. leveling off of maternal and perinatal mortality rates
   b. recognition that a high percentage of deaths are preventable with current diagnostic and therapeutic measures
   c. realization that it is not possible or necessary to provide every pregnant woman and newborn the depth of care required to prevent deaths
   d. understand that most problem pregnancies can be identified prior to labor and delivery
2. Toxemia, hemorrhage, sepsis
3. a. race—nonwhite
   b. age—over 35
   c. past obstetric history—toxemia, hemorrhage, cesarean section, etc.
   d. past medical history—vascular, renal, or cardiac disease
   e. physical examination—positive findings of hypertension, cardiac or pulmonary disease, overdistended uterus
4. a. maternal toxemia
   b. maternal diabetes mellitus
   c. placental disorders
   d. respiratory distress syndrome
   e. malformation
   f. infection
   g. complications of labor

5. a. age—under 15 or over 35
   b. parity—zero or greater than five
   c. past obstetric history—stillbirth, prematurity
   d. past medical history—diabetes mellitus, renal disease
   e. family history—twins, diabetes mellitus, malformations
   f. physical examination—hypertension, failure of uterine growth, excessive uterine growth
   g. laboratory—positive serology, Rh antibodies, proteinuria, abnormal blood sugar

chapter nineteen

pregnancy complications

OBJECTIVES

1. State the risk of sensitization to an Rh negative mother from a term pregnancy with an Rh positive offspring.
2. Identify how indirect Coombs titers are used to detect fetuses at risk of death or damage from Rh isoimmunization.
3. Identify reasons for using amniocentesis to predict fetal outcome in pregnancies complicated by Rh isoimmunization.
4. Identify two methods of preventing intrauterine death from erythroblastosis fetalis.
5. Identify candidates to receive Rh_0 (D) immune globulin at the termination of pregnancy.
6. List two possible mechanisms by which Rh_0 (D) immune globulin might fail to prevent sensitization.
7. Define placenta previa and placental abruption.
8. State the incidence of placenta previa and abruptio placentae and factors that increase the incidence.
9. Identify classic symptoms of placenta previa and abruptio placentae.
10. Identify causes of maternal death with placenta previa and placental abruption.
11. Name direct and indirect methods of diagnosis of placenta previa.
12. Discuss the rationale for "expectant management" of patients with placenta previa and state contraindications to it.

13. Define the hazard of pelvic examination in patients with placenta previa.
14. List common complications of abruptio placentae and relate them to the management of a patient with this condition.
15. Know the three main causes of postpartum hemorrhage and how to predict them by history and physical examination.
16. Know pharmacologic agents that are of value in treatment of uterine atony.
17. List surgical procedures available to treat postpartum hemorrhage resistant to medical management.
18. Describe racial, familial, and pharmacologic factors that influence the occurrence of multiple gestation.
19. Contrast the placenta and membrane structures in monozygotic and dizygotic twinning and give two reasons why monozygotic twins may be at a higher risk than dizygotic twins.
20. Describe maternal changes that make the antepartum diagnosis of multiple gestation possible and list three possible benefits if such diagnosis is made by the 28th week of gestation.
21. List two major complications of twin pregnancy to:
    a. the infants
    b. the mother during the pregnancy
    c. the mother immediately postpartum
22. Know the causes of death for women with amniotic fluid embolus.
23. Know how to diagnose intrauterine fetal death (IUFD).
24. State how to avoid coagulopathy in women with an IUFD.

This chapter discusses the diagnosis and management of conditions that occur only in pregnant women.

RH ISOIMMUNIZATION

Erythroblastosis fetalis or hemolytic disease of the fetus or infant is an immunologic condition in which the mother produces antibodies to fetal erythrocytes. The antibodies cross the placenta and cause hemolysis in the fetus. Three forms of the disease are recognized:

1. Anemia—the mildest form, which results in an anemic newborn
2. Icterus gravis—a more severe form in which the neonate develops high bilirubin levels and has deposits of bilirubin in the basal ganglia causing death or severe retardation

3. Hydrops fetalis—the severe intrauterine form of the disease in which the fetus dies in utero or is moribund at birth from severe anemia and congestive heart failure

Most cases of hemolytic disease of the newborn are caused by ABO incompatibilities between mother and fetus, but these are generally of little concern to the obstetrician because the incompatibility rarely causes IUFD or severe anemia and is readily managed in the nursery. Severe hemolytic disease of the fetus and newborn can be caused by many different incompatible blood group antigens, but 95% are due to incompatibilities in D (Rh_0) antigen of the Rhesus (Rh) group.

Discovery of the Rhesus antigen by Landsteiner and Weiner in 1940 led to a rapid increase in knowledge about erythroblastosis fetalis. In less than 30 years after the etiology of the condition was discovered, ways of detecting and treating the condition in utero and in the nursery were evolved, and a means to prevent the disease was established.

The distribution of the Rh antigen, used here synonymously with Rh_0 or D, varies greatly in different populations. The highest incidence of Rh positive individuals occurs in Asiatics (99%). American Negroes are 92% Rh positive; American Caucasians, 85% Rh positive; and Basques, 66% Rh positive.

The most important factor in managing Rh isoimmunization is diagnosis. This is done on the first prenatal visit. If the patient is Rh negative, an indirect Coombs test is done to detect the presence of antibodies; if present, an antibody titer is done. Obviously, if the patient and the child's father are both Rh negative, no possibility of the disease exists. It can only occur when an Rh negative woman has an Rh positive fetus.

In a prospective study of Rh negative women who had never received blood products, only 1 in 500 developed antibodies in her first pregnancy prior to delivery. In retrospective analyses of Rh negative women who have delivered one Rh positive infant, it has been found that about 10% demonstrated antibodies. This also holds in future pregnancies, so that an Rh negative woman has about a 10% chance of sensitization with each term pregnancy in which she carries an Rh positive infant. Sensitization can occur whenever Rh positive cells enter the circulation of an Rh negative person. This usually occurs at delivery but can also occur with abortion, ectopic pregnancy, amniocentesis, a fetal-maternal transfusion during pregnancy, or following an incorrectly cross-matched blood transfusion.

Once sensitization has occurred, it has been traditional to follow such patients with antibody titers; in general, this is fairly safe. It is extremely unusual for a baby to be born compromised if the titer is less than 1:32 in the first sensitized pregnancy. When titers were the only way of following pregnancies complicated by Rh isoimmunization, it was thought that a two-tube rise in antibody titer during the pregnancy was indicative of hemolytic disease, and preterm delivery was planned. If the titer was unchanged during the pregnancy, it was assumed the fetus was Rh negative and therefore in no danger. There were problems with the reliance on titers, however, for some pregnancies with an Rh negative fetus demonstrated a rising antibody titer (anamnestic response), and the infant was delivered early only to die in the nursery from the complications of prematurity. Other fetuses who were assumed to be Rh negative because the antibody titer did not rise were sometimes stillborn or moribund because their mother had a "fixed titer." At present, the indirect Coombs test and titers are performed only to identify a fetus at risk, and once the titer becomes 1:32 in the first sensitized pregnancy or 1:16 in any subsequent pregnancy, amniocentesis is used to evaluate the risk to the fetus. The risk of fetal death or damage is sufficiently high to warrant amniocentesis when antibody titers reach these levels.

Amniocenteses were done by Bevis in the 1950s. He discovered that the amniotic fluid of sensitized pregnancies contained a large amount of bilirubin that could be quantitated by spectrophotometric analysis. Between 375 and 525 mμ, the optical density of amniotic fluid produces a straight line. Any bilirubin present produces a change in optical density (ΔOD) at 450 mμ. Although it is usual to refer to the ΔOD when discussing Rh sensitized pregnancies, the reader must remember that we are actually discussing the amount of bilirubin present in amniotic fluid. The amount of bilirubin present is an index of the degree of hemolysis occurring in the fetus.

In 1961, Liley described spectrophotometric findings in the amniotic fluid from 101 Rh sensitized pregnancies and suggested how these findings could be used in selecting the optimal time for delivery of affected fetuses. His findings fell into three zones. Zone I, or bottom zone, contained unaffected or mildly affected infants. All changes in optical density in this zone decreased in size as pregnancy progressed. Zone III, or high zone, contained only severely affected infants. Perinatal deaths were usually imminent when the ΔOD fell in this zone. Zone II, or middle zone, was more complex and could be associated with mildly or severely affected infants. Liley's original article, subject to many refinements over the years, is still the basis for managing pregnancies complicated by Rh isoimmunization.

Prior to 1963, there was only one way to prevent intrauterine death in severely affected fetuses—by delivery. This was often equally hazardous for the infant, many of whom died in the nursery from complications of prematurity. In 1963, however, Liley was confronted with a patient whose fetus was virtually without hope. The pregnancy was only about 28 weeks, and the ΔOD of the amniotic fluid was in the high zone, indicating imminent fetal death. If the infant were to be delivered, it would surely have died of prematurity, so Liley performed the first intrauterine transfusion in an attempt to keep the fetus alive until it was mature enough to survive in an extrauterine environment. He succeeded. Following his report, a rash of intrauterine transfusions were performed around the world for many indications. There are now guidelines establishing which patients are candidates for transfusion therapy. It is usually performed between 26 and 32 weeks of pregnancy on a fetus who is certain to die in utero without treatment and who is almost certain to die in the nursery if delivered. After 33 weeks, the fetus generally has a better chance for life if delivered than if transfused. Follow-up examinations to 5 years of age on children who survived intrauterine transfusions have revealed normal development. Intrauterine transfusion is still a heroic form of treatment, and fetal death from the procedure itself is a distinct possibility. Repeated transfusions every two weeks are generally necessary until survival in the nursery is possible.

In the 1960s, a new approach was used in the management of Rh sensitization—prevention. It had long been known that Rh negative mothers with Rh positive fetuses had a much lower incidence of sensitization if there was also an ABO incompatibility between mother and fetus. This was thought to be due to the major blood group incompatibility destroying fetal cells before sensitization to the Rh antigen could occur. This observation led to the idea that Rh sensitization could be prevented by giving Rh_0 (D) immune globulin to a patient who had received Rh positive cells soon after the transfusion, whether it was fetomaternal hemorrhage or a mismatched blood transfusion. It worked. D immune globulin has been on the market since 1968 and is highly efficient in preventing sensitization to the Rh_0 (D) antigen.

In large studies, the incidence of sensitization after administration of D immune globulin is less than 1%. These failures are accounted for by suboptimal doses of D immune globulin being given or to undetected sensitization having occurred before the D immune globulin was given.

The goal of D immune globulin therapy is to prevent sensitization. Candidates to receive it would include all Rh negative patients with no evidence of sensitization by indirect Coombs test who have:

1. delivered an Rh positive infant;
2. received an Rh positive blood transfusion;
3. had ectopic pregnancy or an abortion either spontaneous or, more

especially, induced (here the blood type and Rh of the fetus is not known and is difficult to determine, so the woman is treated);

4. amniocentesis in the second trimester if fetal-maternal bleeding occurs.

Rh_0 (D) immune globulin provides only short-term passive immunity, so patients who are Rh negative, unsensitized, must receive it after each potentially sensitizing event.

BLEEDING IN THE SECOND HALF OF PREGNANCY

The incidence of significant vaginal bleeding in the second half of pregnancy is approximately 3%. Conditions that produce bleeding carry significant risk for the mother and fetus.

Placenta Previa

Placenta previa is defined as implantation of the placenta low in the uterus either overlying or reaching the vicinity of the internal os. Such an implantation causes what the English term "unavoidable hemorrhage." In the course of pregnancy or in early labor, as the cervix effaces, separation of the placenta from its uterine attachment occurs, and maternal blood is lost from venous sinuses. The fetus also has to go through or around the placenta in order to be born.

There are three degrees of severity of placenta previa, which are judged by how much of the internal os is covered by placental tissue. The central or total previa is the most complete form and the most severe for mother and fetus, causing earlier and more profuse hemorrhage.

Placenta previa occurs in about 0.5% of pregnancies (1 per 200). This incidence is fairly consistent throughout the world. The incidence is increased as the age and parity of the patient increases, but parity seems to be the most important factor. The incidence of placenta previa also increases if the patient has a past history of a "low segment" cesarean section or a past history of placenta previa.

The most common (over 80%) patient complaint is painless vaginal bleeding. To be sure, the patient may also complain of labor pains, or placenta previa may be diagnosed fortuitously before bleeding occurs, but painless bleeding is classic. There is also a high percentage of abnormal fetal presentations since no pole of the baby can enter the pelvis obstructed by a placenta.

The most accurate way to diagnose placenta previa is the direct method, by which a finger is introduced through or near the cervix and the placenta is palpated. Unfortunately, this can cause torrential hemorrhage and is *NEVER* to be done unless the patient is in an operating room with blood ready and all preparations made for immediate cesarean section.

Speculum examination of the vagina and cervix, however, is not known to provoke hemorrhage and can be diagnostic of other causes of painless bleeding, such as cervical lesions, vaginal trauma, and fetal blood loss.

Since the direct method of diagnosis is so hazardous for the mother, many indirect methods of diagnosis have been devised. Each attempts to localize the placenta in the uterus and determine whether a previa is possible or not. Each method has a number of false positive and false negative results. X-ray of the abdomen with and without contrast medium in the bladder may localize the placenta. Amniography, where contrast material is placed in the amniotic sac and an x-ray then obtained, may be helpful; retrograde femoral arteriograms are highly accurate but are not often done. Thermography has been abandoned. The most common indirect methods of diagnosis in use today are radioisotope scanning, to look for "hot spots" that occur where the placenta is assumed to be located, and sonography. Ultrasonic localization of the placenta is 95% accurate, and the technique is not invasive. It has replaced almost all other indirect methods of diagnosis.

In the 1920s, Bill noted maternal deaths from placenta previa almost always occurred after delivery because of postpartum hemorrhage due to lacerations at the friable placental implantation site in the lower uterine segment. He advanced the concept of treating patients by blood transfusion and cesarean section to minimize trauma. Immediately after that, institutions that followed his advice reduced maternal mortality from over 10% to much less than 1%.

Once placenta previa became a relatively safe condition for the mother, attention was directed to improving fetal survival, for the perinatal mortality was over 50%. Most of these deaths were from prematurity since a diagnosis of placenta previa was equated with prompt pregnancy termination for maternal reasons. In the 1940s, the concept of expectant management for patients with placenta previa began to appear in an effort to forestall delivery without increasing maternal risk. Expectant management is based on three observations:

1. The first bleeding episode from placenta previa is virtually never fatal unless the patient is examined by a finger in the cervix.

2. Subsequent hemorrhages are rarely fatal unless the patient is already anemic or unless she is examined.
3. The main cause of perinatal loss is premature delivery.

The goal of expectant management is to delay delivery until the fetus is mature enough to survive in an extrauterine environment. Delay of delivery after maturity may result in more hemorrhages and intrauterine death. While awaiting maturity, the patient is usually kept in the hospital. Blood transfusions are given as necessary to maintain a safe maternal blood volume and hemogram.

Expectant management is contraindicated or terminated if:

1. the patient is in labor;
2. the infant is mature;
3. the infant is dead;
4. the membranes are ruptured;
5. there is intrauterine infection;
6. the bleeding is excessive.

Approximately half the patients with placenta previa are near term when the first bleeding episode occurs. Of the remainder, some go into labor or have profuse hemorrhages so that expectant management cannot be continued. Approximately one-fourth to one-third of patients with placenta previa can be managed expectantly; in the best of reports, the perinatal mortality is still over 5%, and in the worst it is 25%. Today, maternal deaths from placenta previa are rare, but the perinatal deaths are still too prevalent.

Abruptio Placentae

Abruptio placentae is defined as separation of the normally implanted placenta between 20 weeks of pregnancy and birth of the infant. The condition is termed "accidental hemorrhage" in the English literature. The incidence of placental abruption varies greatly from one report to another with a range from 0.7% to almost 2%.

Etiologic factors in the genesis of abruptio placentae were investigated by Pritchard in 201 patients with abruptions so severe as to kill the fetus. He found no relationship to short umbilical cord, inferior vena cava occlusion, or maternal folate deficiency. Less than 1% of his cases were associated with external trauma, sudden decompression of an overdistended uterus, or uterine anomalies. Positive associations he found were:

1. maternal hypertension (47% had diastolic blood pressures over 90);
2. high parity (a para 7 had three times the incidence of a para 3 and six times the incidence of a nullipara);
3. past history of an abruption (there was a 30-fold increased chance of a severe abruption occurring in a patient who had had one previously).

The clinical manifestations of abruptio placentae are not as definite as with placenta previa and consist of vaginal bleeding and abdominal pain. Not all patients with abruptio placentae have visible bleeding. In up to 20% of patients, all bleeding may be concealed within the uterus, and in the remaining 80% of patients with visible bleeding, there may be more concealed than apparent blood loss. Concealed blood loss led to the false concept that shock with abruptio placentae is out of proportion to blood loss. The abdominal pain is quite variable. It may be manifested as "labor pains," uterine irritability, uterine tenderness, or uterine tetany.

Placental abruptions may be termed mild when there is external bleeding, no fetal distress, no maternal shock, and mild uterine hypertonus. Moderate abruptions are accompanied by fetal distress or death, uterine tetany and tenderness, and external or concealed bleeding. Severe abruptions are manifested by maternal shock or coagulopathy, uterine tetany, and fetal death.

In a study of maternal deaths from abruptio placentae, 6 of 10 women died from renal cortical necrosis. This was due to severe prolonged hypovolemia. It became apparent in the 1960s that the amount of blood lost by patients with abruptions was usually grossly underestimated, that the blood pressure and pulse were useless indicators of blood loss unless abnormal, and that central venous pressure monitoring and/or urinary output monitoring were valuable guides to correcting hypovelemia. Pritchard found the average blood loss in abruptions severe enough to kill the fetus (moderate or severe) was 2,500 ml, and these patients often did not manifest hypotension or tachycardia on admission. The least blood lost in Pritchard's study was 1,100 ml.

The cause of death in three of the other patients mentioned above was probably related to blood loss—dead on arrival, cardiac arrest, and coagulopathy. Thus, it can be seen that replacement of blood is extremely important in the management of women with placental abruption, for it can prevent most maternal mortality from this condition.

Fetal loss with abruptio placentae remains quite high. The fetus is often dead, severely depressed, or premature. The overall perinatal loss is usually around 50% in most series.

Coagulopathy is another complication of abruptio placentae. It can develop with a live fetus, although this is very rare. Usually, it develops only with abruptions severe enough to kill the fetus, and it complicates around 25% of these cases. Studies done by Pritchard reveal the coagulopathy is usually present at admission, will not develop if it has not done so by eight hours, and, if present, will not improve until after delivery. Many therapeutic regimens have been developed to treat coagulopathy, but the most important thing to remember is to replace blood loss and treat the basic pathologic process by emptying the uterus.

Fortunately, most women with placental abruptions deliver within several hours after its occurrence. Their labors are usually rapid and hypertonic. Maternal morbidity and mortality generally increase with the time from onset of symptoms to delivery. This can be offset by liberal transfusions, but it is generally thought best to have a patient with an abruption delivered within eight hours. Recent studies, however, indicate delivery can safely be awaited if the mother is adequately transfused.

Management of women with placental abruption includes immediate treatment with blood and intravenous fluids to restore the circulating blood volume. If the fetus is alive and in distress, immediate delivery should be accomplished; this usually means cesarean section. If the fetus is dead, vaginal delivery is preferred unless there is an obstetric contraindication. Live fetuses with no indication of distress by constant monitoring may be allowed to deliver vaginally. A method of monitoring blood loss must be utilized. This may be as simple as monitoring hourly urine volumes and giving sufficient blood and fluids to assure a minimum urinary output of 30 ml/hr or may involve central venous pressure or balloon flotation catheter monitoring. Usually, coagulopathy is first detected by monitoring blood clotting on an hourly basis; coagulation studies are rarely reported in time to be of clinical utility. Should a coagulopathy develop, it may require treatment with fibrinogen or fresh frozen plasma if surgery is necessary; otherwise, no specific treatment is needed unless the woman sustains lacerations at the time of delivery. The coagulopathy rapidly disappears once delivery is accomplished. Oxytocin is rarely necessary to initiate or augment uterine contractions, and uterine pressure recordings are usually of value in assessing the contractile pattern since there is often a hypertonic contractile response. Amniotomy to hasten labor and perhaps to decrease bleeding is desirable; it is necessary if bioelectric fetal monitoring is carried out.

The outlook for women with abruptio placentae is not as good as for women with placenta previa. Abruptions are much more sudden and dramatic, more lethal for the fetus, and much more likely to recur in future pregnancies.

POSTPARTUM HEMORRHAGE

Hemorrhage is one of the three leading causes of maternal mortality. Although placenta previa and abruptio placentae contribute to this cause of mortality, postpartum hemorrhage occurs much more frequently (5% of deliveries) and causes many deaths.

Postpartum hemorrhage is defined as an estimated blood loss of over 500 ml from the time of delivery until 24 hours postpartum. We know, however, that estimates of blood loss are notoriously low. Actual blood loss is often twice the estimated loss, and accurate measurements after normal delivery often record an average loss of over 500 ml of blood. It is apparent that the definition should be changed, but so far it has not been.

The three main causes of postpartum hemorrhage, uterine atony, maternal lacerations, and retained placental fragments, account for 98% of all postpartum hemorrhages. Minor causes include coagulopathies, inverted uterus, uterine rupture, and vaginal vault rupture.

Ninety percent of hemorrhages are due to uterine atony (failure of the uterus to contract). This may occur immediately after delivery of the placenta but rarely occurs later than two hours postpartum. Bleeding from uterine atony is rarely explosive, resulting in immediate exsanguination, but usually is an insidious flow of a moderate amount of blood until the patient is "suddenly" found in shock. The average interval between delivery and death in one series was five and one-half hours. Patients rarely die within one and one-half hours of delivery. This means there is ample time to recognize postpartum hemorrhage and treat it vigorously. The physician can predict postpartum hemorrhage in most patients prior to delivery by simple data obtained on history and physical examination. With this ability to predict, he can order blood cross-matched on admission to the labor room and avoid being told later, at a critical time, that appropriate blood is unavailable.

Factors that predispose to atony and, therefore, postpartum hemorrhage include:

1. overdistention of the uterus—from polyhydramnios, twins, or simply large babies. If an infant weighs over 4,000 gm, the mother's chance is one in four of having uterine atony.
2. past history of a uterine atony—75% of women with a history of atony repeat in a future pregnancy.
3. grand multiparity—patients of higher parity are at increased risk of developing uterine atony.
4. prolonged labor—produces maternal exhaustion and with this goes "uterine exhaustion," making atony likely.
5. toxemia of pregnancy—patients with toxemia of pregnancy are more prone to have uterine atony. The cause is unknown.
6. general anesthesia, especially if the patient is deeply anesthetized or is anesthetized with halothane or ether—causes uterine atony and hemorrhage.

Typically, uterine atony is manifested by a "boggy" uterus as one would suspect, but if the uterus is firm and the patient is bleeding, one should suspect maternal lacerations or retained placental fragments. These patients must be examined carefully in the delivery room, preferably with an assistant and certainly with an intravenous infusion running, blood available, and the patient anesthetized.

The uterine cavity must be explored manually to check for retained placenta or membranes and to determine whether or not it is ruptured. The entire cervix and vagina must then be visualized and any lacerations repaired. The presence of an assistant is necessary.

Uterine atony is usually diagnosed in the delivery room or the recovery room by continued bleeding of a moderate degree. Since the *normal* pregnant woman has increased her blood volume by almost 50% near term, most women can tolerate a blood loss of 1,000 ml in the postpartum period with little change in vital signs. Shock is a late sign of hemorrhage. Exceptions to this general rule are patients who are dehydrated and exhausted after a prolonged labor, who are febrile while in labor, or patients with toxemia who do not increase their blood volume as do normal pregnant women. These patients exhibit shock much earlier.

In treating uterine atony, the blood loss should never exceed 1,000 ml without blood replacement. When this is approached, administration of blood should be started.

Uterine massage may be of some value in maintaining uterine tone and bimanual compression may be life saving during an acute hemorrhage from atony.

The main treatment of uterine atony is with oxytocin and/or methylergonovine. Both these agents are powerful stimulators of the myometrium. Oxytocin causes rhythmic contractions and methylergonovine causes sustained tetanic contractions. In many institutions, these agents are administered routinely after delivery in an attempt to maximize uterine tone and minimize uterine bleeding. Ten units of oxytocin given intramuscularly or added to an intravenous solution usually assures good uterine tone. If the patient develops atony while on this regimen, more oxytocin (20, 30, or 40 units) can be added to the solution, or 0.2 mg of methylergonovine can be given intramuscularly. Since methylergonovine acts just as well and just as rapidly whether given orally, intramuscularly or intravenously, the intramuscular route is chosen because it is safer. It is rarely necessary to continue these agents for more than eight hours.

Rarely, uterine atony is severe and unresponsive to pharmacologic agents. In this event, bimanual uterine compression can be lifesaving until more blood can be obtained, an operating room prepared, and assistants summoned.

The procedures available for surgical treatment of postpartum hemorrhage include:

1. hysterectomy;
2. uterine artery ligation (with or without ovarian artery ligation)
3. hypogastric artery ligation (with or without ovarian artery ligation).

Prior to ascribing unresponsive hemorrhage to atony alone, it is necessary to inspect carefully the cervix and vagina. Lacerations in these locations may continue to bleed after a hypogastric or uterine artery ligation. If the patient has unresponsive atony and has another factor that would make hysterectomy desirable (ruptured uterus, fibroids, desire for sterilization, amnionitis, dysplasia), this may be the procedure of choice to control bleeding and prevent future surgery. If, on the other hand, she desires to retain the potential for future childbearing, a uterine artery or hypogastric artery ligation can be done. With the changes of pregnancy, there is a rich anastomosis between the uterine and ovarian arteries, so the ovarian arteries must be ligated to ensure success of the operation.

After arterial ligations, all blood supply to the uterus and ovaries is not shut off. Rather, there is instant anastomosis through other channels to provide blood to the uterus, tubes, and ovaries. The pulse pressure is decreased, however, and this allows the normal clotting mechanism to provide the hemostasis that could not be obtained through uterine muscle tone. Should this not occur, hysterectomy must be done. There are now well-documented cases of uneventful pregnancies after ligation of both ovarian arteries and both hypogastric or uterine arteries.

MULTIPLE GESTATION

Women usually produce one offspring per pregnancy. In the United States, twins occur about once in every 90 births and triplets the square of that number—1 per 8,100 births. Quadruplets occur about the cube of that number, or once every 729,000 births. Multiple pregnancy occurs more often with increasing age and with increasing parity, both operating as independent variables. Black women have a frequency of twins of 1 per 75; Caucasian women, 1 per 105; and Asiatics, 1 per 155. This racial difference also applies to triplets. There is a positive familial influence for dizygotic (fraternal, double ovum) twinning, probably acting only through the mother. Agents to induce ovulation are responsible for an increased number of dizygotic multiple gestations in the past few years, and this is one of the least desirable side effects of these drugs. Oral contraceptives are not known to have any effect on multiple gestation.

Litter mates are not intended for human babies, and multiple pregnancy must be considered pathologic. The risk of death for the infant is increased about three-fold with perinatal mortality reported at 80–100 per 1,000 multiple births. There is significant perinatal morbidity since about half of all twins weigh less than 2,500 gm at birth. Maternal complications also increase. Toxemia and postpartum hemorrhage are encountered more frequently.

Two-thirds of twins are dizygotic. The two ova may come from the same follicle, from two follicles in the same ovary, or from follicles in both ovaries. They may be of the same or different sexes, and the offspring resemble each other as sibs born at different times. When the two zygotes enter the uterine cavity, implantation may occur close together or widely separated. When close, the placentas may fuse, but normally there is no intermingling of the fetal circulation, and the fetal membranes remain separate. All dizygotic twins have two chorions and two amnions, although the placenta may appear single. In multiples of higher order than twins, double-ovum and single-ovum siblings may be present at the same time.

Monozygotic twinning appears to be totally unrelated to race, age, and parity and is not familial. Division of zygotes can be experimentally induced at certain times by temporary retardation of growth using reduced oxygen environment or cold. The cause in humans is not known, but in the armadillo, which normally produces four or five monozygotic offspring, delayed implantation is the rule. Division may occur at any time from morula through primitive streak. Placentation and arrangement of the fetal membranes in monozygotic twinning depends on the state of development at the time of division. If division occurs very early, there will be two placentas, two chorions, and two amnions. Late division produces twins with a single placenta, amnion, and chorion. Very late division produces conjoined twins and other malformations.

There is a relationship between placentation and perinatal mortality, the most hazardous condition existing when two infants are in the same amnion (monoamniotic twins). Monochorionic twins in separate amnions may have anastomoses of their placental circulations. When this occurs, an arteriovenous shunt can result in the overwhelming of one infant's cardiovascular system by the other—the fetal transfusion syndrome. When artery-to-artery and vein-to-vein anastomoses occur, one heart may beat for two and an acardiac monster develop.

Each infant's growth depends somewhat on the quality of its placentation. The best arrangement seems to be a fused or single placenta with a dichorionic membrane arrangement. Such twins may be either monozygotic or dizygotic, but in both circumstances, the total weight of the two infants is usually higher and the differences between the two least. Widely separated placentas or placentas with anastomotic circulations often are associated with smaller infants and greater intrapair weight difference.

Intrauterine growth is reduced in multiple gestation, probably due to maximum utilization of the transfer function of the placenta. In twins, each fetus grows at the same rate as a singleton until about the 28th week of pregnancy. For the next 10 weeks, the rate of growth is lower than a singleton, and each twin commonly weighs about 500 gm less than a singleton at term. Of course, the combined weight of the two will be much greater. Twins usually lie parallel to each other and parallel to the long axis of the mother. Frequently, both infants are cephalic presentation, but any arrangement is possible. Growth retardation occurs earlier in triplets and quadruplets.

The increased volume of the contents of the uterus produces enlargement greater than expected for the period of gestation. The uterus is also larger than expected when first fetal movement is detected or when the fetal heart is first heard. Unexpected uterine enlargement is often the first clinical indication of multiple pregnancy. Different fetal heart beats are not easily distinguished. At times, the very multiplicity of fetal parts on abdominal examination may lead one to suspect twins. Regardless of what led to the suspicion, the diagnosis may be confirmed by sonography or radiography. Under ideal circumstances, the diagnosis of twins should be made by the 28th week of pregnancy.

Low birth weight in twins occurs because of growth retardation in utero and also because the pregnancy terminates prematurely. Preterm labor is probably related to uterine distention. Triplet and quadruplet pregnancies manifest more growth retardation and earlier pregnancy termination.

Preeclampsia occurs about three times more frequently than in single pregnancies and may be more difficult to manage when it does occur. Maternal anemia is also more frequent because of increased fetal and maternal demands for iron. The maternal blood volume increase is greater than in single pregnancies, and this becomes especially relevant in patients with chronic cardiopulmonary disease. Patients carrying twins have more musculoskeletal complaints associated with increased lumbar lordosis, abdominal distention, and edema. These maternal and fetal problems associated with multiple pregnancy call for early diagnosis and management during the prenatal course. Enforced rest, as much as 20 hours a day between 28 and 36 weeks, has been urged, and some physicians recommend hospitalization during this time. One study found that six to eight weeks of hospitalization did not change the number of infants born weighing 2,500 gm or less, but the average weight of these low-birth-weight infants was about 500 gm greater than the control group. The incidence of preeclampsia was also reduced. Nutritional counseling and supplements, weekly office visits, avoidance of laxatives, and reduction in physical work are all measures that give mother and infants a healthier and longer pregnancy.

Labor with multiple pregnancy frequently begins in a desultory fashion but progresses, once underway, as for a singleton. Prolapse of the cord is more common because of the increased frequency of malpresentation. Excessive blood loss at delivery can be anticipated, and blood for transfusion should be made available early in labor. Plans for pain relief in labor are modified by the frequent prematurity and preeclampsia. Hypnotics and sedatives should be avoided. Analgesics should be used sparingly and only when an antidote is available. One is frequently left with small doses of meperidine for analgesia in labor. Epidural anesthesia in the hands of experienced persons is quite satisfactory except in the extremes of blood pressure. (Pain relief in labor is covered in Chapter 23.)

The initial steps of the delivery are conducted as for a singleton, according to the fetal presentation. Following delivery of the first baby, the cord is doubly clamped and cut. The operators gloves are changed, and bimanual examination determines the presentation of the second infant. If the presentation is transverse, external version to cephalic presentation is attempted but sometimes fails. Internal version to breech presentation followed by a total breech extraction, under general anesthesia, is then the usual procedure. If, on the other hand, examination reveals a breech or cephalic presentation, the presenting part is encouraged to enter the inlet, the membranes are ruptured, and uterine contractions will usually ensue. If not, delivery can be hastened by oxytocin, breech extraction, forceps or vacuum extraction. The interval between the birth of twins is usually 10–20 minutes. Vaginal bleeding following the birth of the first twin may mean placental separation or a tear of a velamentous insertion of the cord of the second twin. In either case, immediate delivery is required, and, at times, immediate transfusion of the neonate is needed.

Postpartum, the overdistended uterus may contract poorly. Such uterine atony is often accompanied by postpartum hemorrhage and delayed or incomplete separation of the placenta. Therapy must be aggressive, with manual removal of the placenta, uterine massage, and oxytocic agents.

The accurate and careful determination of the zygosity of the multiple births is of importance. It is of much interest to the parents, and in this era of tissue and organ transplantation, it may later become medically useful information. Following determination of the sex of the infants, further analysis can be made by thorough gross and microscopic study of the fetal membranes. Figure 19–1 shows a practical guide to follow.

AMNIOTIC FLUID EMBOLUS

Large infusions of amniotic fluid into the maternal circulation may occur whenever a rent in the membranes allows the fluid to come in contact with large venous sinuses in the uterus. The infusion usually occurs late in labor and has been reported with cesarean section. Since

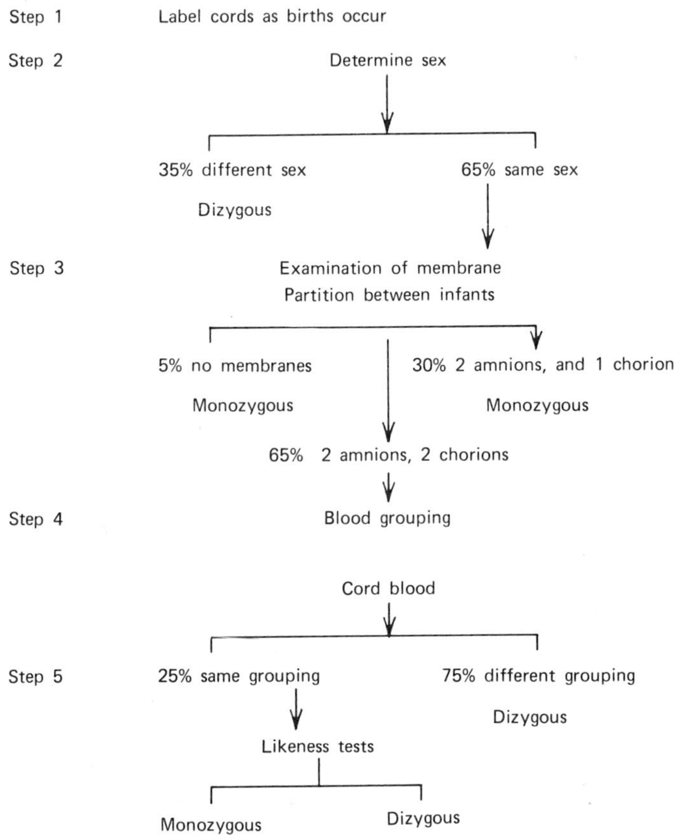

Figure 19–1. Flow chart for the determination of zygosity in multiple gestation.

amniotic fluid contains cellular debris, lanugo, meconium, and vernix as particulate matter, the effect is one of multiple small pulmonary emboli. Patients have sudden cessation of respiration, cyanosis, shock, and coma. Uterine hemorrhage associated with incoagulability of blood may be noted if they survive long enough. There is little data on which to base preventive measures, and death often ensues very rapidly. Sections of lung obtained at autopsy reveal particulate matter in the arterioles of the pulmonary vascular tree. Rarely, the condition may be manifested by sudden hypofibrinogenemia, occurring immediately after delivery, unassociated with the marked systemic changes previously described.

In patients who survive the initial insult, oxygen, peripheral vascular dilators, and possibly heparin could be utilized. In patients whose chief problem is uterine hemorrhage, replacement of blood loss by transfusion and intravenous fibrinogen may be lifesaving. Three to five women die annually in Texas as a result of amniotic fluid embolus.

FETAL DEATH

IUFD occurs in 1–2% of all pregnancies terminating after 20 weeks. The mother commonly volunteers that she no longer feels fetal movement, and the fetal heart, once audible, can no longer be detected. Together, these two pieces of evidence are strongly presumptive of IUFD. Definite diagnosis can be made by real-time ultrasonography, which reveals the absence of cardiac motion and an assymetric fetal skull. Diagnosis can also be made by a roentgenogram, which shows collapse of the fetal skeleton, overlapping of the skull bones, or gas in the vascular system or joints. Roentgenographic changes may take one to two weeks to appear. Doppler devices for detection of the fetal heart may assist in making the diagnosis.

Once IUFD has occurred, three-fourths of the patients will deliver within two weeks. In cases in which an IUFD has remained undelivered for more than a month, a coagulopathy may develop with fibrinogen levels falling below 100 mg/dl. Active intervention to induce labor once the diagnosis of IUFD has been made will prevent this complication and will lessen the psychologic problems of a woman who is carrying a dead fetus.

SUGGESTED READING

ACOG Technical Bulletin: Management of erythroblastosis. No. 17, July 1972.

Bill AH: The treatment of placenta previa by prophylactic blood transfusions and cesarean section. *Am J Obstet Gynecol* 14:523, 1927.

Liley AW: Intra-uterine transfusion of foetus in haemolytic disease. *Br Med J* 2:1107, 1963.

Liley AW: Liquor amnii analysis in the management of the pregnancy complicated by rhesus sensitization. *Am J Obstet Gynecol* 82:1359, 1961.

O'Driscoll K, McCarthy JR: Abruptio placentae and central venous pressure. *J Obstet Gynaecol Br Commonw* 73:923, 1966.

Pritchard JA: Genesis of severe placental abruption. *Am J Obstet Gynecol* 108:22, 1970.

Pritchard JA, Brekken AL: Clinical and laboratory studies on severe abruptio placentae. *Am J Obstet Gynecol* 97:681, 1967.

Queenan JT: *Modern Management of the Rh Problem*, ed. 2. Hagerstown, Md., Harper & Row, Inc., 1977.

QUESTIONS

1. Match the population with the percentage of Rh positive individuals:

(1) American Caucasian	A. 99%
(2) American Black	B. 92%
(3) Asians	C. 85%

2. The risk of sensitization to an Rh negative woman from a term delivery of an Rh positive infant is ______________________.
3. Indirect Coombs titers are currently used to determine which pregnancies are at enough risk of death or damage to the fetus to justify amniocentesis. TRUE FALSE
4. Amniocentesis:
   A. can infer the amount of hemolysis in the fetus by the bilirubin content in the amniotic fluid
   B. can indicate to the physician the optimum time for delivery
   C. both
   D. neither
5. Intrauterine death from erythroblastosis can be prevented by ______________ and ______________.
6. Patients who are candidates to receive Rh_0 (D) immune globulin include all of the following *EXCEPT*:
   A. Rh negative unsensitized women after delivery of an Rh positive infant
   B. Rh negative sensitized women after a therapeutic abortion
   C. Rh negative unsensitized women after receiving an Rh positive unit of blood
   D. Rh negative unsensitized women after a spontaneous abortion
7. Failure of Rh_0 (D) immune globulin to prevent sensitization can occur if:
   A. the dose of Rh_0 (D) immune globulin given is not sufficient to neutralize the amount of Rh positive blood received

B. the patient is already sensitized but has insufficient antibody to be detected by the indirect Coombs test
C. both
D. neither

8. Define placenta previa. ______________________________

____

____

9. The incidence of placenta previa is: ______________________________

10. Match symptoms with diagnosis:

(1) placenta previa	A. painless bleeding
(2) abruptio placentae	B. pain and bleeding

11. Women who die as a consequence of abruptio placentae usually die of:
A. postpartum hemorrhage
B. lacerations
C. complications of blood loss

12. Name three indirect methods of diagnosing placenta previa:
A. ______________________________
B. ______________________________
C. ______________________________

13. List three contraindications to expectant management for placenta previa.
A. ______________________________
B. ______________________________
C. ______________________________

14. Pelvic examination is hazardous in patients with placenta previa because of ______________________________

15. List three common complications of abruptio placentae:
A. ______________________________
B. ______________________________
C. ______________________________

16. Mark the following statements TRUE or FALSE:
A. Multiple gestation occurs less often with increasing age and increasing parity.
B. There is a racial difference in the frequency of multiple gestation, with Black women experiencing twins most frequently, Asian women least frequently.
C. Familial influences for multiple gestation operate both in dizygotic and monozygotic twinning.

D. Oral contraceptives are the probable cause for an increased frequency of multiple gestation.

17. Listed below are eight features of placentation and membrane arrangement for twinning. Those that are characteristic of monozygotic twinning, identify with the letter "M." For those characteristics of dizygotic twinning, identify with the letter "D." For those characteristics found in both types, identify with both letters.
    A. two-thirds of all twins
    B. same sex
    C. two chorions
    D. monoamniotic
    E. cord entanglement
    F. single placenta
    G. fetal transfusion syndrome
    H. same blood group

18. Which of the following maternal changes make the antepartum diagnosis of multiple gestation possible?
    A. increased maternal blood volume
    B. increased lumbar lordosis
    C. greater uterine enlargement than expected for the period of gestation
    D. peripheral edema

19. List three possible benefits of early diagnosis of multiple gestation.
    A. ______________________________
    B. ______________________________
    C. ______________________________

20. List two major complications of twin pregnancy for each of the following:
    A. infants
    B. mother during pregnancy
    C. mother immediately postpartum

21. The most common cause of postpartum hemorrhage is
    A. cervical laceration
    B. uterine atony
    C. retained placental fragments
    D. coagulopathy

22. List four factors known to predispose to uterine atony
    A. ______________________________
    B. ______________________________
    C. ______________________________
    D. ______________________________

23. Drugs used to treat uterine atony include
    A. oxytocin
    B. methylergonovine
    C. both
    D. neither
24. Women who have had hypogastric ovarian artery ligation to treat postpartum hemorrhage cannot become pregnant.
    TRUE FALSE
25. What are the autopsy findings of women dying of amniotic fluid embolus?

    ____

    ____

    ____
26. Which of the following prove the diagnosis of IUFD?
    A. collapsed skull on x-ray or ultrasonography
    B. gas seen in the fetal vasculature on x-ray
    C. absence of cardiac motion by real-time ultrasonography
    D. all of the above
27. Prompt pregnancy termination after IUFD may prevent:
    A. psychological problems
    B. coagulopathy
    C. both
    D. neither

ANSWERS

1. (1) C
   (2) B
   (3) A
2. 10%
3. TRUE
4. C
5. Delivery and intrauterine transfusion
6. B
7. C
8. See text
9. 0.5%

10. (1) A
    (2) B
11. C
12. any three: sonar, radioisotopes, any x-ray method
13. any three: labor, excessive hemorrhage, infection, mature baby, dead baby, ruptured membranes
14. potentially fatal hemorrhage
15. any three: shock, anuria, fetal distress, coagulopathy
16. A. FALSE
    B. TRUE
    C. FALSE
    D. FALSE
17. A. D
    B. M-D
    C. M-D
    D. M
    E. M
    F. M-D
    G. M
    H. M
18. C
19. A. treatment for the prevention of toxemia
    B. treatment to prevent premature labor
    C. treatment to prevent anemia
20. A. prematurity, abnormal presentation, RDS
    B. toxemia, anemia, polyhydramnios
    C. postpartum hemorrhage, retained placenta
21. B
22. prolonged labor, overdistended uterus, high parity, previous history of postpartum hemorrhage, deep general anesthesia (halothane and ether)
23. C
24. FALSE.
25. Vernix, fetal cells, hair, and other amniotic fluid debris in the pulmonary vasculature
26. D
27. C

chapter twenty

medical problems in pregnancy

OBJECTIVES

1. Identify when the fetal thyroid begins to function.
2. Identify which agents commonly used to treat thyroid diseases cross the placenta.
3. Identify how different tests of thyroid function are altered by pregnancy.
4. Identify drugs and tests of thyroid function that are contraindicated by pregnancy.
5. Establish fetal risks from the treatment of thyroid disease in pregnancy.
6. Identify risks from not treating hyperthyroidism in pregnancy.
7. State the chief concern in the management of diabetic pregnancies.
8. List major causes of intrauterine and nursery death in the infants of diabetics.
9. Identify the diabetogenic effect of pregnancy.
10. Identify pregnant patients at risk for diabetes mellitus by data obtained on history, physical examination, and laboratory tests.
11. Classify diabetics by their need for insulin, age of onset, duration of disease, and organ changes (White's classification).
12. List the prognostically bad signs in diabetic pregnancies.

13. Identify the two main concerns in timing the delivery of a diabetic patient.
14. Identify tests of value in determining fetuses at risk from intrauterine death or nursery death in pregnancies complicated by diabetes.
15. Identify signs of cardiac disease in a pregnant patient.
16. Classify patients with cardiac disease in pregnancy by the New York Heart Association classification.
17. List medical management for pregnant women with cardiac disease.
18. Identify salient points in labor and delivery management of pregnant cardiac patients.
19. List and define the classification of hypertensive diseases of pregnancy.
20. List and define the pathophysiologic changes in latent preeclampsia.
21. List and define the pathophysiologic changes in apparent preeclampsia/eclampsia.
22. State the incidence of population at risk for developing preeclampsia/eclampsia.
23. State the principles of therapy and drugs used to treat preeclampsia/eclampsia.
24. List three factors associated with pregnancy that may result in intravascular clotting.
25. Contrast superficial thrombophlebitis in the saphenous vein with deep thrombophlebitis in the femoral vein with regard to symptoms, physical signs, time of occurrence in pregnancy, prognosis, and therapeutic goals.
26. Discuss pelvic thrombophlebitis from the standpoint of etiology, diagnosis, and treatment.

This chapter deals with medical problems commonly encountered during pregnancy and with pregnancy-induced hypertension (preeclampsia). In dealing with these diseases, one must always consider how pregnancy affects the diagnosis, management, and course of the disease as well as how the disease affects pregnancy. Anemias were discussed in Chapter 16, and infections will be discussed in Chapter 21. We will consider thyroid disease, diabetes mellitus, heart disease, hypertensive disorders, and thrombophlebitis in this chapter.

THYROID DISEASE

Normal pregnant women demonstrate changes in thyroid function that could be interpreted as evidence of hyperthyroidism. The diagnosis of thyroid disease is altered by pregnancy, and treatment is adjusted because of the fetus.

The fetal thyroid begins to function at approximately 12 weeks. Prior to 10 weeks' gestation, there is no central colloid in the fetal thyroid and no increased concentration of iodine. Between 10 and 11 weeks, colloid is present, and iodotyrosines and iodothyronines can be found. Fetal thyroid stimulating hormone (TSH) has been identified as early as 11 weeks, and at about 12 weeks the fetal thyroid begins to concentrate iodine.

Thyroid binding globulin (TBG) can be detected at 12 weeks, and tetraiodothyronine (T_4) can be detected about the same time. T_4 levels in the fetus remain in the range of 3–4 μg/dl from the 16th week of pregnancy until delivery. After delivery, T_4 rapidly rises to about 11 μg/dl for several days and then slowly drops to a more normal range.

All TSH in the fetus is of fetal origin because maternal TSH does not cross the placenta. Maternal triiodothyronine (T_3) and T_4 cross the placenta very slowly, but some gets across because infants who are born without thyroid glands are euthyroid at birth.

In contrast, long-acting thyroid stimulator (LATS) from a mother with Grave's disease readily crosses the placenta, and infants born to these patients can be thyrotoxic at birth. Iodides and thioureas (antithyroid drugs) also cross the placenta easily and can profoundly affect the fetal thyroid.

In normal pregnancy, the maternal thyroid gland enlarges due to hyperplasia and increased vascularity. The basal metabolic rate (BMR) increases by 25% as a result of the metabolic activity of the fetus and placenta. There is also an increase in the amount of TBG and a concomitant rise in protein bound iodine (PBI) and bound T_4. These findings are compatible with a diagnosis of hyperthyroidism, but the normal pregnant woman is not hyperthyroid. The amount of unbound, and therefore "active," T_4 is not increased in pregnancy as it is in hyperthyroidism. Resin uptake of T_3 is actually decreased because of increased protein binding. Pregnancy increases the amount of TBG, which, in turn, raises the bound T_4 determination and lowers the T_3 resin uptake without affecting the amount of metabolically active free hormone.

The diagnosis of thyroid disease in pregnancy is clinical, with laboratory tests playing only a confirming role.

Hypothyroidism

Hypothyroidism is uncommon in pregnancy, with an incidence of about 1 in 20,000 pregnancies. Seventy percent of hypothyroid women are anovulatory, and if they do conceive, there is a high incidence of abortion. Most patients with hypothyroidism in pregnancy are hypothyroid following radioactive iodine treatment or surgical removal of the gland. The treatment for this condition is thyroid replacement.

The physician often encounters a woman who is taking a "low dose" of thyroid hormone and who is pregnant. In this circumstance, one should increase the therapy to full replacement for the remainder of her pregnancy and then evaluate her need for thyroid hormone after delivery.

Hyperthyroidism

Hyperthyroidism poses a threat to the life of the mother and the fetus. The incidence of this condition ranges from 1 in 500 pregnancies to 1 in 2,500.

The difficulties in diagnosing hyperthyroidism in pregnancy have already been discussed. One would not do radioactive iodine uptake studies on a pregnant woman since iodides cross the placenta freely and can be picked up by the fetal thyroid. The signs and symptoms of thyrotoxicosis are the same in pregnancy as they are in the nonpregnant state. If the resting heart rate is below 80 beats per minute, the diagnosis is virtually excluded, and if it is over 100 beats per minute while the patient is asleep, this is presumptive evidence of thyrotoxicosis. Eye findings such as "staring," lid lag, and exophthalmos are helpful in making the diagnosis, as are nervousness, palpitations, tremor, heat intolerance, diarrhea, muscle weakness, and personality changes. Weight loss is an impressive finding during pregnancy, and weight gain can be used as a guide to the adequacy of treatment. No specific criteria can be given for T_3 resin uptake or T_4 levels that are diagnostic of hyperthyroidism in pregnancy, but usually the T_3 resin uptake is greater than 35%, and the T_4 level is greater than 15 μg/dl. The free thyroxine index, obtained by multiplying the T_4 by the T_3 resin uptake, should be in the normal range during pregnancy. If it is elevated, hyperthyroidism is suggested.

If the pregnant hyperthyroid patient is left untreated, she has a much higher incidence of toxemia of pregnancy—up to 75% in some series. She also has an increased risk of premature labor and a chance of developing high output heart failure or thyroid storm. Thyroid storm carries a 25% mortality. Intrauterine growth retardation of the fetus is also quite common, and perinatal mortality is increased.

In treating hyperthyroidism during pregnancy, the effect of the treatment on the fetus must be considered. Radioactive iodine cannot be used, and the prolonged use of iodine compounds is also contraindicated because they may cause fetal goiter.

The proper treatment has been in debate for years because there is some fetal and/or maternal risk in any management program. Thioureas (propylthiouracil and methimazole) have been the mainstay of medical management. They act by inhibiting the iodination of tyrosine and have no effect on stored thyroid hormone in the gland. Their therapeutic effects may not be noted for four to six weeks. Thioureas, however, easily cross the placenta and affect the fetal thyroid. This can lead to the development of fetal goiter that can be large enough to cause airway obstruction in the newborn. Children may also be born hypothyroid and may be mentally retarded. In treating patients with hyperthyroidism, it is important to use the lowest effective dose of thioureas to minimize effects on the fetal thyroid. Approximately 10% of patients treated with thioureas also develop drug reactions, which range from relatively minor rashes to agranulocytosis, which is serious and may be fatal.

Other drugs may be used to decrease the symptoms of thyrotoxicosis during the four to six weeks it takes thioureas to work. These include reserpine, guanethidine, and propranolol. Propranolol is probably the best of the three.

Surgical treatment is also advocated for pregnant women with thyrotoxicosis. This is usually done in the first or second trimester after the patient has been controlled with thioureas. With surgery, the effects on the fetal thyroid are eliminated, but the patient has a 0.7% chance of dying as a result of surgery, a 5% chance of recurrent disease, a 3.5% chance of permanent hypothyroidism, a 1% chance of hypoparathyroidism, and a 1% chance of vocal cord problems due to surgical injury to the recurrent laryngeal nerve.

The problem in managing hyperthyroidism during pregnancy is that some form of treatment is necessary but no treatment is ideal. Most authors now suggest medical management with thioureas at the lowest possible dose to control hyperthyroidism and surgical management only if the patient cannot be satisfactorily controlled medically and if there is an excellent thyroid surgeon available.

DIABETES MELLITUS

Diabetes mellitus is the most commonly encountered metabolic disease in pregnant women. It is of importance because of the deleterious effects of the disease on the fetus. Nationally, 1% of pregnancies are complicated by diabetes; however, in some institutions, the incidence is 2–3%. The detection of pregnant diabetes and their management is a constant source of concern in obstetrics.

The chief concern in the management of pregnancies complicated by diabetes mellitus is the prevention of perinatal mortality. Perinatal mortality arises from intrauterine fetal demise and deaths in the nursery. With good obstetric and nursery care, perinatal mortality is less than 10%. In the absence of prenatal care, it is 35–40%.

Toxemia and maternal ketoacidosis are often listed as causes of intrauterine deaths, but the etiology of most intrauterine deaths is unknown. The tendency toward intrauterine death is increased as the pregnancy continues, reaching approximately 20% by 40 weeks' gestation. This fact argues for preterm delivery of infants of diabetic mothers, but too often deaths in the nursery occur because of respiratory distress in the neonate or from birth trauma due to attempts at vaginal delivery of a large infant. There is also an increased incidence of congenital anomalies in infants of diabetic mothers that accounts for some nursery deaths. Metabolic causes, such as hypoglycemia in the first few days of life, have almost been eliminated as reasons for nursery death.

Pregnancy is diabetogenic, and some women with normal glucose metabolism in the nonpregnant state become diabetic during pregnancy. Overt diabetics almost always require more insulin while pregnant. This tendency toward diabetes is produced by placental insulinase, which leads to accelerated degradation of insulin by the pregnant women. Human placental lactogen (chorionic somatomammotropin) is present in maternal serum in increasing amounts as pregnancy continues. It induces lipolysis, increases free fatty acids, and inhibits cellular uptake of glucose and gluconeogenesis. These placental products combine to antagonize the action of insulin. A normal pregnant woman responds by increasing insulin synthesis and release, but women with marginal pancreatic reserve exhibit glucose intolerance and, therefore, diabetes mellitus during pregnancy. Immediately after delivery, the insulin antagonists vanish, and in the puerperium, the patient's ability to handle glucose quickly reverts to prepregnant levels.

There are two types of diabetes in pregnancy. First is the previously diagnosed overt diabetic who happens to become pregnant. She can be expected to have a higher perinatal mortality and an increasing insulin requirement as pregnancy progresses. Second is the gestational diabetic, the woman with normal glucose tolerance in the nonpregnant state who develops glucose intolerance during pregnancy, usually in the last trimester. Gestational diabetics also have increased perinatal mortality, but of a much lower magnitude than overt diabetics, perhaps double the normal perinatal mortality rate. To be sure, some gestational diabetics are previously undiagnosed overt diabetics, but this cannot be determined until the pregnancy is over.

The detection of gestational diabetics is approached in many different ways, but the final diagnosis is usually based on an oral glucose tolerance test (O-GTT). It is not feasible to perform this test on every pregnant woman, so women at risk for gestational diabetes are usually first identified on the basis of historical data, physical examination, and simple laboratory evaluation. At the Robert B. Green Memorial Hospital, a woman is evaluated for gestational diabetes if she meets one or more of the following criteria:

1. Past obstetric history: a history of unexplained stillbirths, nursery death, children with anomalies, recurrent preeclampsia, an infant born weighing over 9 lb, or a past history of gestational diabetes
2. Family history: diabetes in one close relative (parent or sibling) or two distant relatives
3. Physical examination: obesity, 20-lb weight gain in the first half of pregnancy, polyhydramnios, or recurrent monilial vaginitis
4. Laboratory: Glycosuria on two or more occasions or an elevated postprandial blood sugar

Table 20–1 summarizes the frequency with which these findings were associated with gestational diabetes.

This method of selecting patients for an O-GTT is not used universally. O'Sullivan believes that every pregnant patient over age 25 should have a blood sugar test one hour after ingesting 50 gm of glucose as a screening test for gestational diabetes. If that value is above 130 mg/dl, the woman should have an O-GTT.

Once patients are selected for an O/GTT, best results will be obtained after 28 weeks' gestation. Many clinicians prefer to obtain fasting and two-hour postprandial blood sugar values prior to 28 weeks. If normal, an O-GTT is then done after 28 weeks; if abnormal, an O-GTT is done then.

Table 20–1. Reasons for Suspecting Gestational Diabetes (85 patients)

	Number	*Percentage*
History		
Previous 9-lb infant	48	56
Family history	30	35
Previous perinatal loss	21	25
Previous gestational diabetes (O-GTT not needed)	12	14
Child with congenital anomalies	5	6
Physical Examination		
Obesity (all suspected if over 200 lb)	29	34
Polyhydramnios	2	2
Recurrent monilia	2	2
Laboratory		
Glycosuria (in three women, this was the only reason)	13	15

For two or three days prior to an oral GTT, the woman should ingest a high-carbohydrate diet. This can be achieved by having the woman eat one or two candy bars a day during this time. On the morning of the test, a fasting blood sugar is drawn; if this is above 140 mg/dl, the remainder of the test is cancelled because it is not needed for the diagnosis, and it may be dangerous. If the fasting value is below 140, the test is usually continued by giving the patient a 100-gm oral load of glucose. Blood sugar determinations are then obtained at one, two, and three hours. Most institutions use O'Sullivan's criteria to diagnose gestational diabetes. These are based on blood sugar. If plasma sugar is determined, the values are approximately 15% higher. Two of the four values must be elevated in order to diagnose diabetes mellitus. The values are 90 mg/dl fasting, 165 mg/dl at one hour, 145 mg/dl at two hours, and 125 mg/dl at three hours. If a patient is suspected of having diabetes mellitus, and the GTT does not diagnose it early in pregnancy, then the O-GTT should be repeated after the 28th week of pregnancy since the diabetogenic effect of pregnancy increases as pregnancy continues.

Once a diagnosis of diabetes mellitus has been made, patients are generally classified according to the duration and severity of the disease and their need for insulin. Below is an adaptation of the classification proposed by Priscilla White in 1949.

- Class A—blood sugar controlled with diet alone. This group contains mainly gestational diabetics and rarely some overt diabetics whose blood sugar can be controlled by diet. The best perinatal survival—around 95%.
- Class B—this class and all below it require hypoglycemics (almost always insulin during pregnancy) to control blood sugar. These patients have had diabetes less than 10 years, had the diagnosis made after age 20, and have no vascular disease. Some gestational diabetics require insulin and are included here.
- Class C—these patients have been diabetic for 10–19 years or had the diagnosis made between age 10 and 19, or they have minimal vascular disease in the eyes or legs.
- Class D—these patients have been diabetic for more than 20 years or had the diagnosis made before age 10. The presence of retinitis, transient hypertension, or transient albuminuria also puts a patient in this class.
- Class F—these patients have diabetic nephropathy with hypertension and proteinuria. This class has the worst perinatal mortality, 35–50%.

In addition to White's classification, Pedersen described four prognostically bad signs in diabetic pregnancy. The presence of any one of these four factors is associated with increased perinatal mortality regardless of classification, whereas the absence of all of these factors is associated with a great decrease in perinatal mortality. Pedersen's Prognostically Bad Signs of Pregnancy (PBSP) are:

1. Pyelonephritis. This exerts an adverse effect through premature labor as well as through possible development of ketoacidosis. Its development can usually be predicted by finding a stage of asymptomatic bacteriuria and can then be prevented by appropriate antibiotic therapy. In the prenatal care of diabetic patients, a check for bacteriuria should be done at least once a month.
2. Ketoacidosis. This can cause intrauterine deaths and can be prevented by frequent hospitalization for control of diabetes as well as to treat "minor" illness. In the clinic, weekly blood sugars as well as home testing of urine could alert the physician to impending acidosis.
3. Toxemia. Diabetics have a 25% incidence of toxemia, and approximately 25% of diabetic toxemic gravidas sustain fetal loss. We do not know how to prevent toxemia, but it should be detected early in clinic and treated in the hospital.

4. "Neglectors." This is a term used to encompass a rather nebulous group of patients who can not, or will not, follow the physician's advice. Typically, they are of low intelligence, appear for prenatal care after 32 weeks of pregnancy, or are sociopathic. Often such patients are unwilling to accept a diagnosis of diabetes mellitus. Rapport with the physician and nursing personnel is essential to motivate such patients.

Blood sugar control in pregnant diabetics is of utmost importance. Dietary changes and insulin dosages must be managed by blood sugar determinations since urinary sugar determinations are altered by pregnancy-induced changes in renal function. The incidence of excessively large infants and nursery problems can be minimized by adequate blood sugar control. No two authors agree on a definition of adequate control, but a minimum definition would seem to be maintenance of the fasting and two-hour postprandial values in the normal range, less than 90 and 145 mg/dl, respectively.

No area of diabetic management arouses more discussion than timing delivery of the diabetic patient, for one is constantly balancing the risk of intrauterine death against the risk of nursery death. If diabetics are allowed to deliver at term, there will be a high incidence of intrauterine deaths that could possibly be prevented by "preterm" delivery. In many situations, however, a baby delivered "preterm" to prevent intrauterine death dies in the nursery from the complications of prematurity, usually respiratory distress syndrome.

With modern technology, we are better able to predict which infants are at risk of dying in utero and which infants are at risk of developing respiratory distress in the nursery.

Twenty-four hour collections of urine can be submitted for determination of urinary estriol values. As long as these values are stable or rising, the fetus is at little risk of intrauterine death. Such tests should be done daily, near term, which usually means the patient must be hospitalized. If a urinary estriol value shows a 35% drop from the average of the three previous days, the infant is at risk of dying in utero, and delivery should be considered.

Contraction stress tests have also been used to select fetuses at risk. This test uses an external labor monitor and intravenous oxytocin of a dose sufficient to produce three uterine contractions, each 10 minutes. If there are no fetal heart rate patterns suggestive of placental insufficiency, one can assume that the fetus will not die in utero within the next week. If there are changes in fetal heart rate indicative of placental insufficiency, it is assumed that the fetus is at risk and delivery should be considered.

Amniocentesis is used to obtain amniotic fluid for evaluation of fetal maturity. Most of the studies have been replaced by one study, the lecithin/sphingomyelin (L/S) ratio. This test measures the ratio of these phospholipids, and when the ratio is 2:1 or greater, it is assumed that the fetal lung is mature and extrauterine life is possible without the complication of respiratory distress. Those tests are described in greater detail in Chapter 22.

Proper use of these tests can allow the physician to delay delivery of a diabetic woman whose fetus is at no risk of intrauterine death and deliver fetuses in jeopardy after the risk of respiratory distress syndrome in the nursery has passed. Hopefully, this will lead to a further decrease in perinatal mortality for the infants of diabetics.

HEART DISEASE

Heart disease is the fourth leading cause of maternal mortality. (In some series, it is the leading cause.) In normal pregnancy, signs and symptoms develop that may mimic cardiac disease. The physician must be aware of these in order to evaluate properly the 1% of pregnant women who actually have cardiac disease. Pregnancy-induced changes that mimic cardiac disease include:

1. Easy fatigability. This is a common complaint of many pregnant women, and in some it can seem almost pathological. In fact, it is listed as an early presumptive sign of pregnancy. The cause is unknown, but it often decreases after the first trimester, only to reappear near term.
2. Edema. Edema is common in the second half of pregnancy. Dependent edema can be detected in most pregnant women, probably associated with increased venous pressure in the lower extremities. Approximately 15% of normal pregnant women exhibit generalized edema.
3. Dyspnea. Pregnant women often get dyspneic, especially near term, supposedly just from walking around with 20–25 lb of extra weight gained during pregnancy.
4. Cardiac enlargement. The gravid uterus rises progressively during pregnancy, elevating the diaphragm and displacing the heart upward and rotating it to the left. This can make the heart appear enlarged on roentgenograms as well as on physical examination. Whether the heart actually enlarges or not during pregnancy has been debated. It probably does enlarge, but only about 10%. Such

enlargement is not detected clinically, so what is often interpreted as cardiac enlargement is due to elevation and rotation of the heart.

5. EKG changes. The most common electrocardiographic change in pregnancy is left axis deviation, which is due to rotation of the heart. Inversion of the T waves can also be seen at times in normal pregnant women.
6. Murmurs. Approximately 50% of pregnant women have systolic ejection murmurs. These murmurs are thought to be due to lowered blood viscosity or possibly to torsion on the great vessels. They are usually grade I or II/VI and heard at the apex or along the left sternal border. Diastolic murmurs are not a feature of pregnancy.

Ideally, the diagnosis of heart disease would be made prior to pregnancy, and the patient would be in good condition. Since this frequently does not happen, the obstetrician must always be alert to the possibility of undiagnosed cardiac disease complicating pregnancy. Burwell and Metcalfe established criteria to diagnose heart disease in pregnancy that include: diastolic murmur, holosystolic murmur, grade III or higher (thrill) systolic ejection murmur, arrhythmia and unequivocal cardiac enlargement, or electrocardiographic changes. Once a woman is diagnosed as having cardiac disease in pregnancy, her care is optimized if a cardiologist and an anesthesiologist are consulted.

The greatest danger to the pregnant woman with heart disease is the development of congestive heart failure. The prognosis for pregnancy correlates best with her functional classification according to the New York Heart Association criteria. This classification is:

- Class I—patients with cardiac disease and no limitation of physical activity. There are no symptoms of cardiac insufficiency and no anginal pain.
- Class II—slight limitation of physical activity. Patients are comfortable at rest, but ordinary activity causes discomfort in the form of excessive fatigue, dyspnea, palpitation, or anginal pain.
- Class III—marked limitation of physical acitvity. Comfortable at rest but less than ordinary activity causes discomfort in the form of excessive fatigue, palpitation, dyspnea, or anginal pain.
- Class IV—unable to perform any physical activity without discomfort. Symptoms of cardiac insufficiency or angina may occur at rest, and any physical activity increases this discomfort.

In pregnancy, cardiac output rises. The increase begins as early as eight weeks and reaches a maximum at appropriately 28–30 weeks of gestation. The actual increase may vary from 27 to 64%, but in most women, the increase is in the range of 30–40%. After 30 weeks of pregnancy, the cardiac output levels off until term, only to rise again in labor and postpartum as fluids are mobilized from the extremities and the blood volume rapidly changes toward normal.

There are only two mechanisms for increasing cardiac output: by increasing heart rate or by increasing stroke volume. In pregnancy, the heart rate usually stays the same or only increases modestly (10%), so most of the increased cardiac output is accounted for by increased stroke volume.

The blood pressure in normal pregnancy stays nearly the same as prepregnant or may decrease slightly. Therefore, in pregnancy, peripheral vascular resistance decreases since blood pressure is determined by cardiac output and peripheral resistance.

The medical management of heart disease in pregnancy is dictated by the functional capacity of the heart, as defined above. In general, one cannot lessen the increase in cardiac output due to pregnancy, so the physician can only try to decrease the cardiac work load due to other causes. Management is aimed at preventing heart failure and includes:

1. Rest. Limiting activity decreases excessive work by the heart.
2. Allay anxiety. Anxiety increases the work of the heart and can often be allayed by discussion or mild sedatives.
3. Diet. Excessive weight gain and abnormal fluid retention should be prevented.
4. Correct anemia. Anemia increases cardiac output and should be prevented or corrected.
5. Treat infection vigorously. This is especially true of febrile illnesses and includes even upper respiratory infections.

Class I and II cardiacs usually do well if they follow several specific recommendations. They should rest in bed 10 hours nightly and should lie down after each meal. They should do only light housekeeping and spare themselves all unnecessary effort.

Outpatient follow-up is aimed at detecting cardiac failure early and treating it. Failure can be suspected early by finding decreased vital capacity, rales at the lung bases that do not clear after deep breathing, easy fatigability, orthopnea, resting tachypnea, resting tachycardia, and vascular engorgement in the upper lung fields on chest x-ray.

Class III cardiacs should be kept in bed throughout pregnancy, preferably in the hospital. Approximately one-third of these women go into heart failure during pregnancy. Patients who have been in congestive failure prior to pregnancy should be considered class III.

Class IV cardiacs are essentially in cardiac failure at rest, so maternal and fetal mortality are high.

Cardiac disease in pregnancy is usually the result of rheumatic fever. Mitral stenosis seems to be the most lethal lesion. Recently, congenital heart disease has become relatively more common because more children born with congenital defects are surviving into the reproductive age. Almost any type of heart disease can be encountered during pregnancy. The prognosis depends on the functional classification rather than the etiology of the disease.

Hospitalization should be accomplished a week or two prior to labor and delivery to ensure that the patient is in the best possible condition. Delivery should be vaginal unless there are obstetric indications for cesarean section. Cardiac patients generally tolerate the physical effort of labor better than they do cesarean section.

Relief of pain and apprehension are important to cardiac patients in labor. Epidural or caudal anesthesia are valuable except in patients who have shunts. The transient hypotension that may occur with conduction anesthesia can be rapidly fatal if the shunt reverses and blood bypasses the lungs. Narcotics and pudendal block are also quite effective agents for anesthesia in cardiac patients.

During labor, the patient's head should be elevated. If congestive failure seems to be developing, it should be treated vigorously by medical management. Once the cervix is completely dilated, the physician should effect delivery with as little delay as is safe in order to minimize the bearing-down efforts of the patient.

Cardiac patients should continue to be monitored closely for several days in the puerperium. During this time, women who have previously done well may go into congestive failure due to another rise in cardiac output.

HYPERTENSIVE DISORDERS

Hypertensive diseases are common in pregnancy. They constitute the leading cause of maternal death and account for a large number of stillborns and neonatal deaths.

Hypertensive disorders can arise before, during, or immediately after pregnancy. The signs and symptoms of these disorders are nonspecific, yet their consequences to the pregnancy and overall prognosis differ. Some texts refer to these disorders collectively as toxemias of pregnancy. The American College of Obstetricians and Gynecologists has proposed the following classification of hypertensive disorders in pregnancy:

1. Preeclampsia and eclampsia
2. Chronic hypertension of whatever cause
   a. Essential
   b. Secondary
3. Chronic hypertension with superimposed preeclampsia/eclampsia
4. Late or transient hypertension

Preeclampsia/Eclampsia

Preeclampsia is a clinical syndrome of unknown etiology that develops after the 20th week of gestation with the onset of acute hypertension (arbitrarily defined as a rise in blood pressure of at least 30 mm Hg systolic and 15 mm Hg diastolic) associated with either proteinuria and/or generalized edema. Preeclampsia can progress to the more severe syndrome of eclampsia by the development of generalized tonic-clonic convulsions followed by coma. Eclampsia may occur before labor (50%), during labor (25%), or in the early postpartum period (25%). Preeclampsia/eclampsia occurs more frequently in the first pregnancy and only in humans. Factors predisposing to the development of preeclampsia include diabetes mellitus, multiple gestation, hydatidiform mole, renovascular disease, preexisting hypertension, and malnutrition.

Preeclampsia is divided into mild and severe forms. Mild preeclampsia becomes severe when the blood pressure is 160 mm Hg systolic or 110 mm Hg diastolic or more, proteinuria increases (5 gm/24 hr), oliguria develops (less than 400 ml/24 hr), cerebral or visual disturbances occur, or pulmonary edema develops.

Characteristically, preeclampsia progresses in the following fashion; edema of the hands and face as well as the lower extremities first appears, and the patient begins to gain weight excessively (over 2 lb per week). The next sign is an increase in blood pressure of greater than 30/15 mm Hg above baseline values on two occasions six or more hours apart. Proteinuria (at least 0.3 gm/24 hr) is usually the last major sign to develop. In mild preeclampsia, these last two signs fluctuate from hour to hour. As the syndrome progresses to its severe form, the blood pressure remains elevated, proteinuria increases, and the patient may develop symptoms of cerebral ischemia and edema—blurred vision, a severe generalized headache, and hyperactive deep tendon reflexes. Epigastric pain may develop as a result of stretching of the liver capsule by tissue edema. These latter symptoms often precede the onset of eclampsia.

Recent investigations have detected alterations in the early midtrimester of pregnancy, long before the woman exhibits the clinically evident syndrome of preeclampsia. These alterations are listed below:

1. Uterine ischemia. Histopathologic studies of the decidual and myometrial spiral arteries of normal and preeclamptic pregnant human uteri at the placental implantation site reveal significant differences. Normal tissues reveal cytotrophoblastic cells infiltrating the myometrial and decidual spiral arteries. Much of the elastic tissue is replaced by fibrous and fibrinoid changes. These vessels are widely dilated. The study of the preeclamptic pregnant tissue reveals lack of cytotrophoblastic cells infiltrating myometrial vessels and the presence of nonpregnant amounts of fibroelastic tissues in the undilated spiral artery walls. These findings suggest that placentation and subsequent alterations of placental blood flow are abnormal early in the pregnancy destined to develop preeclampsia. Preeclampsia accounts for a high percentage of small-for-dates infants and asphyxial fetal deaths. These findings suggest a cause-and-effect relationship.
2. Vascular sensitivity to pressor substances. The amount of angiotensin infusion necessary to elicit a 20 mm Hg diastolic pressure rise is increased two and one-half times in normal pregnant women when compared to nonpregnant women. Pregnancy thus blunts the pressor effect of angiotensin. This blunting is diminished in preeclamptics and also in patients in the midtrimester destined to develop preeclampsia later in gestation.
3. Metabolic alteration. The fetoplacental unit is intimately involved in the conversion of estrogen precursors (dehydroepiandrosterone sulfate) to estrogens. The metabolic conversion rate (MCR) of this unit can be measured by infusion of labeled dehydroepiandrosterone sulfate and measurement of labeled estrogens excreted in the maternal urine. Studies have shown the MCR progressively increases as pregnancy progresses. In patients destined to develop preeclampsia, the MCR initially increases above expected values, then decreases below expected values and remains low. These changes precede the development of overt preeclampsia by several weeks and suggest a decrease in placental blood flow or altered placental metabolism.

The following alterations usually occur with development and progression of preeclampsia:

1. Arteriolar spasm. Early in the development of the clinical manifestations of preeclampsia, blood pressure increases because of generalized intermittent arteriolar vasospasm. This can be seen by visualizing the retinal vessels that exhibit segmental arteriolar narrowing occurring intermittently. As the syndrome develops, the segmental narrowing becomes more generalized and sustained, resulting in increased peripheral resistance and decreased blood flow to such vital organs as kidney, uteroplacental bed, and central nervous system.
2. Major fluid shift. As preeclampsia develops, there is an unexplained major shift of intravascular fluid to the extravascular space. This shift results in generalized tissue and organ edema and hemoconcentration, as evidenced by a rising hematocrit and rapid weight gain. Although aldosterone blood levels in normal pregnancy progressively rise to 100 times nonpregnant levels by term, the levels are lower in preeclamptics. Aldosterone apparently does not play a major role in the fluid shift.
3. Sodium retention. Although some controversy exists, most studies show that the total exchangeable sodium of the body is increased in preeclampsia. This accompanies the increase in total body water; however, the increase in sodium is proportionately less than the water increase. Preeclamptic patients excrete less sodium in the urine after an intravenous load than do normal pregnant patients.
4. Decreased renal and uterine blood flow. Sustained arteriolar vasospasm produces a decrease of renal blood flow, and the glomerular filtration rate drops. This is reflected by decreased creatinine clearance, decreased urine output, and increased serum creatinine. Serum uric acid levels increase because of altered tubular function, and albuminuria occurs because of altered glomerular membrane function. These changes are reversible. In the uterus, blood flow to the intervillous space is compromised early leading to smaller products of conception. In addition, acute vasospasm compromises uterine blood flow further and accounts for the higher intrauterine fetal death rate and higher incidence of fetal distress during labor.
5. Decreased renin and aldosterone levels. Serial measurements of renin activity and aldosterone have shown lower levels in preeclamptic patients than normal pregnant women. Angiotensin and catecholamine infusions have shown increased pressor response by preeclamptic patients. Renal biopsies have shown degranulation of the juxtaglomerular apparatus (JGA) in preeclamptic patients. This

would suggest exhaustion of the renin-angiotensin-aldosterone system, yet the patient is now hypertensive. The pathophysiologic role of this system in preeclampsia is unclear.

6. Decreased estrogen and progesterone. The lower metabolic clearance rate for steroids is reflected by lower endogenous estrogen and progesterone production by the placenta of the preeclamptic patient. Measurements of these hormones as a predictive function test lose their value in preeclamptic patients. The role of decreased hormonal output from the fetoplacental unit has yet to be defined in the pathophysiology of preeclampsia.
7. DIC. In severe preeclampsia and eclampsia, there is evidence that disseminated intravascular coagulation (DIC) occurs. Measurements of increased fibrin monomers, fibrin split products, and decreased fibrinogen and platelets have been made. The characteristic renal lesion of preeclampsia is glomerular endotheliosis or thickening of the glomerular capillary endothelium by deposition of fibrinoid material. Recently, this material has been shown to stain with immunofluorescent antifibrinogen antibodies. The DIC is apparently low grade since a bleeding diathesis is uncommon in preeclamptic/eclamptic patients.

Chronic Hypertension

Chronic hypertension (blood pressure above 140/90 mm Hg) may be diagnosed by the history of hypertension prior to pregnancy or by detecting elevated blood pressure in the first half of pregnancy. Frequently, chronic hypertensives have a decline in blood pressure in the midtrimester of pregnancy; their blood pressure may remain low throughout pregnancy, increase to the nonpregnant level, or become severely elevated. If prepregnancy blood pressures are unknown, the woman's exact diagnosis (chronic hypertension or preeclampsia) is often in doubt when she is seen late in pregnancy and is noted to be hypertensive. In this case, elevation of blood pressure after delivery is consistent with chronic hypertension. Funduscopic findings of retinal arteriolar narrowing and tortuosity as well as hemorrhages or exudates are much more characteristic of chronic hypertension than of preeclampsia.

Most chronic hypertensive patients are older than preeclamptic patients, and they are more frequently parous. Usually, the cause of chronic hypertension in pregnancy is unknown (essential), but other causes include chronic pyelonephritis, chronic glomerulonephritis, lupus nephritis, and diabetic nephropathy. Less common causes are coarctation of the aorta, pheochromocytoma, and primary aldosteronism.

Eighty-five percent of chronic hypertensives do well in pregnancy, though the fetus may suffer intrauterine growth retardation. Perinatal mortality is increased.

Chronic Hypertension with Superimposed Preeclampsia/Eclampsia

In perhaps 15% of pregnant women with chronic hypertension, a worsening of their condition occurs during pregnancy; this is termed superimposition of preeclampsia. Superimposed preeclampsia tends to occur earlier in pregnancy than preeclampsia and progresses more rapidly. Eclampsia occurs in almost 1% of chronic hypertension patients. Diagnosis of superimposed preeclampsia is made by finding an aggravation of hypertension (at least 30 mm Hg systolic and 15 mm Hg diastolic), sustained proteinuria, and edema. Chronic hypertension with superimposed preeclampsia is a more lethal disease for the mother than preeclampsia. This is because the patients are older and have not been in good health previously.

Late or Transient Hypertension

Transient hypertension is defined as the occurrence of hypertension alone late in pregnancy, in labor, or in the first 24 hours postpartum in a previously normotensive woman. Hypertension disappears within 10 days of delivery. Proteinuria and generalized edema are absent. Some of these women probably have early, mild preeclampsia and others latent essential hypertension.

Incidence and Prognosis

United States statistics indicate that 7% of primigravidas develop preeclampsia and that the overall incidence of hypertensive disorders in pregnancy is 6–7%. The incidence of hypertensive disorders varies with the population studies; at the Robert B. Green Memorial Hospital in San Antonio, the incidence is 10%. Eclampsia develops once in

1,000–1,500 deliveries. Maternal mortality from hypertensive disorders is 3.5 per 100,000 live births. Maternal deaths usually occur when the woman is eclamptic or has chronic hypertension with superimposed preeclampsia. The cause of death is usually from pulmonary edema, cerebrovascular accident, or hepatic rupture.

Overall, perinatal mortality is increased three to four times. The causes of death are usually stillbirth, intrapartum asphyxia, or complications after premature delivery.

Preeclampsia is a reversible condition and does not cause chronic hypertension. However, latent hypertension and preexisting hypertension predispose to preeclampsia. Recurrent preeclampsia in several pregnancies or preeclampsia in pregnancies after the first should make the physician suspicious that he is dealing with a woman with latent hypertension that is being unmasked during pregnancy.

Principles of Treatment

Salt restriction, weight restriction, and the use of diuretic medication do *not* prevent preeclampsia and may have deleterious effects on the pregnancy (maternal hyponatremia, hypokalemia, pancreatitis or fetal hyponatremia, thrombocytopenia). Early detection of preeclampsia is one of the reasons for increasing the frequency of prenatal visits as term is approached. Good dietary habits (high protein and no salt restriction) are thought to be important in the prevention of preeclampsia.

Once the diagnosis of preeclampsia is made, hospitalization is mandatory. In the hospital, bedrest in the lateral recumbent position is one of the most important therapeutic measures. Renal blood flow and naturesis are maximal in this position, and uterine blood flow is increased. The hospitalized patient responds to therapy by improvement (decrease in blood pressure, hematocrit, and edema), remaining stable, or progressing to severe preeclampsia.

The only cure for preeclampsia is delivery. If the patient with mild preeclampsia improves or remains stable, spontaneous labor is awaited as long as the tests of fetal health remain good. Should she progress to severe preeclampsia or eclampsia, the pregnancy should be terminated for maternal and fetal reasons. Vaginal delivery is usually achieved after amniotomy and induction of labor with oxytocin. The mother and fetus both deserve intensive monitoring during labor, delivery, and the early puerperium.

Magnesium sulfate is the drug of choice to prevent or treat eclamptic convulsions, and it may be administered intravenously or intramuscularly. Pritchard has reported excellent results with magnesium sulfate in the treatment of eclampsia. Therapeutic blood levels prevent convulsions by blocking release of acetylcholine at the motor end plate of skeletal muscle. Magnesium ions are excreted in urine. Though the drug crosses the placenta, there is usually no effect on the fetus or newborn unless a large intravenous dose is administered just prior to delivery.

Antihypertensive medications have generally been of little value in treating preeclampsia. Rapid lowering of the maternal blood pressure creates the risk of decreasing uterine blood flow and placing the fetus in more jeopardy. If hypertension is so high that the mother is judged to be in danger of a cerebrovascular accident (diastolic pressure above 115 mm Hg), antihypertensives should be used. Hydralazine is generally thought to be the drug of choice since there is some evidence that it increases uterine blood flow. The diastolic pressure should not be lowered below 90 mm Hg.

Chronic hypertensives should be carefully followed in an outpatient setting. Tests of fetal health should be performed after 34 weeks of gestation. Induction of labor may be indicated when the fetus is mature and the cervix is favorable. Hospitalization is necessary whenever superimposition of preeclampsia is suspected (proteinuria or blood pressure elevation). Superimposition of preeclampsia on chronic hypertension is usually treated by delivery whenever it occurs. This aggressive management has resulted in increased survival of mothers and infants.

THROMBOPHLEBITIS

Several changes occur associated with pregnancy that increase the likelihood for intravascular clotting. Changes in coagulation factors include increased amounts of factors VII, VIII, IX, and an increase in plasma fibrinogen by 50%. There is also increased platelet adhesiveness. Stasis occurs in the lower extremities and pelvis because of mechanical pressure from the gravid uterus, positioning during delivery, and immobility of the puerperal woman. Stasis may enhance clotting. Complications of pregnancy such as pelvic infection or operative delivery augment the tendency for intravascular clotting. The most common sites for vascular thrombosis associated with pregnancy are the venous system in the lower extremities and pelvis.

Thromboembolic phenomena may occur at any time in pregnancy but are most common in the third trimester and in the immediate puerperium. With proper management, they play a very small role in perinatal mortality and morbidity, but pulmonary embolism is the most common nonobstertic cause of postpartum death.

Thrombophlebitis of the Saphenous and Ileofemoral Systems

Thrombophlebitis of the saphenous vein occurs in pregnant and puerperal patients who have large varicosities. The process is usually restricted to one or more segments of the vein. The patient commonly complains of localized pain. On examination a 4–10-cm venous segment is red, warm, tender, and firm. Occasionally, varicose veins of the vulva develop thrombophlebitis. When this happens, the vulva and inguinal areas may exhibit firm, tender, warm, indurated vessels and swollen labia majora. The process may be unilateral or bilateral.

The systemic response to segmental thrombophlebitis in the saphenous system is usually slight with little or no elevation of temperature and pulse. Cultures from these veins are negative, showing no growth of pathogenic organisms. The cause of the inflammatory response is not understood. Uncomplicated, superficial thrombophlebitis in varicose veins runs a self-limited course of 7–14 days. Treatment is designed to relieve discomfort and prevent extension of the process into the femoral system.

Careful observation of the involved extremity, local application of heat, and systemic analgesics are utilized. Bed rest with the leg elevated may be employed for comfort and observation, but after 24–48 hours, an elastic dressing is placed on the extremity, and the patient is allowed to walk. Care is taken to avoid flexion at the knee when the patient is at rest. Anticoagulants and antibiotics are not necessary for thrombophlebitis of the superficial veins. Prenatal patients with varicose veins should be advised to wear elastic stockings, encouraged to sit with uncrossed legs, and urged to report possible inflammatory changes in the veins. Surgical removal of varicosities may be performed three to six months following a pregnancy, at which time maximum regression would have occurred. Previous saphenous stripping and ligation is not a contraindication for pregnancy.

Thrombophlebitis of the femoral veins is a serious but very infrequent complication prior to delivery and is encountered most frequently in the puerperium. It is more frequently noted in patients who have operative deliveries. Regardless of when it occurs, thrombophlebitis of the lower extremities may present many or few signs and symptoms. Patients may complain of all degrees of pain and/or swelling. They may or may not have systemic symptoms of elevated temperature and malaise. Symptoms associated with pulmonary embolism such as chest pain and hemoptysis may be the initial complaint. On physical examination, the leg is nearly always swollen. There may be discoloration (cyanosis or pallor), but unless there is associated involvement of the saphenous system, there is rarely redness. Tenderness may be diffuse of follow the course of the vein, especially in the calf. The swelling may be pitting or nonpitting and may be most marked in the thigh. Careful examination and measurements of the circumference of the legs both below and above the knee may reveal as much as 10 cm enlargement of one leg as compared to the other. The temperature may be elevated as high as 101F, and the pulse be more rapid than the temperature would indicate. Phlebography may be helpful in identifying deep venous thrombosis. A high level of suspicion must be maintained.

Treatment is designed to prevent pulmonary embolus, reduce the intensity of the phlebitis, and prevent permanent damage to the peripheral vascular system. At times, associated arterial spasm exists, and treatment is necessary in order to relieve the resulting ischemia. Anticoagulant therapy decreases coagulation, shortens the course of the disease, and should prevent complications.

In pregnancy, coumarin derivatives are contraindicated because they cross the placenta, producing a defect in the fetal coagulation system of sufficient severity to result in intracranial, pulmonary, and hepatic bleeding, especially during labor and delivery. Heparin does not cross the placenta, and for this reason it is the anticoagulant of choice during pregnancy. It may be used during labor and vaginal delivery, although operative delivery may require reversal of the heparin effect. The goal of heparin administration is to produce a clotting time two to three times the normal control. This usually requires intravenous administration of heparin. Treatment is continued for weeks after the signs and symptoms have disappeared. When femoral thrombophlebitis is encountered postpartum, heparin and a coumarin derivative may be started at the same time, with discontinuation of the heparin as soon as the elevated prothrombin time indicates effective coumarin control.

Occasionally, the first evidence of thrombophlebitis is the sudden catastrophe of a pulmonary embolus, which may result in immediate death. Postpartum patients must be observed carefully and symptoms of a pulmonary nature carefully investigated. The most accurate way to diagnose a large pulmonary embolus is a pulmonary arteriogram, and even this may fail to detect multiple small pulmonary emboli. If pulmonary embolus is suspected, an arterial blood gas with a PO_2 below 80 mm Hg or a lung scan showing a perfusion defect are compatible with the diagnosis. A clear lung scan or an arterial PO_2 above 90 mm Hg virtually exclude the possibility of a pulmonary embolism. Treatment of pulmonary embolization involves anticoagulation with heparin, oxygen, and possibly venous ligation if repeat embolization occurs while the patient is anticoagulated.

Pelvic Thrombophlebitis

Pelvic veins may be involved during ileofemoral thrombophlebitis; however, pelvic thrombophlebitis is seen most often as a complication of severe pelvic infection. Under these circumstances, it is a part of a generalized pelvic cellulitis, which involves the generative organs, their supporting soft tissue, and the vessels themselves. In the usual situation, the patient has developed postoperative or postpartum pelvic peritonitis and cellulitis that has failed to respond completely to antibiotic therapy. Pain, ileus, and prostration may have improved, but the temperature elevation continues in a spiking pattern. There is rarely evidence of deep vein thrombosis of the legs.

Occasionally, the first evidence of septic pelvic thrombophlebitis is a chest film revealing one or more small areas of density—small pulmonary emboli from the pelvic veins. These emboli are rarely massive but threaten the patient primarily by spreading the infection and reducing pulmonary function.

Treatment is designed to prevent further extension of the thrombophlebitis and initial or continuing embolic phenomena and to allow reestablishment of normal circulation within the pelvic veins. Intravenous heparin is effective and is given along with the antibiotics employed in the treatment of the primary infection. The response to heparin is ordinarily prompt. Evidence of continued pulmonary embolization indicates failure of anticoagulant therapy, requiring prompt surgical ligation of both ovarian veins as well as the inferior vena cava.

Acute postpartum thrombophlebitis of the ovarian vein has been described by several authors. The disorder occurs infrequently and presents as unilateral abdominal pain associated with fever, abdominal tenderness, and leukocytosis. The differential diagnosis most frequently involved is acute appendicitis or urinary lithiasis. The diagnosis is usually made at laparotomy. Suggested management includes ligation of the ovarian vessels above the point of the thrombosis followed by postoperative anticoagulants.

SUGGESTED READING

Burrow GN: *The Thyroid Gland in Pregnancy*, Philadelphia, WB Saunders Co, 1972.

Burwell CS, Metcalfe J: *Heart Disease and Pregnancy*, Boston, Little, Brown and Co, 1958.

Gant NF, Hutchinson HT, Siiteri PK, et al: Study of the metabolic clearance rate of dehydroepiandrosterone sulfate in pregnancy. *Am J Obstet Gynecol* 111:555, 1975.

Gant NF, Paley GL, Chand S, et al: A study of angiotensin II pressor response throughout primigravid pregnancy. *J Clin Invest* 52:2682, 1973.

Lindheimer MD, Katz AJ: Sodium and diuretics in pregnancy. *N Engl J Med* 288:891, 1973.

Mestman JH, Manning PR, Hodgman J: Hyperthyroidism and pregnancy. *Arch Intern Med* 134:434, 1974.

Pedersen J: *The Pregnant Diabetic and Her Newborn*, ed. 2. Baltimore, Williams & Wilkins, 1977.

Pritchard JA, MacDonald PC: *Williams Obstetrics*, ed. 15. New York, Appleton-Century-Crofts, 1976, pp 551–581.

Pritchard JA, Pritchard SA: Standardized treatment of 154 consecutive cases of eclampsia. *Am J Obstet Gynecol* 123:543, 1975.

O'Sullivan JB, Mahan CM: Criteria for the oral glucose tolerance test in pregnancy. *Diabetes* 13:278, 1964.

O'Sullivan JB, Mahan CM, Charles D, et al: Screening criteria for high-risk gestational diabetic patients. *Am J Obstet Gynecol* 116:895, 1973.

Talbert LM, Thomas GG Jr, Holt WA, et al: Hyperthyroidism during pregnancy. *Obstet Gynecol* 36:779, 1970.

White P: Pregnancy complicating diabetes. *Am J Med* 7:609, 1949.

QUESTIONS

1. The fetal thyroid begins to function at approximately_____weeks.
2. Circle the following agents that cross the placenta easily from mother to fetus:
   A. T_4
   B. TSH

C. LATS
D. Iodides
E. Thioureas

3. Identify as increased or decreased how each of the following tests of thyroid function are altered by pregnancy:
A. BMR__________ C. T_4__________
B. PBI__________ D. T_3 resin uptake______
4. The diagnosis of hyperthyroidism in pregnancy is a (clinical, biochemical) diagnosis.
5. Which test of thyroid function is contraindicated by pregnancy?
____
6. Which treatments for hyperthyroidism are contraindicated by pregnancy? ______________________________
7. What effects can thioureas have on a fetus? ______________________
8. What are maternal effects of untreated hyperthyroidism?
____
9. What is the chief concern in managing diabetic pregnancies?
____
10. List three causes of intrauterine fetal death in diabetes:
A. ______________________________
B. ______________________________
C. ______________________________
11. List four causes of neonatal death in infants of diabetic mothers:
A. ______________________________
B. ______________________________
C. ______________________________
D. ______________________________
12. In pregnancy there is an increase in circulating insulin.
TRUE FALSE
13. Plasma insulin is antagonized by chorionic somatomammotropin (human placental lactogen) and placental insulinase.
TRUE FALSE
14. Pregnancy patients to be screened for diabetes mellitus with a glucose tolerance test can be identified by history, physical exam, and lab work. List two items in history, three on physical, and two on lab.

History	*Physical*	*Lab*
A. ____________	A. ____________	A. ____________
B. ____________	B. ____________	B. ____________
	C. ____________	

15. Using White's classification for diabetes, match the class with the statement:

A. _____ (1) onset over age 20 or duration less than 10 years
B. _____ (2) diabetic nephropathy
C. _____ (3) onset age, 10–19; duration, 10–19 years
D. _____ (4) diet controlled
F. _____ (5) onset before age 10 or duration over 20 years, retinitis, or hypertension (transient)

16. List the prognostically bad signs of diabetic pregnancy of Pedersen:

A. ______________________________
B. ______________________________
C. ______________________________
D. ______________________________

17. In timing the delivery, the obstetrician is constantly trying to balance the chance of _________ against _________.

18. Which two tests are of value in predicting intrauterine death?

A. ______________________________
B. ______________________________

19. Which test is of value in predicting possible problems from respiratory distress syndrome in the nursery?

____

20. List four normal findings of pregnancy that may mimic cardiac disease:

A. _______________ C. _______________
B. _______________ D. _______________

21. The heart appears larger in pregnancy because of elevation of the diaphragm and some rotation of the apex of the heart.
TRUE FALSE

22. Cardiac output begins to rise in the (first, second, third) trimester and increases to (10–20%, 30–40%, 50–60%) above the prepregnant level by (14–18 weeks, 28–30 weeks, 36–38 weeks).

23. List four signs of cardiac disease in pregnancy:

A. _______________ C. _______________
B. _______________ D. _______________

24. A cardiac patient who is comfortable at rest but with minor physical activity develops fatigue, dyspnea, palpitation, or angina is classified as class _____.

25. List four common causes of cardiac disease in pregnancy—list the *most* common as A.

A. ______________________ C. ______________________
B. ______________________ D. ______________________

26. Cardiac work in pregnancy is determined mainly by cardiac output. The physician cannot limit the increase in cardiac output due to pregnancy, but he can relieve any undue increases in output due to other causes. List four ways this can be done.

____

____

27. Most cardiac patients should be delivered by cesarean section to decrease the strain of labor. TRUE FALSE
28. Patients with congenital heart disease with shunts should not be allowed to become hypotensive, or they may have reversal of the shunt. TRUE FALSE
29. Cardiac patients should have their second stage of labor shortened in order to decrease the strain of expelling the fetus. TRUE FALSE
30. Which of the following does NOT fit the syndrome of preeclampsia:
    A. usually in primigravida
    B. associated with edema and weight gain
    C. associated only with pregnancy in humans
    D. can progress to a convulsive disease
    E. has a normal creatinine clearance
31. Which of the following is NOT true of preeclampsia:
    A. associated with a mortality rate of 25%
    B. associated with a high perinatal loss
    C. begins as an intermittent generalized arteriolar vasospasm
    D. proteinuria is not an early manifestation
    E. associated with a major fluid shift
32. Treatment for preeclampsia/eclampsia includes all of the following *EXCEPT*:
    A. a salt-restricted diet
    B. bedrest and sedation
    C. magnesium sulfate if preconvulsive
    D. evaluation of fetal maturity
    E. induction of labor
33. The differential diagnosis of hypertension in a pregnant patient includes:
    A. coarctation of the aorta
    B. pheochromocytoma

C. lupus nephritis
D. preeclampsia
E. all of the above

34. Answer the following true or false:
A. fibrin split products can be seen in preeclampsia
B. hydralazine is the treatment of choice for preeclampsia to prevent convulsions
C. the early diagnosis of preeclampsia will not alter the final outcome

35. Answer the following true or false
A. transient hypertension is classified with chronic hypertension
B. severe preeclampsia is defined as a blood pressure of > 160/110 mm Hg, proteinuria 3 +, and oliguria
C. spiral arterioles in preeclampsia are normally altered by cytotrophoblastic cells
D. vascular sensitivity to angiotensin is decreased in preeclampsia-prone patients

36. The most important initial therapeutic measure to treat preeclampsia is ______

37. To prevent eclampsia or to treat it the drug of choice is
______

38. List three predisposing factors in pregnancy that may result in an increased likelihood of intravascular coagulation.
______
______
______
______

39. Label the following TRUE or FALSE:
A. Thromboembolic disease is a significant cause of maternal mortality in the United States. TRUE FALSE
B. Maternal thromboembolic disease is a significant cause of perinatal mortality in the United States.
TRUE FALSE

40. Listed below are six features of thrombophlebitis of the lower extremities in pregnancy. For those that are characteristics of disease in the saphenous system, identify with the letter S. For those characteristics of the ileofemoral system, identify with the letter I. When characteristic of both, identify with both letters.
A. occurs in varicosities ____
B. few systemic symptoms ____
C. generalized edema of the extremity ____
D. embolization ____

E. segmental redness and heat ____
F. most commonly encountered postpartum ____

41. List three significant clinical manifestations of pelvic thrombophlebitis.

____

____

____

ANSWERS

1. 12
2. C, D, E
3. A. increased; B. increased; C. increased; D. decreased
4. clinical
5. RAI uptake
6. iodides and radioactive iodine
7. goiter, airway obstruction, mental retardation
8. toxemia, cardiac failure, thyroid storm
9. prevention of perinatal mortality
10. unknown, acidosis, toxemia
11. hypoglycemia, birth trauma, anomalies, respiratory distress syndrome
12. TRUE
13. TRUE
14. Hx—family hx., OB hx.; Physical—obesity, 20 lb. wt. gain in first 20 weeks, recurrent monilia, polyhydramnios; Lab—glycosuria, 1 hr pp blood sugar over 130 mg/dl
15. A. (4); B. (1); C. (3); D. (5); F. (2)
16. toxemia, pyelonephritis, acidosis, "neglectors"
17. intrauterine death—neonatal death
18. contraction stress test, 24-hour urinary estriol
19. lecithin/sphingomyelin ratio
20. dyspnea; easy fatigability; edema; murmur; "cardiac enlargement" on x-ray or physical exam; inverted T waves on EKG
21. TRUE

22. First, 30–40%, 28–30 weeks
23. diastolic murmur; loud systolic murmur with a thrill; cardiac enlargement, holosystolic murmur, arrhythmia
24. III
25. A. rheumatic; B. congential; C. hypertensive; D. thyroid; E. lupus erythematosus; F. coronary artery disease, etc.
26. bed rest; relieve anxiety; correct anemia; treat infections vigorously; control obesity
27. FALSE
28. TRUE
29. TRUE
30. E
31. A
32. A
33. E
34. A. TRUE; B. FALSE; C. FALSE
35. A. FALSE; B. TRUE; C. FALSE; D. FALSE
36. Bed rest—lateral recumbent position
37. $MgSO_4$
38. Changes in clotting factors; stasis of the circulation; complications of pregnancy
39. A. TRUE; B. FALSE.
40. A. S; B. SI; C. I; D. I; E. S; F. I
41. 1. associated with pelvic cellulitis
    2. marked temperature elevation
    3. no evidence of femoral involvement
    4. multiple small pulmonary emboli

chapter twenty-one
infections in pregnancy

OBJECTIVES

1. Describe pathogenesis of puerperal infection and chorioamnionitis.
2. Describe fetal and maternal complications of acute bacterial chorioamnionitis.
3. Discuss the treatment for herpes simplex genital infections in early and late pregnancy.
4. Discuss the features of pregnancy that predispose to pyelonephritis. List maternal and fetal complications of pyelonephritis.
5. Describe the fetal complications of maternal rubella acquired during pregnancy. Describe the conditions that should be present prior to immunizing a woman of the childbearing years against rubella.
6. Describe the fetal complications of congenital cytomegalovirus infection, toxoplasmosis, tuberculosis, syphilis, listeria, and varicella (chicken pox).

PUERPERAL INFECTION (CHILDBED FEVER)

Infectious diseases have always threatened the pregnant woman. For most of recorded history, the great infectious threat to the gravid woman has been postpartum genital tract infection, commonly called "childbed fever." Epidemics swept through maternity wards, leaving grief and death in their paths. Nearly a century and a half ago, two great figures, Semmelweis, in Vienna, and Oliver Wendell Holmes, in Boston, observed independently that it was the physician himself who was guilty of transmitting infection from patient to patient. Both urged the simple practice of washing the bare hands between pelvic examinations. Initially, this practice was greeted with ridicule by the medical establishment, but its lifesaving value was gradually recognized. A few years later, Pasteur isolated the group A streptococcus as the pathogen in these epidemics. Throughout the 19th and well into the 20th century, streptococcal epidemics were active, but with the spontaneous decrease in the virulence of the organism and the introduction of sulfonilamide, control was eventually achieved. Other organisms, however, quickly rose to prominence in causing puerperal infections.

Despite the introduction of newer and more potent antibiotics, there has been no panacea. Indeed, since the early 1950s, the incidence of puerperal infection has remained at 2–8%, and in selected groups it may be as high as 40–50%. Furthermore, severe puerperal infections continue to demand our attention, and infection persists as a leading cause of maternal death. Patients at high risk for genital infection are those with cesarean section, with prolonged rupture of the membranes, and with trauma and anemia.

With uncommon exception, puerperal infection is a mixed infection involving aerobic and anaerobic bacteria that ascend from the patient's vagina and cervix. The most common organisms are anaerobic streptococci, aerobic streptococci, *Bacteroides* species, and *E. coli*. Occasionally, *Staph. aureus*, *Clostridia*, and Gram negative organisms may be involved.

In view of this wide range of possible pathogens, initial treatment of postpartum infection should consist of broad spectrum antimicrobial agents. Few critical studies have been performed, but ampicillin appears to be effective in most mild cases. In more severe infections, combination antibiotic therapy is appropriate. Usually, intravenous penicillin G plus an aminoglycoside antibiotic (kanamycin or gentamicin) are employed for the more serious infections. Clindamycin may be used in place of penicillin to treat serious pelvic infections when *Bacteriodes fragilis* is suspected.

A few infected postpartum patients (perhaps 2–5%) do not respond to proper antibiotic therapy alone. In some of these cases, surgical intervention is needed to drain an abscess or remove retained, necrotic placental fragments. At other times, septic pelvic thrombophlebitis may be the cause of a persistent spiking fever. The treatment for this entity is most often heparin as well as antibiotics.

CHORIOAMNIONITIS

Clinical infection of the fetal membranes and amniotic fluid occurs in approximately 0.5–1% of all pregnancies. Leukocytic infiltration of the amnion can be found much more frequently (in about 10% of all placentas). Most cases of chorioamnionitis are caused by the ascent of bacteria from the cervix and vagina following spontaneous rupture of the membranes, and its frequency appears to be directly related to the elapsed time following this event. Chorioamnionitis may rarely precede rupture of the membranes. Under most conditions, the cervix, its tenacious mucus "plug," and the intact membranes are an effective barrier to ascending infection. However, certain organisms, particularly *Listeria monocytogenes*, may cause a fulminant chorioamnionitis and even fetal death with intact membranes. In addition, maternal blood-borne infections may rarely produce chorioamnionitis, or it may occasionally occur following amniocentesis or intrauterine transfusion. Microorganisms are the same as those causing puerperal infection.

Clinically, chorioamnionitis is an intrauterine exudative infection that, in its fulminant form, presents with a foul, purulent cervical discharge, a tender uterus, maternal fever, and fetal tachycardia. In its early stages, though, the diagnosis requires a high index of suspicion. If the patient is not already in labor, uterine contractions will begin, leading to expulsion of its contents. Since the uterus is so vascular, systemic spread develops very early. Microabscesses in the myometrium, pelvic cellulitis, peritonitis, thrombophlebitis, septic shock, and death may occur as a result of spread of the uterine infection. The infection commonly gains entry into the fetus through swallowing, intrauterine respirations, invasion of the umbilical cord, or through the skin. Fetal involvement is almost always present when the mother has clinical symptoms. The fetus may die prior to birth or in the nursery of sepsis, pneumonitis, or meningitis.

Management of chorioamnionitis consists of obtaining cultures of the amniotic fluid and maternal blood, administering broad spectrum antibiotics, and terminating the pregnancy. Since this infection is a potentially serious one, it is appropriate to employ a combination of antibiotics such as penicillin-aminoglycoside or clindamycin-aminoglycoside. However, it is essential to effect delivery in order to drain the infected uterus and treat the infant. Often, delivery can be achieved by inducing labor. If a vaginal delivery can not be anticipated within a period of four to six hours, then it is best to achieve delivery by cesarean section or cesarean hysterectomy.

Prevention of chorioamnionitis is best achieved by delivering the mature fetus within approximately 24 hours of rupture of the membranes. For premature pregnancies, the risk of infection to the mother and fetus must be weighed against the maturity of the fetus. No standard policy exists to resolve this dilemma.

URINARY TRACT INFECTION

Pregnancy predisposes to ascending urinary tract infection because of stasis in the collecting system induced by progesterone and mechanical pressure. The urine of perhaps 5% of asymptomatic pregnant women contains pathologic numbers of bacteria (greater than 10^5 organisms per milliliter). This condition is known as asymptomatic bacteriuria and predisposes to clinical infection. Symptomatic urinary tract infections occur in perhaps 1% of pregnant and 1% of postpartum patients. *E. coli* is the organism usually encountered, but other Gram negative organisms (such as *Klebsiella* species or *Proteus* species) and *Streptococcus faecalis* (enterococci) may also be the cause. Catheterization predisposes to the development of infection, especially when performed during labor or in the postpartum period.

Pyelonephritis is a clinical diagnosis based on the presence of chills, fever, back pain, dysuria, and frequency. It is confirmed by finding significant bacteruria upon culture. Pyelonephritis may occur at any time during pregnancy but especially in the second half of gestation. Patients with asymptomatic bacteruria are much more likely to develop clinical pyelonephritis, but it is very doubtful that other problems, such as premature labor and growth retardation, are more frequently seen in association with asymptomatic bacteriuria.

Pyelonephritis may affect the outcome of pregnancy in several ways. It may initiate premature labor, or it may cause septic shock. When diabetes is present, the infection may lead to acidosis and intrauterine death. Recurrent infections are also more likely to occur in pregnancy. Chronic pyelonephritis manifested by diminished renal function or hypertension may produce intrauterine fetal growth retardation. Chronic renal disease may worsen during pregnancy. Renal calculi or congenital obstructions are occasionally present when pyelonephritis is recurrent during pregnancy. Proper management consists of screening selected patients for asymptomatic bacteriuria. Depending on the population, as many as 35% of all antepartum patients might be tested. In addition, there may be active treatment of patients with either asymptomatic bacteriuria or symptomatic infection.

For patients with asymptomatic bacteriuria or with cystitis, treatment should consist of an oral drug such as sulfonamide (except in the last two months of pregnancy), ampicillin, or a cephalosporin. The antimicrobial sensitivities of the bacteria should be obtained, and therapy should be modified accordingly. Pregnant patients with pyelonephritis should be hospitalized in order to observe for sepsis or premature labor. The patient should be given adequate glucose and electrolyte solutions and parenteral antibiotics. For the first or second episode of pyelonephritis, ampicillin, or a cephalosporin is indicated. In recurrent infections, when more resistant organisms become likely, initial therapy with gentamicin should be considered. When sensitivities are available, antibiotic therapy may be changed to less toxic agents. Urine culture should be obtained prior to treatment and repeated two weeks after cessation of treatment and every four weeks therafter to identify recurrent infection. Finally, catheterization should be avoided unless necessary for safety, and patients with persistent or recurrent infections should have a complete postpartum evaluation, including intravenous pyelography.

HERPES VIRUS HOMINIS TYPE II (HVH II) (HERPES SIMPLEX)

Herpes type II produces a painful but self-limited vulvovaginitis and cervicitis in the adult female. It is characterized by typical vesicles and ulcers. There is no specific treatment, and therapy is aimed at symptomatic relief of discomfort. The disease usually lasts two to three weeks. When infection occurs in late pregnancy, the fetus is threatened by ascending infection. Transplacental infection is very rare, and the virus does not seem to produce chorioamnionitis as long as the membranes are intact. Very rapidly after the rupture of the membranes, however, the virus can invade the uterine cavity and infant. In addition, descent of the fetus into a birth canal infected with herpes virus results in the acquisition of the disease by the infant in 20–40% of cases. Neonatal herpes rarely is present at birth, as it usually takes several days to develop. It is highly contagious to other neonates, and the disease has a mortality rate in the newborn approaching 50%. Thus, abdominal delivery is indicated in the fetal interest if the membranes are intact or have been ruptured for less than approximately four hours. If herpes occurs in early pregnancy but is cleared virologically and cytologically by term, then a vaginal delivery may be allowed.

RUBELLA, CYTOMEGALOVIRUS, AND TOXOPLASMOSIS

A number of maternal infections are known to be transmitted across the placenta to the fetus. The severity of the maternal illness is not an accurate index of the degree or the intensity of the fetal involvement. The effects on the conceptus are extremely variable, depending on the type of infection and the period of gestation. Abortion, malformation, mental retardation, stunted growth, and intrauterine death are all seen. Specific lesions of the central nervous system, heart, and organs of special sense are common in surviving children.

Rubella was the first disease to be recognized as a cause of congenital malformation. Perhaps 35% of susceptible pregnant women, when exposed, may acquire and transmit the virus to the fetus. The Center for Disease Control in Atlanta estimated that there were over 6,000 perinatal deaths and 20,000 abnormal children born as a result of the 1964 epidemic, and the estimated economic cost alone was in excess of one billion dollars. The rubella virus produces fetal malformations of various kinds—congenital heart disease, cataracts, deafness, thrombocytopenia, and chorioretinitis. They occur in about 25% of fetuses when the mother has a primary infection in the first trimester. When rubella infection occurs in the second trimester, some infants, who are apparently normal at birth, will have defects manifested by age 6.

Susceptibility to rubella should be determined serologically by a hemagglutination inhibition test (HAI). It is useful when patients are exposed to rubella because the presence of antibodies can give assurance of protection. If antibodies were not present initially, their presence in a convalescent titer (two to three weeks later) will indicate the mother has has an infection and the fetus is at risk. From studies during epidemics, many women will have viremia without clinical evidence of the infection. The HAI test is also used in screening populations and is required in some states for premarital and prenatal patients. If antibodies are absent, an attenuated live virus vaccine can be offered to the nonpregnant woman if she will avoid pregnancy for 90 days. Approximately 80% of the adult population is immune to rubella.

Cytomegalovirus (CMV) is found in the cervix or urine of perhaps 5% of pregnant women and may be transmitted across the placenta, especially when there is a primary maternal infection with viremia. However, the maternal infection is usually subclinical. Pregnant women employed in neonatal nurseries or in renal transplant units are at particularly high risk. The clinical lesions in infants may be similar to those produced by the rubella virus, but most cases of congenital CMV infection are subclinical. It has been estimated that 50,000 infected infants are born annually in the United States, and 5,000 of these have suffered major brain damage. At present, there is no definitive therapy for CMV, but research to develop a vaccine is in progress.

Toxoplasmosis is a protozoan infection caused by *Toxoplasma gondii*. The definitive host is the house cat, while intermediate hosts may be humans, rodents, and domestic animals. Humans may become infected by consuming undercooked meat of infected animals or by contacting feces of an infected pet cat. Approximately 30–50% of adults in the United States have had toxoplasmosis.

When a pregnant women becomes infected, a transplacental infection may affect her fetus. From recent studies in France (where toxoplasmosis is a more common disease), approximately one-third of all fetuses will be affected with either clinical or subclinical disease. Subclinical disease is one that is diagnosed solely on serologic critieria at birth. Clinically, the characteristics of the fulminant disease are chorioretinitis, cerebral calcifications, and microcephaly. Owing to French gastronomic pleasures, including a penchant for undercooked meat, the incidence of congenital toxoplasmosis has been estimated at 1% of all pregnancies, but in the U.S. it is approximately 0.1–0.2%. Prophylaxis for congenital toxoplasmosis consists of avoiding undercooked meat, keeping pet cats indoors and away from wild rodents, and having kitty litter changed daily before the excreted cysts become infective.

OTHER INFECTIONS

Syphilis is still a common and lethal problem. Intrauterine syphilis produces fetal death by tissue necrosis. Surviving children may have central nervous system and musculoskeletal abnormalities. Premarital and prenatal tests for syphilis have been very useful in preventing the congenital variety of the disease. (For treatment of syphilis, see Chapter 7.)

Neisseria gonorrhoeae is present in the uterine cervix in 0.2–5% of pregnant women and may be entirely asymptomatic. The organism does not commonly ascend into the uterine cavity during pregnancy. Dissemination can occur, however, with an infectious arthritis as an expression. Asymptomatic carriers of the gonococcus and their sexual consorts should be treated with appropriate antibiotic regimens. During passage through the birth canal, the newborn may acquire a gonococcal infection of the cornea and conjunctiva, which will result in blindness without appropriate treatment. Silver nitrate or an antibiotic ointment is instilled into the eyes of all newborns in order to prevent this disease—ophthalmia neonatorum.

Monilial vaginitis due to *Candida albicans* is a frequent infection of pregnancy. Although often asymptomatic, it may cause itching, burning, dysuria, and dyspareunia. When symptomatic, it may be treated satisfactorily by vaginal medications. Neonatal moniliasis (thrush) is an infection of the oral cavity probably acquired during labor. Moniliasis is not regarded as a cause of intrauterine infection, antepartum or postpartum infection.

Some infectious diseases seem to be more hazardous to the patient when acquired during pregnancy. The Asian flu epidemic of 1957–1958 was associated with many maternal deaths. Since then, immunization with the killed virus vaccine against influenza has been recommended for women during epidemic years. Chicken pox (varicella) may occur as a coincidental infection in pregnancy, but it is not clear whether the risk is greater to pregnant women than to other adults. If mothers develop varicella within four days prior to delivery, the neonate may develop varicella in the immediate neonatal period. This neonatal infection may have a high mortality (approximately 20%).

In the pregnant patient, coccidioidomycosis is more likely to become disseminated; if so, the mortality is extremely high (90%).

Tuberculosis and pregnancy often have coexisted, and pregnant patients with pulmonary tuberculosis usually do well. Tuberculosis of the reproductive organs usually results in sterility, so that genital tuberculosis generally does not coexist with pregnancy. Isoniazid (INH) may be administered in pregnancy but should be given with vitamin B_6 to counteract possible central nervous system side effects in the fetus. Long-term streptomycin therapy is curtailed because of potential VIIIth cranial nerve damage to the fetus. Infants born to mothers with active tuberculosis require isolation from the mother until her sputum cultures are negative. When there is maternal miliary tuberculosis or tuberculous meningitis, there is a 30% risk of congenital infection, and these infants are usually treated immediately. Infants born of mothers with active tuberculosis who do not manifest miliary spread or tuberculous meningitis may be followed with skin tests or may be immunized with bacillus Calmette-Guerin (BCG).

Vaccination in pregnancy should not be carried out if it involves live or attenuated virus vaccine preparation. Rubella vaccination is, of course, contraindicated in pregnancy, and smallpox vaccination should not be carried out unless absolutely required.

SUGGESTED READING

Charles D, Finland M: *Obstetric and Perinatal Infections.* Philadelphia, Lea & Febiger, 1973.

Gibbs RS, Weinstein AJ: Puerperal infection in the antibiotic era. *Am J Obstet Gynecol* 124:769, 1976.

Monif G: *Infectious Diseases in Obstetrics and Gynecology.* Hagerstown, Md., Harper & Row, Inc, 1974.

Nahmias AJ: "The TORCH complex," *Hospital Practice*, May 1974, p 65.

QUESTIONS

Pick appropriate answers; one or more may be correct.

1. In pathogenesis of puerperal infection and chorioamnionitis:
   A. there is usually an ascending infection with the patient's own flora
   B. there is often an exogenous infection with group A streptococci
   C. hospital acquired organisms such as *Serratia* and *Pseudomonas* are commonly found
   D. prolonged rupture of the membranes is an important predisposing factor
2. Chorioamnionitis may cause:
   A. maternal sepsis and shock
   B. neonatal sepsis
   C. neonatal pneumonia
   D. neonatal meningitis
3. Herpes simplex genital infection in pregnancy
   A. is often fulminant, leading to disseminated maternal herpes
   B. commonly causes transplacental infection
   C. may be an indication for cesarean section to avoid neonatal herpes
   D. is adequately treated in pregnancy with specific antiviral agents
4. Complications of acute pyelonephritis in pregnancy are:
   A. maternal septic shock
   B. premature labor
   C. recurrent bacteruria
   D. adverse drug reactions to antibiotics in fetus and mother

5. Which of the following statements are correct about rubella and pregnancy?
   A. If a susceptible pregnant woman acquires rubella in the first 13 weeks of pregnancy, the fetus may have significant abnormalities at birth.
   B. When rubella is acquired in the second trimester, fetal abnormalities frequently are not manifested until after the second, third, or fourth year of life.
   C. Women in the reproductive years should be tested for susceptibility to rubella in order that they may be immunized if susceptible.
   D. Prior to immunizing a woman in the childbearing years, it should be ascertained that she is not pregnant and that she will avoid pregnancy for 90 days thereafter.
6. Which of the following maternal diseases can cause definite transplacental infection of the fetus?
   A. cytomegalovirus and toxoplasmosis
   B. syphilis and tuberculosis
   C. *Listeria monocytogenes*
   D. monilia (*Candida albicans*)
7. Which of the following diseases has a worsened maternal prognosis when it occurs in pregnancy?
   A. tuberculosis
   B. coccidioidomycosis
   C. rubella
   D. all of the above

ANSWERS

1. A, D
2. A, B, C, D (all are correct)
3. C
4. A, B, C, D
5. A, B, C, D
6. A, B, C
7. B

chapter twenty-two

fetal growth and maturity

OBJECTIVES

1. Identify the ideal birth weight and duration of gestation for human pregnancy.
2. Define immaturity, prematurity, term and postterm pregnancies.
3. Differentiate between the effects of deviation from ideal birth weight and pregnancy duration on neonatal health (fetus too big, too small, pregnancy too short, too long).
4. List the determinants of normal fetal growth.
5. Define dysmaturity and list its causes.
6. Describe fetal growth by history and physical measurements and assess their reliability.
7. Describe the principle of diagnostic ultrasound and its modes of display.
8. State the relative limitations of certain commonly used measurements of fetal monitoring.
9. List tests that can be used to assess fetal health.

As advances in the field of perinatology are made, tools used to assess the intrauterine patient are clinically useful. These tools allow the obstetrician to assess the growth and maturity of the fetus, which, in turn, aids him in making clinical decisions. In this chapter, both older and newer concepts will be reviewed.

A newborn infant weighing 3,300 gm (7 lb 5 oz) and delivered between 37 and 42 weeks of gestation has the highest survival rate

Table 22–1.

Definition	*Completed Weeks Gestation*[a]	*Birth Weight (gms)*
Immature birth	20–28	500–999
Premature birth	29–36	1,000–2,499
Term or mature birth	37–42	> 2500
Postmature birth	> 42	> 2500

[a]Dated from first day of last normal menstrual period (LMP).

in the neonatal period. Deviations from this optimal circumstance (shorter or longer duration of pregnancy, larger or smaller birth weights) are associated with increased perinatal loss and damage.

Table 22–1 lists the definitions for the time of birth as it relates to gestational age (maturity) and range of birth weights. These data assume normal fetal growth.

The assessment of fetal growth is difficult because of the inherent biologic variability in the natural termination of pregnancy and natural intrauterine growth patterns. For example, a study of 2,614 pregnancies where women kept a menstrual diary and the pregnancy terminated after 20 weeks provided the following observations:

1. 4.3% of pregnancies terminated prior to 252 days (36 weeks) from the last menstrual period (LMP).
2. About 50% of the pregnancies terminated prior to 280 days (40 weeks) from the LMP—before the expected date of delivery (EDD).
3. 3.4% of pregnancies terminated after 301 days (43 weeks) from the LMP.

Varying lengths of individual menstrual cycles may contribute to this variability in gestational length. A recent evaluation of a large number of menstrual cycles from a number of different studies reveals that normal cycles range from 25 to 30 days. The mean cycle length was 29 days, with a standard deviation of ± 7 days. Over 50% of women had cycle lengths less than 30 days.

The older method for defining fetal maturity is birth weight. We consider the normal, average, mature newborn to weigh 3,250–3,350 gm, but the 10th percentile weight is about 2,700 gm and the 90th percentile weight about 3,900 gm. Prematurity is defined as either a birth weight less than 2,500 gm or duration of gestation of less than 36 weeks. However, at 34 weeks, the 90th percentile baby weighs about 2,900 gm. On the other hand, at least 25% of newborn infants weighing less than 2,500 gm make the physiologic adjustments required for intact survival in the extrauterine environment as mature infants would. Thus, fetal weight is frequently not a reliable guide to fetal maturity. This concept is clinically important since the true premature infant cannot adjust easily to extrauterine environment. The premature may develop progressive respiratory insufficiency, a major cause of perinatal death, and it is also immature in its ability to resist and handle bacterial infection, maintain its body temperature, and concentrate its renal excretory products. Thus, a fetus who is large for gestational age (LGA) and is thought to be mature on the basis of an estimated fetal weight of over 2,500 gm may be delivered by an elective repeat cesarean section only to succumb in the neonatal period due to its premature state. Likewise, in a situation where the fetus is small for gestational age (SGA) and is estimated to be less than 2,500 gm, the physician may elect to temporize termination of pregnancy in a chronic hypertensive mother, trying to balance the risk of prematurity against the increasing risk of fetal and maternal damage as gestation advances. This discrepancy between presumed maturity according to fetal weight and the actual maturity behavior is called dysmaturity. The challenge to the obstetrician is to determine the true functional maturity of the fetus and to reliably date the gestation. Before describing the tools that are available, let us review the determinants of normal fetal growth and the alterations that result in abnormal fetal growth (Fig. 22–1).

The fetal growth pattern in the first trimester tends to be linear, but the rate of increase in absolute weight of the fetus is relatively small until early in the second trimester when there is an accelerated linear growth pattern that persists until about 36 weeks when the slope of the fetal growth curve is reduced. This plateauing of the growth curve suggests a rate limiting factor since infant growth after delivery proceeds at an increased rate, parallel to that in the second trimester. This rate limiting factor is the placenta, which progressively restricts the transfer of essential substrates to support fetal growth.

Determinants of fetal growth include genetic factors, availability of growth substances, normal level of growth-promoting hormones, and vascular support of the uterus and placenta.

In general, the ultimate size of the fetus at term is predetermined by the individual genetic compliment; for example, male fetuses are usually larger than female fetuses. Events that result in chromosomal abnormalities often are associated with a decreased number of cells in

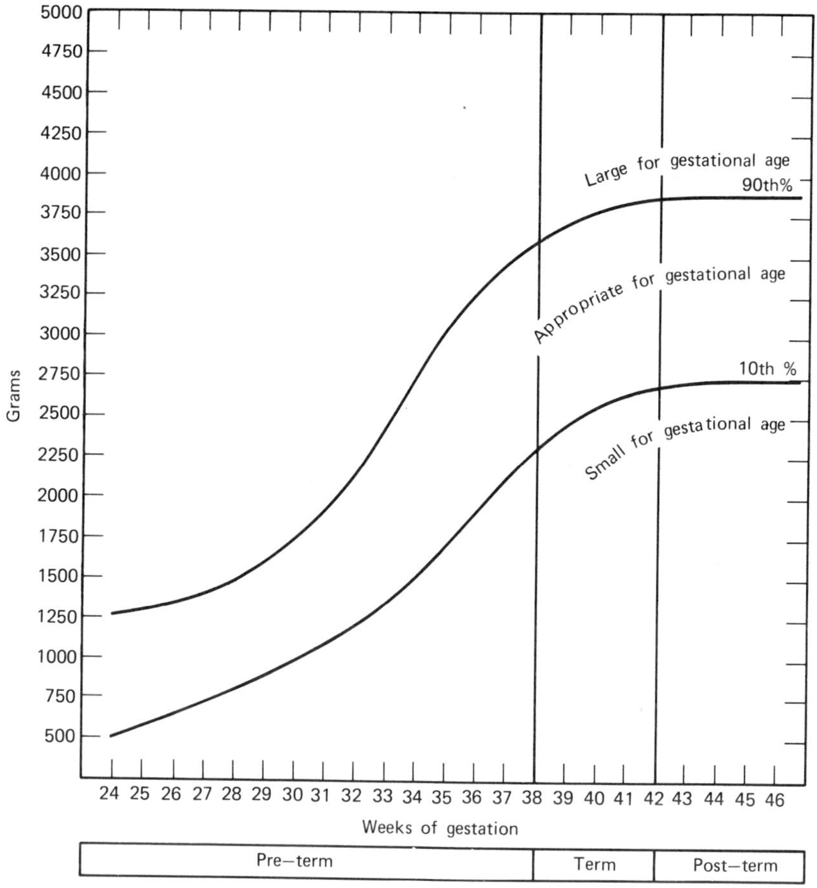

Figure 22–1. A graphic representation of birth weight and gestational age correlation throughout pregnancy (University of Colorado Medical Center). (From Battaglia FC, Lubchenco LO. A practical classification of newborn infants by weight and gestational age. *J Pediatr* 71:159, 1967.)

fetal organs, producing a growth-retarded fetus. Significant events occurring in early pregnancy, such as a teratogenic insult resulting in a gross congenital anomaly or chronic intrauterine infection with rubella, toxoplasmosis or cytomegalic viruses, syphilis, or Listeria, can result in a reduction in total number of cells in the fetus. An exception to this concept is that transposition of the great vessels may be associated with a fetus large for dates. Reduction in the number of fetal cells results in a symmetrical type of intrauterine growth retardation in which the fetal growth curve deviates from the norm early and progressively. The overall size of the fetus (head and body) is smaller than a normal fetus at equivalent gestation ages.

The maternal organism converts energy generation from carbohydrate metabolism to fatty acid metabolism in order to ensure the fetus a steady supply of glucose, since it is wholly dependent on glucose for energy production. During early pregnancy, the mother also supplies the fat requirements for the growing fetus. As pregnancy advances, the fetus and/or placenta develop enzyme systems to synthesize the more complex lipids and derive from the mother only simple precursor compounds such as acetate and ketones. Under normal conditions, the fetus probably does not mobilize its own nutritional lipids to provide energy but continues to build up its fat stores. Under conditions of intrauterine malnutrition, these fat stores are depleted, resulting in a decreased body size due to loss of subcutaneous fat, smaller fetal organs, and diminished muscle mass. The result is an asymmetrical growth retardation pattern in which the fetal body is considerably smaller than the fetal head. The fetal central nervous system is least affected by intrauterine malnutrition.

The concentration of essential amino acids in the plasma of the normal fetus is greater than in maternal plasma because of an active placental transport system. Intrauterine growth rates can be correlated with a reduction in maternal plasma amino acids. Maternal protein restriction can result in failure of cell replication in many viscera, including the brain. Such deficiency, if severe enough, can result in symmetrical fetal growth retardation pattern.

There is little evidence to suggest that maternal hormones regulate fetal growth directly. Growth-promoting hormones of fetal origin probably represent a principle mechanism for the regulation of fetal growth. Increasing evidence supports the following concept: blood glucose perfusing the fetal pancreas stimulates insulin secretion, which has as its primary effect the promotion of growth rather than regulation of fetal blood glucose, as in adults. Fetuses of diabetic mothers are large for gestational age because of high levels of glucose being delivered to the fetus. The fetus responds with a relative hyperinsulinemia and grows larger. (In tissue culture studies, insulin antagonizes the cellular effects of glucocorticoids. In vivo, elevated insulin levels may relate to the relative immaturity of many enzyme systems in the infant of the diabetic mother.)

The trophoblastic alteration of the placental bed arterioles allows maximal blood flow. (See Chapter 16.) Hypertensive diseases in pregnancy alter the placental arterioles, resulting in impaired intervillous blood flow. Furthermore, it has been observed that over 50% of stillborn infants who were small for gestational age were born to hypertensive mothers. Progressive vascular disease, associated with the more advanced stages of diabetes, can result in impairment of fetal growth. Growth retardation resulting from insufficient uteroplacental blood supply is of the asymmetrical type and operates principally in the third trimester. The deviation from normal fetal growth curve occurs in the third trimester. This type of growth retardation occurs much more commonly than the symmetrical type. Investigation of mothers who deliver growth-retarded infants demonstrates that, as a group, the mothers weighed less at birth, and they never have given birth to large children in previous pregnancies. Dysmature fetuses do not tolerate labor as well as normally grown fetuses, and the development of fetal distress during labor is sufficiently common to warrant careful bioelectronic and biochemical monitoring.

The postdate pregnancy represents a different problem. In such patients, the fetus appears to be of adequate size ($>$ 2500 gm), but the pregnancy continues. From the end of the 43rd week on in primigravidas, the perinatal mortality increases dramatically, and this is true for women of all parities if hypertension is present. Women with postdate pregnancies come in two varieties. In one, the placenta seems adequate, supports the fetus well, and the pregnancy terminates after 43 weeks without apparent physical difficulty. In the other variety, the placenta begins to fail, and if the fetus is not delivered, intrauterine fetal death may occur. Determining the status of fetal well-being in utero may be helpful in differentiating these two varieties. However, the presence of a maternal disease such as preeclampsia, chronic hypertension, or diabetes is a clear-cut indication for termination of the postdate pregnancy. The remainder of this chapter will discuss tools the obstetrician has to aid in the determination of fetal growth and maturity and techniques used to evaluate fetal health. This information is clinically useful when an elective cesarean section or induction of labor is contemplated or when there is an indication for premature termination of pregnancy, such as maternal diabetes, hemolytic disease of the fetus, hypertensive complication in the mother, or intrauterine growth retardation with chronic fetal distress.

A history and physical exam of the mother early in pregnancy yields base line information. The date of the last menstrual period (properly defined as the beginning date of the menstrual flow) was maligned as an indicator of the length of gestation before it was realized how frequently low-birth-weight infants may be born at term. When, in the woman's opinion, the last menses was normal, and the date is known, the probability of error is less than 5%. An exception would be women who conceive during the first ovulation following discontinuance of oral contraceptives. The interval between stopping the "pills" and ovulation is highly variable. Another exception is the woman who experiences postconceptional vaginal bleeding.

Inasmuch as intrauterine growth retardation or hydramnios (excessive amniotic fluid) does not ordinarily occur before 28 weeks, estimates of uterine size in terms of gestational weeks during the first two trimesters are reasonably reliable unless one is dealing with multiple gestation. A bimanual pelvic examination in the first trimester by an experienced observer can give a good estimation of gestational age. Once the uterus rises out of the pelvis to become an abdominal organ, serial measurements of the fundal height (FH) are made by performing McDonald's measurements (determining with centimeter tape measure the distance along the curvature of the lower abdomen from the top of the symphysis pubis to the umbilicus, then along a horizontal plane to the top of the uterine fundus). This distance usually increases somewhat linearly until the last few weeks of gestation (Fig. 22–2). Such serial measurements point out the possibility of SGA or LGA fetuses if the measurements deviate from the normal shaded areas. (Note that the FH in centimeters [± 2] corresponds to the numerical week of gestation until around 32 weeks.)

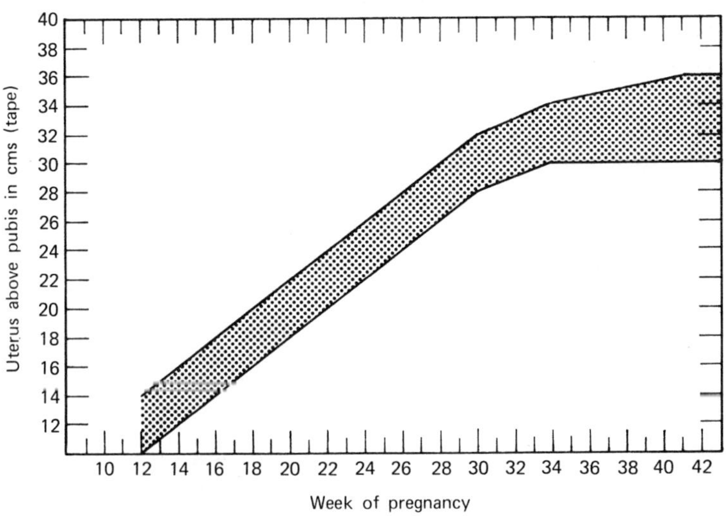

Figure 22–2. A graphic representation of the normal fundal height measurement (stippled area) at different times in gestation.

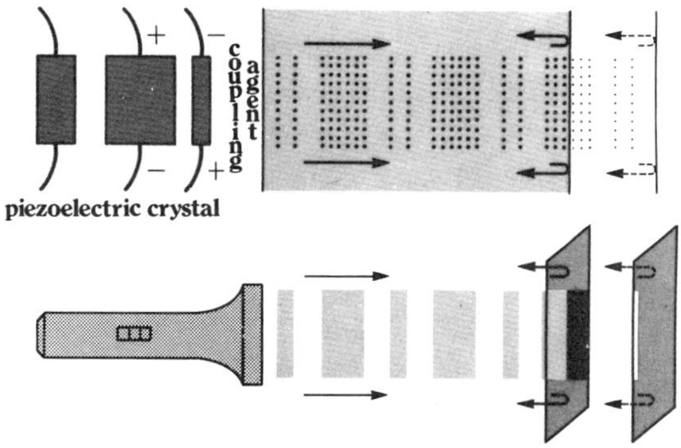

Figure 22–3. Comparison of a piezoelectric crystal generating ultra-high-frequency sound waves that are partially reflected by interfaces of different density to a blinking flashlight beamed through partial mirrors.

When the fetal heart rate can be counted with the stethoscope, it may be assumed that the gestation is 20 or more weeks along. When the patient first detects fetal movement (quickening), the mean time to date of delivery is 147 days but with a standard deviation of 15 days. The variability is too great to inspire confidence.

A relatively recent development that has become an extremely helpful tool in the management of obstetric problems is diagnostic ultrasound. The principle of ultrasonography is relatively simple. Ultra-high sound frequencies of 10–20 million cycles per second are generated by the reverse piezoelectric effect. (The piezoelectric effect is that rapid contraction and expansion of a certain type of crystal generates an electrical current.) This ultra-high-frequency sound wave is essentially a single beam and can penetrate tissues by a series of rarefactions and condensations of tissue particles. As the sound beam penetrates tissues, some of it is reflected as an echo from a plane or interface between tissues of different density; the rest of the sound energy penetrates further only to echo back from a density interface deeper in the body (Fig. 22–3). Using a probe containing a piezoelectric crystal capable of emitting pulsed sound waves, 400 pulses per second, and capable of recording the echoes coming back as electric impulses between the pulsed emittence of ultrasound, one can record the reflected echo on an oscilloscope to reproduce a picture of the different tissue density surfaces. The probe behaves like a blinking flashlight passing through partial mirrors. This information is displayed on an oscilloscope by different modes (Fig. 22–4). The A-mode (amplitude)

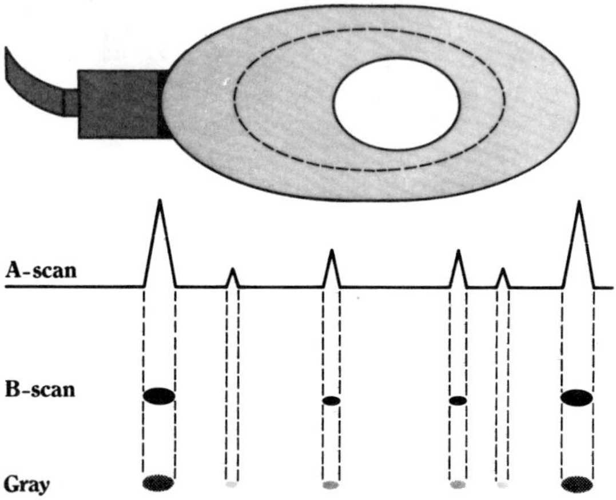

Figure 22–4. Schematic representation of different modes of display for ultrasonography.

displays the echoes as a positive deflection of a horizontal line. The B-mode (brightness) records the reflected echoes as dots at varying distances from the probe, depending on the time it took to leave and return to the probe. It does have a threshold for weaker echoes. The gray-scale mode shades the dots, darker dots for stronger echoes and lighter gray for weaker ones. It picks up the weaker echoes to give a texture to the soft tissue being scanned. As the probe moves across the abdominal wall, the dots coalesce, as in a drawing, depicting a two-dimensional anatomic cross section (Fig. 22–5). By using this method, one can outline the contents of the pregnant uterus and accurately measure diameters of fetal structures. Table 22–2 depicts the time-table for ultrasonographic evaluation of various fetal structures during pregnancy.

Measurement of the biparietal diameter of the fetal head (BPD) correlates closely with gestational age (Fig. 22–6). Since the central nervous system is least affected by nutritional deprivation, the BPD is a reliable indicator of fetal growth when measured serially. A BPD measured between 20 and 30 weeks gives a reliable estimate of the EDD (± 11 days) due to less variability during this time. Later, the variability increases, yielding less reliable results. Serial readings of BPD will also increase at a weekly rate that progressively decreases as gestation advances. When following fetal growth by this method, readings should be obtained at a minimum interval of two weeks.

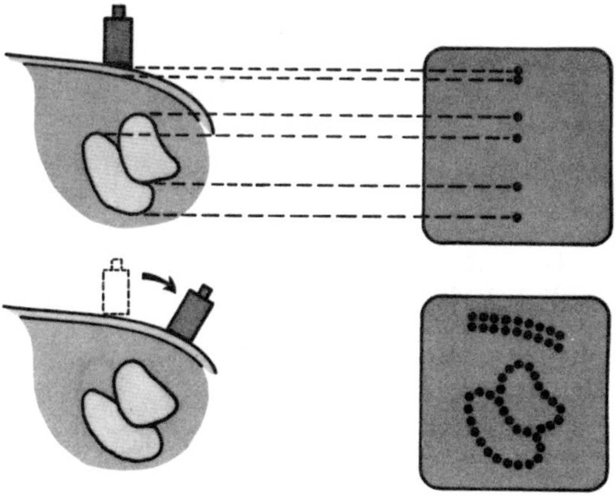

Figure 22–5. Representation of B-mode ultrasonography displayed on a storage oscillograph.

Table 22–2. Timetable for Ultrasound Evaluation of Pregnancy Structures

Gestation sac	5–10 weeks
Fetus	12 weeks to term
Biparietal diameter of fetal head	15 weeks to term
Placenta	9 weeks to term

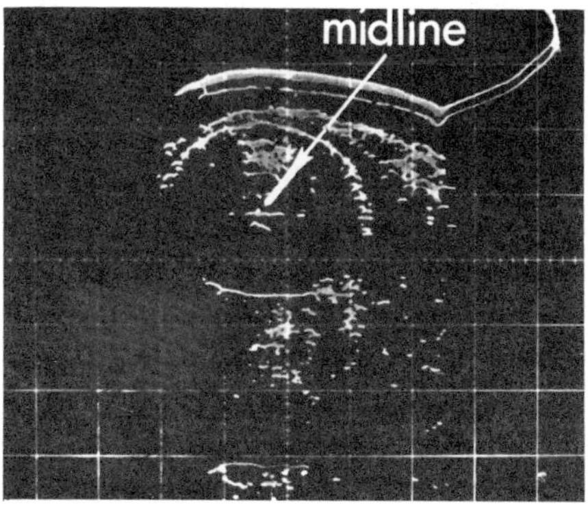

Figure 22–6. Actual B-mode ultrasonographic display of a fetal head from which the BPD may be measured.

Serial BPD readings show deviation of fetal growth of both the symmetrical and asymmetrical pattern types. Asymmetrical growth retardation can be confirmed by measuring the anteroposterior (AP) diameter of the lower fetal chest. As pregnancy advances, a malnourished fetus has an AP chest diameter that becomes progressively smaller than the BPD. A BPD reading indicating a mature infant will generally be highly reliable except in a diabetic pregnancy where the BPD may indicate a less mature LGA infant. In normal pregnancies, the BPD increases slightly beyond 37 weeks, whereas it continues to increase almost linearly in many diabetic patients carrying LGA fetuses. Serial BPD readings of a large number of dated pregnancies has yielded a percentile chart of BPD for gestational age. In a normal pregnancy, a BPD reading early in gestation in the 90th percentile will result in a newborn weighing in the 90th percentile. A fetus has a high probability of remaining in its percentile growth pattern, supporting the concept of inherent biologic variability. A 10th percentile growth pattern fetus need not necessarily demonstrate the findings of asymmetrical growth retardation in the newborn period. Ultrasound is useful in early detection of gross morphologic congenital defects such as hydrocephaly, anencephaly, spina bifida, and meningomyelocele. Sonar can detect multiple gestation, fetal death, and erythroblastosis. Sonar can locate the placenta, which is important in pregnancies complicated by bleeding or prior to doing an amniotic fluid tap (amniocentesis). The use of diagnostic ultrasound is thought to be safe, with no proven damage to growing tissues or genetic material.

The following paragraphs discuss tests of functional fetal maturity. The use of radiology as a test of fetal maturity is limited. The appearance of certain anatomic epiphyseal areas (distal femoral) on films is a fairly accurate estimate of fetal maturity in a normal infant. However, female fetuses seem to precede the male fetus by two weeks in the appearance of these landmarks. In fetuses of diabetics, the appearance of epiphyseal signs may be delayed. Because of exposure to ionizing radiation, these examinations cannot be repeated.

Presently, the most reliable tests of fetal maturity utilize an analysis of the constituents of amniotic fluid. Amniotic fluid is a vital material in constant dynamic relationship to the membranes, fetus, and maternal circulation. It is the only part of the conceptus that can be sampled reliably, with relative impunity and, if necessary, repeatedly. The sources of amniotic fluid appear to be multiple, including secretion from the fetal membranes, especially covering the placenta and cord, fetal kidneys, and fetal tracheobronchial tree. The constituents include desquamated cells, excretion products from fetal urine and gastroin-

Table 22–3. Commonly Used Amniotic Fluid Tests of Fetal Maturity

Test	*Organ(s) Tested*
1. Lecithin: sphingomyelin (> 2.0)	Fetal lung
2. Creatinine concentration (> 2.0 mg/dl)	Fetal kidney
3. % Fat-laden epithelial cells (> 20%)	Fetal skin
4. Bilirubin pigments (ΔO.D. 450 μm = 0.000)	Fetal liver
5. Osmolality (< 260 mOsm)	Fetal kidney

testinal tract (if the fetus is stressed), secretory products from fetal lungs, hormones, salts, protein, and amino acids. Table 22–3 lists amniotic fluid tests of fetal maturity in order of decreasing reliability. The procedure for obtaining the amniotic fluid sample is by transabdominal needle aspiration. It is called amniocentesis. The procedure is not without risk to the fetus, with a complication rate just under 1%. Placental localization by ultrasound minimizes the risk of placental puncture. The clinical situation should have a risk factor to mother and/or fetus greater than 1% in order to justify amniocentesis.

The most reliable and important amniotic fluid test is the measurement of the lecithin to sphingomyelin ratio (L/S ratio). Respiratory distress syndrome or hyaline membrane disease of the newborn is responsible for 40% of infant mortality in preterm deliveries. The L/S ratio can be used to predict risk to the infant. Respiratory distress syndrome (RDS) occurs when the newborn has insufficient amounts of a phospholipid material called surfactant in the alveolus. This material lowers surface tension in the pulmonary alveoli, preventing collapse during expiration. The phospholipid surfactant is a lecithin. Lecithin is produced in the type II alveolar cell by two major pathways, the CPD-choline pathway and the CPD-ethanolamine pathway. High surface activity requires the presence of two saturated fatty acids at the α and β positions of the lecithin molecule. Most lung lecithins contain two molecules of palmitic acid. Some contain myristic acid at the β site. The concentration of surface active lecithin is low until the 24th week of gestation, when it begins to increase. Palmitoylmyristoyl lecithin first appears at this time. At 35–36 weeks' gestation, large quantities of dipalmitoyl lecithin appear in amniotic fluid. The appearance of dipalmitoyl lecithin is associated with a surge of surface active lecithin in the amniotic fluid. Increased activity in the CDP-choline pathway is responsible for this surge in lecithin and is thought to be the major pathway for synthesis of surface active lecithin in the term infant and in the adult.

The biosynthesis of lecithin in the fetal lung may be enhanced by glucocorticoids. Preliminary clinical studies have demonstrated that maternal injections of steroids reduce the incidence of RDS in newborns of mothers in premature labor before 32 weeks' gestation. These studies indicate that it may be possible to prevent hyaline membrane disease, presumably by accelerating lung maturation, but further investigative work and corroborative evidence are needed.

In 1971, Gluck introduced the L/S ratio. Since sphingomyelin concentration in amniotic fluid remains relatively constant and even decreases slightly near term, the ratio of the relative concentration of lecithin to sphingomyelin, as measured by paper chromatography, reflects the alveolar production of lecithin. An L/S ratio of 2:1 occurs after 35 weeks' gestation and indicates the production of enough surfactant to prevent the development of hyaline membrane disease. A L/S ratio of less than 2:1 indicates that the fetus, if delivered, would be at risk of developing hyaline membrane disease.

The creatinine content of amniotic fluid increases as pregnancy advances. This represents a progressive increase in the excretion of creatinine by the maturing fetal kidneys. There is a high correlation with fetal maturity when the creatinine concentration in amniotic fluid reaches 2.0 mg/dl. It also appears that the test is reliable in many high-risk pregnancies such as Rh disease, diabetes, toxemia, and even in the presence of meconium. There are two exceptions. In a pregnancy with intrauterine growth retardation, the creatinine levels may be lower for the degree of fetal maturity attained, probably due to a smaller total body mass. Also, elevated maternal serum creatinine gives a falsely elevated amniotic fluid (AF) creatinine level.

As gestation advances, the amount of exfoliated cells in the AF increases, and the percentage of these cells containing lipid granules in the cytoplasm increases. These lipid-containing cells, "fetal fat cells," stain orange with 0.1% nile blue sulfate stain, whereas the non-fat containing epithelial cells stain pale blue. No premature infants have been delivered if lipid cells constitute more than 20% of the exfoliated cell count in several drops of amniotic fluid. Disconcerting, however, is that several authors have found no fat stained cells in up to 40% of known term pregnancies. The source of these lipid-laden cells is not definitely known.

In midgestation, there normally appears a certain amount of indirect bilirubin bound to albumin in the AF. Provided there is no fetal hemolysis or maternal hyperbilirubinemia or meconium in the AF, bilirubin disappears before 36 weeks of gestation. This presumably occurs as the fetal liver enzyme system matures enough to conjugate all bilirubin. Many authors question the reliability of this test since there are reports documenting disappearance of bilirubin in AF in less than 40% of mature infants, especially in diabetic and toxemic mothers. Bilirubin is measured by a change in optical density at 450 $m\mu$ (ΔOD) in amniotic fluid.

As pregnancy advances, the amniotic fluid becomes progressively hypotonic, mainly due to decreased sodium concentration. At term, the AF osmolarity decreases below 260 mOsm. The decline is quite gradual, and the difference between different gestational age groups is too small to make this test reliable as a test of fetal maturity.

Each of these tests depends on the maturation of a specific fetal organ system; whether maturity of an individual system is predictive of functional maturation of the entire fetus is not yet determinable. Nonetheless, the pulmonary system is the most important in terms of neonatal survival.

For the obstetrician managing a high-risk pregnancy, tests of fetal well-being provide an expanding data base from which intelligent decisions can be made. These tests, however, are not 100% predictive of fetal outcome and must be used in the context of the clinical situation.

The pregnant woman is able to perceive movements of her fetus from midgestation on to term. Daily fetal movements have been documented to occur up to 60 times per hour at midgestation and decrease slightly near term. When the mother perceives sluggish or decreased fetal movement, the fetus is usually in jeopardy. This simple observation may provide a "red flag" in the surveillance of a high-risk pregnancy.

A normal healthy fetus demonstrates a temporary elevation of baseline fetal heart rate when actively moving or sometimes when the uterus contracts. Absence of this fetal heart rate acceleration with fetal movement is usually associated with chronic fetal distress. A clinical test using the Doppler ultrasound recording of the fetal heart rate on a monitor and marking fetal movements on the strip chart read-out is called a nonstress test of fetal well-being. The results are either reassuring (presence of acceleration with fetal movement) or nonreassuring (absence of acceleration with fetal movement).

A fetus in chronic distress or with compromised reserve may respond to a temporary impairment of intervillous blood flow by developing an acute bradycardia that begins during the cessation of blood flow and continues after return of intervillous blood flow for a short period (a late deceleration). Intermittent cessation of intervillous blood flow can be accomplished by administering enough intravenous oxytocin by infusion to attain three uterine contractions in 10 minutes. If the fetal heart rate does not respond with intermittent bradycardia of the late deceleration pattern, the fetus is most likely healthy; whereas, if the late deceleration pattern is present, then the fetal environment is such that it may threaten fetal well-being.

Twenty-four-hour urinary estrogen output or plasma estriol levels (see Chapter 14) slowly increase as gestation advances. When these levels drop significantly (usually 50% of previous levels), either rapidly or gradually, there is indication that the fetus is in jeopardy. The more rapid and profound the decline, the greater the jeopardy for the fetus.

As with tests of fetal maturity, these tests of fetal well-being are used as a battery of tests in most medical centers with the hopes of not allowing a false negative result of one test to cause a catastrophe. No one test is infallible.

SUGGESTED READING

Begneaud WP Jr, Hawes TP Jr, Mickal A, et al: Amniotic fluid creatinine for predication of fetal maturity. *Obstet Gynecol* 34:7, 1969.

Doran TA, Bjerre S, Porter CJ: Creatinine, uric acid, and electrolytes in amniotic fluid. *Am J Obstet Gynecol* 106:325, 1970.

Gluck L, Kulovich MV, Borer RC Jr, et al: Diagnosis of the respiratory distress syndrome by amniocentesis. *Am J Obstet Gynecol* 109:440, 1971.

Gluck L, Kulovich MV, Eidelman AI, et al: Biochemical development of surface activity in mammalian lung. IV. Pulmonary lecithin synthesis in the human fetus and newborn and etiology of the respiratory distress syndrome. *Pediatr Res* 6:81, 1972.

Moore WMO, Murphy PJ, Davis JA: Creatinine content of amniotic fluid in cases of retarded fetal growth. *Am J Obstet Gynecol* 110:908, 1971.

Murata Y, Martin CB Jr: Growth of the biparietal diameter of the fetal head in diabetic pregnancy. *Am J Obstet Gynecol* 115:252, 1973.

QUESTIONS

1. The ideal birth weight and duration of gestation are ________ gm and ________ weeks.

2. Immature births range from ________ gm to ________ gm in weight and ________ weeks to ________ weeks in duration
3. Dysmature term births account for about one-third of all "premature newborns." TRUE FALSE
4. The diagnostic terms, dysmature and postmature, are incompatible and cannot be used to describe the same fetus. TRUE FALSE
5. Approximately how many women whose pregnancies pass beyond 20 weeks will have delivered by 40 weeks' gestation (from LMP)?
6. The prematurity rate for all live births is ________% for a U.S. nonindigent population.
7. Premature infants who are not dysmature tend to have ________.
8. Dysmature infants tend to become ________.
9. The three mechanisms for dysmaturity are ________, ________, and ________.
10. Placental insufficiency usually has its origin in some maternal disease complicating pregnancy or in pregnancy toxemia. TRUE FALSE
11. The dysmature infant is defined as being too small or too large for its gestational age. TRUE FALSE
12. Which of the following statements is *FALSE*:
    A. the probability of correctly dating a pregnancy by the last menstrual period is good except if conception occurred shortly after stopping oral contraceptives.
    B. the fundal height corresponds closely to the gestational week at midgestation.
    C. quickening is a reliable sign of gestational age.
    D. the gestation is over 20 weeks if fetal heart rate is counted.
13. The following tests of fetal maturity were done on a patient. Which would you be *least* willing to accept as valid?
    A. bimanual exam of uterus done 18 weeks after the LMP
    B. BPD measurement 26 weeks after the LMP
    C. amniotic fluid creatinine level with maternal creatinine elevated due to renal disease
    D. amniotic fluid L/S ratio
    E. menstrual history
14. Which of the following is true about ultrasound use in obstetrics:
    A. if used repeatedly the effects of high frequency sound waves are deleterious to fetal cells

B. the BPD measurement is most reliable for estimating gestational age after 34 weeks' gestation
C. the localization of the placenta is reliable with ultrasonography
D. the B-scan is an unidirectional technique
E. normally the BPD increases linearly after 36 weeks

15. The L/S ratio:
A. is grossly affected by fluid volume changes
B. is a reflection of the amount of central nervous system development
C. correlates well with lung maturity
D. is associated with high levels of palmitoyl-myristoyl lecithin rather than dipalmitoyl lecithin
E. rises rapidly at 36 weeks due to phosphatidyl ethanolamine pathway induction

16. Match the following with regard to the establishment of fetal maturity or fetal size:

(1) AF—creatinine	a. 2:1
(2) AF—L/S ratio	b. decreases to zero
(3) AF—fetal fat cells	c. 20%
(4) AF—bilirubin	d. 2.0 mg/dl
(5) AF—osmolarity	e. unreliable but decreases

17. Match the following with regard to fetal growth and nutrition:

(1) Fetal insulin	a. growth hormone
(2) Fetal energy metabolism	b. rubella
(3) Asymmetric growth retardation	c. maternal hypertension
(4) Symmetric growth retardation	d. glucose

18. Which of the following statements is *FALSE*:
A. Fetal movements as noted by the mother may herald impending disaster if increased.
B. Acceleration of baseline fetal heart rate in conjunction with fetal movement indicates fetal good health.
C. The apparent rate limiting factor in fetal growth is the placenta.
D. Fetal bradycardia during and following a uterine contraction induced by oxytocin is not a reassuring event.
E. Increasing values of plasma estriol correlate with decreasing perinatal loss.

19. Match the following:

(1) L/S ratio over 2:1	a. 35 weeks
(2) LGA infant	b. 26 weeks

(3) SGA infant	c. hyperglycemia
(4) McDonald measurement of 25 cm.	d. hypoglycemia
(5) amino acid concentration	e. active transport

ANSWERS

1. 3,300 gm and 40 weeks
2. 500 to 999 gm   20 weeks to 28 weeks
3. TRUE
4. FALSE
5. 50%
6. 6%
7. Respiratory disease
8. Hypoglycemic
9. Placental insufficiency, congenital anomalies, and chronic intrauterine infections.
10. TRUE
11. TRUE
12. C
13. C
14. C
15. C
16. (1) d
    (2) a
    (3) c
    (4) b
    (5) e
17. (1) a
    (2) d
    (3) c
    (4) b
18. A
19. (1) a
    (2) c
    (3) d
    (4) b
    (5) e

chapter twenty-three
labor

OBJECTIVES

1. Describe structural and functional changes in the uterine muscle and collagen during pregnancy.
2. Describe the functional role of the cervix.
3. Describe the evolution of myometrial activity during pregnancy.
4. Describe changes in the cervix during pregnancy.
5. Identify the time limits of normal first and second stage labor.
6. List the forces necessary to effect the second stage.
7. List the structures cut in an episiotomy.
8. Know the cellular phenomena that occur during myocyte contractions.
9. Know the control systems for prelabor and labor.
10. Identify etiologic factors in dysfunctional labor.
11. Describe the actions of anesthetics on the uterus.
12. Identify the origin and pathways of pain in the first and second stages of labor.
13. List indications for pain relief in labor.
14. Identify three preparatory techniques for labor.
15. Identify the mechanism of pain relief for paracervical block, caudal anesthesia, lumbar epidural block, pudendal block, and spinal anesthesia.
16. Identify the potential fetal or neonatal harm secondary to the administration of opiates, barbiturates, hypnotics, and tranquilizing drugs during labor.

17. Identify the fetal cost of labor from asphyxia, trauma, poisoning, and infection.

PHYSIOLOGY OF NORMAL LABOR

Labor is the process that empties the uterus and vagina of the conceptus during the last half of pregnancy. It is a physical contest between the forces of extrusion provided by the uterine myocytes and the increase in intraabdominal pressure provided by voluntary maternal striated muscle and the forces of resistance of the connective tissue and muscle of the cervix and the pelvic floor.

Nonpregnant Uterine Anatomy Relevant to Uterine Emptying

The uterus is a hollow organ (100 μl) comprised essentially of myometrial smooth muscle cells, collagen, vasculature, and a mucosal lining. The upper portion of the mature uterus is called the corpus and contains about 35% muscle (Fig. 23–1A). The nonvascular collagen to muscle ratio is about 1:1. Muscle cell distribution is fairly random, although the cells are in bundles. Individual cells may pass from bundle to bundle, so that the whole corpus is an interlaced network of collagen and muscle fibers from top to bottom and bottom to top and from outside in and inside out. The cervix, the inferior portion of the uterus, is comprised primarily of collagen (Fig. 23–1A, B). This is the truly unique feature of the uterus when compared to other hollow organs, for the uterus has a collagenous, not a muscular, sphincter. The junction of the corpus and cervix is the anatomic internal os, identified as a constriction of the upper end of the cervical canal (Fig. 23–1A, B). Microscopic examination does not identify any significant difference between the myometrial tissue just above or just below the internal os. This area is a zone of transition from tissue with much muscle content above to tissue with much collagen content below.

Adaptations of the Uterine Cavity to Distention

When a mass that maintains its shape and volume under pressure is created within the uterine cavity, be it a myoma, large polyp, or gestation, the muscle cells respond with hvpertrophy; the individual

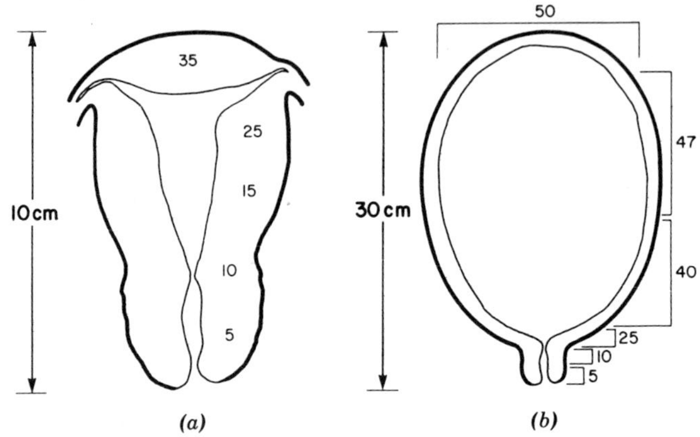

Figure 23–1. Size and percent of muscle content for various areas is shown for (*a*) nonpregnant adult uterus and (*b*) term pregnant uterus.

muscle cells become larger. This response occurs even in the absence of estrogenic stimulation and is a characteristic, intrinsic response of uterine muscle cells to chronic stretch. When the levels of circulating estrogen are high, as in pregnancy, the uterus increases in weight from an organ of approximately 90 gm to an organ weighing about 1,100 gm. This kilogram increase in mass occurs in the corpus, with insignificant change in the mass of the cervix. Both muscle cells and collagen increase in content in the corpus, but the percentage increase in muscle mass is greater (Fig. 23–1B). In late pregnancy, the corpus is about 45–50% muscle (Fig. 23–1B).

The corpus of the uterus during gestation is an organ of accommodation, permitting the expansion of its volume to approximately 4,500 ml. The cavity changes from a flat inverted triangle in the nonpregnant state to a spherical shape that continually expands until approximately 24 weeks of gestation. From that point on, the uterus becomes ever more cylindroid or pyriform (Fig. 23–1B).

The cervix serves as a sphincter, keeping the pregnancy in situ until labor begins. There are two evidences for this sphincteric action in clinical experience. If the cervix is amputated prior to pregnancy, the uterus will extrude the conceptus between the 16th and 24th week of gestation. Patients with a poorly identified pathophysiologic entity called the "incompetent cervical os syndrome" behave in pregnancy as if the cervix were amputated. The cervix dilates painlessly, and the amniotic sac containing the fetus slides out of the uterus into the vagina, usually between the 16th and 24th week of gestation.

The collagen of the cervix is in the form of fibrils with no distinct or specific orientation. Stiffness or noncompliance is the result of mucopolysaccharides binding the proteins of collagen fibrils into large filamentous structures. During the latter half of pregnancy and especially as the onset of labor is approached, the cervix becomes progressively more compliant. This change is effected by loss of binding, permitting the filaments to slide apart under physical stress. This quality of compliance, which permits the mechanical forces of uterine contraction to dilate the cervix, is the result of a biochemical change in the forces of adhesion between collagen fibrils and is not associated with significant change in the collagen content of the cervix in rodents. In humans, the collagen concentration in terms of dry weight and the collagen to protein ratio decrease in association with the development of cervical compliance.

Control systems permitting or effecting these cervical changes are unknown, but studies in ungulates suggest that prostaglandins may be responsible for the development of cervical compliance, independent of their ecbolic action on uterine musculature. Cervical shortening and dilatation occur in postmenopausal women with the passage of a pedunculated submucous myoma from the uterine cavity into the vagina. This result is obtained in the total absence of a pregnancy and suggests that the contractions of the uterus pressing the tumor against the cervix eventually result in the development of compliance. Obviously, there is much to be learned about the systems that control these phenomena.

Gross Anatomic Changes in the Cervix during Pregnancy

Starting at various times after the twenty-fourth week of pregnancy, women manifest changes in the cervix, consisting of shortening of the canal (effacement) and dilatation of the internal and the external os (Fig. 23–2). Nulliparous women differ somewhat from parous women, and within each group the patterns of change during the latter half of pregnancy are highly variable. The average nullipara tends to achieve more effacement but has less dilatation than the average multipara on arrival at the hospital in labor (Table 23–1). Effacement is measured in terms of percentage of shortening of the canal, assuming that the uneffaced cervix is 3 cm in length (0% effaced). Dilatation is measured in centimeters of the diameter of the orifice of the external os.

Figure 23–2. Effacement in a primigravida. Shaded area is the cervix, comprised primarily of collagen: (*a*) uneffaced, (*b*) beginning effacement, (*c*) completely effaced.

Table 23–1. State of Cervix on Arrival in Hospital in Labor

Mean	*Nullipara*	*Multipara*
Effacement (%)	80–90	70–80
Dilatation (cm)	2–3	3–4

Myometrial Contractile Activity

At all times during adult life, some myometrial cells in every uterus are contracting, but the contractile pattern in respect to strength and repetition frequency differs according to the clinical state, such as between menses, during menstruation, or during pregnancy and labor. The most useful technique for assessing myometrial activity in humans is measurement of the pressure within the uterine cavity (the intrauterine pressure, IUP), which reflects myometrial activity. An increase in IUP above the resting pressure followed by a decrease in IUP to the resting pressure is called a "uterine contraction." The IUP in this semielastic closed hydrostatic system is proportional to the cross-sectional tension developed in the wall of the uterus and inversely proportional to the radii of the cavity (LaPlace). Thus, the mean amplitudes of the IUP are lower in *normal* term labors with twins than with singleton fetuses because the radii are larger in multiple pregnancy.

Uterine Contractility in the Menstrual Cycle

Between menstrual episodes, the uterine contractions are characterized by high frequency (20–30/10 min) and low amplitude (Fig.

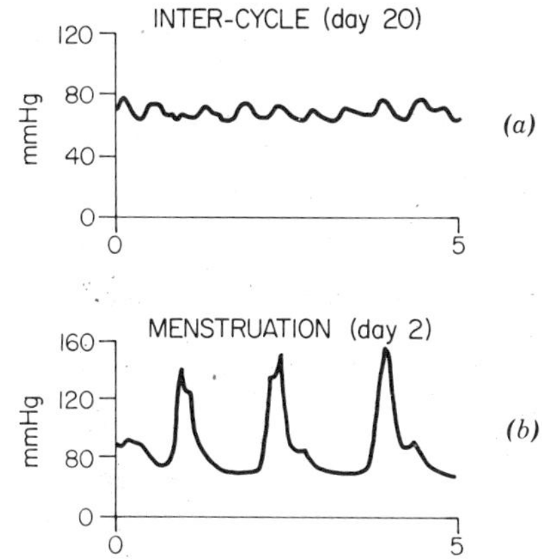

Figure 23–3. Nonpregnant intrauterine pressure recorded for five minutes: (*a*) day 20 of a 30-day menstrual cycle; (*b*) second day of menstruation.

23–3A). There is practically no difference in this activity before and after ovulation, suggesting that the variations in tissue levels of sex steroids occurring in the nonpregnant state between episodes of menstruation have little effect on myometrial function. Just prior to menstruation, the contractions become less frequent but stronger, and at menstruation they occur every two to three minutes and are of considerable strength (Fig. 23–3B). This suggests a myometrial cycle that seems to produce patterns of coordinated myometrial activity at menstruation and is imitative of the events during pregnancy and labor. The myometrium throughout the menstrual cycle is insensitive to oxytocin but is sensitive to vasopressin. The cramping suprapubic abdominal pains experienced by women with primary dysmenorrhea are synchronous with menstrual contractions, but the contractions of women without dysmenorrhea can be identical in intensity and frequency.

Uterine Contractility During Pregnancy

The human uterus is fairly quiet during the first 18 weeks of pregnancy. At about midpregnancy (18–22 weeks), some obvious change in the myometrial cells occurs, for the myometrium manifests increasing spontaneous contractility and becomes progressively more sensitive to oxytocin. If one averages the results obtained from studies of the IUP in large numbers of women, spontaneous change in contractility increases from about 20 to 36 weeks of gestation and seems to remain the same or continues to increase from 36 weeks until the initiation of labor. However, the variation from woman to woman of similar length of gestation, or within a single woman from hour to hour, day to day, or week to week, is considerable until labor is well established. By the 30th–32nd week of pregnancy, examination of uterine contractility by external tocography detects an average of two to three contractions per hour. This repetition frequency differs grossly from the usual range of four to six per 10 minutes observed in women with threatened preterm labor.

Since the safety of hospital obstetric delivery has been established, it is the parturient and her family who must make the decision about the onset of labor by leaving home for the hospital. In some women during the last two to three weeks of pregnancy, episodes of spontaneous uterine activity become so strong that early labor is suspected, and the woman and family rush to the hospital only to be disappointed because, on or very shortly after arrival, the contractions cease and the doctors inform them that the woman is in "false labor." While some women make the shift from weak and disorganized prelabor activity to strong active labor in a very brief period of time, other women require many hours or episodes of discernible contractions, during which the myometrium converts from weak, disorganized, irregular activity to integrated, rhythmic contractions. This period has been called the "latent period," "prodromal labor," or "preactive labor." If one of these prelabor bursts of myometrial activity is followed by quiescence, we call it false labor because the patient has sensed labor only to have it go away. Of course, it is not false, only a disappointment for the woman who is anxiously looking forward to labor, to have, see, feel, hold, and examine her newborn infant. If the woman demonstrates cervical effacement and dilatation over a few hours' time but is less than 3–4 cm dilated, she is in active prelabor; if 3–4 cm dilated or more, she is in active labor.

ACTIVE LABOR

Once definitely initiated, labor is characterized by uterine contractions usually occurring at a mode repetition frequency of four per 10 minutes with a mean amplitude of 40–50 mm Hg. A strong uterine contraction can be identified if the corpus of the uterus cannot be indented at the peak of the contraction by an extended index finger. During periods of relaxation, the uterus should be fairly flaccid, and the resting pressure, which averages 8–12 mm Hg, should not exceed 16 mm Hg (Fig. 23–4).

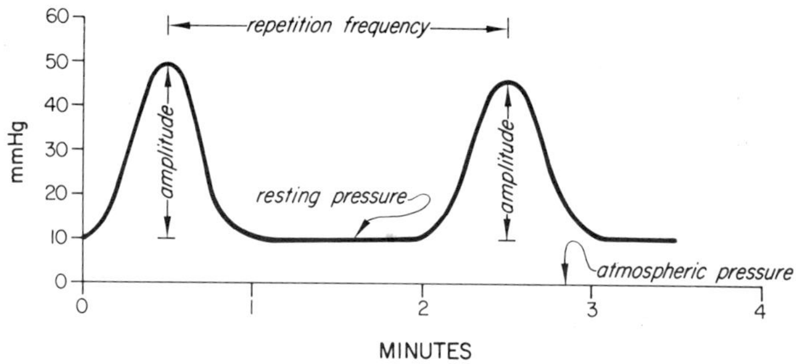

Figure 23–4. Intrauterine pressure during labor.

Cellular Contractile Phenomena

Unlike the myocardium, skeletal muscle, or ciliary muscle, the uterus, a "syncytial" type of smooth muscle organ, has no nervous system integration of activity. We know very little about the mechanisms of coordination that result in contractions. Bits and pieces of evidence do permit construction of a reasonable hypothesis, which will certainly be modified as more information is gained.

The plasma membrane of the myocyte maintains a resting potential of approximately −50 mV. When action potentials are sufficiently repetitive and prolonged (burst discharges), tension results (Fig. 23–5). Every small muscle biopsy (1 × 1 × 10 mm) from the corpus demonstrates many cells that manifest slow depolarizations prior to the burst discharge. These cells are believed to function as pacemaker cells initiating excitation and contraction in adjoining myocytes. No geographic area of the corpus of the uterus dominates pacemaker function. The control system for the repetition frequency of burst discharges is unknown. Physiologic contractions vary from two to three per minute, observed between menses in the nonpregnant woman, to three or four per 10 minutes, observed either during menstruation or labor.

Excitation-contraction coupling and the resultant shortening of actomyosin filaments is effected by the cytosol concentration of calcium ions. It is not known whether the source of Ca^{2+} is external, for the surface to volume ratio of myocytes is great compared to other muscle cells, or internal as suggested by these observations: 1) The quantity of bound calcium is 10 times that required as Ca^{2+} to produce contraction, 2) isolated myocyte sarcoplasmic reticulum binds calcium in large

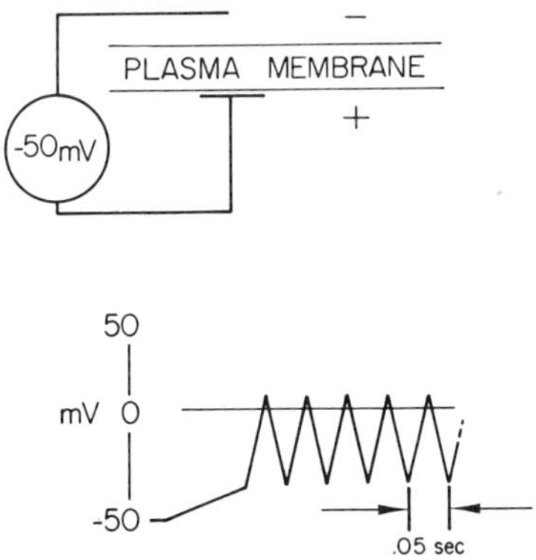

Figure 23–5. Diagrammatic representation of plasma membrane with a resting potential of −50 mV. Slow depolarization is shown followed by burst discharges every 0.05 seconds. These result in the development of tension in the muscle.

quantities in nonionic form, and 3) sarcoplasmic reticulum tubules are infrequent and poorly developed.

Relaxation of spontaneous contractions appears to be the result of the innate biochemical characteristics of the myocyte. Calcium release is associated with membrane dephosphorylation, and at the completion of depolarization, the membranes are repolarized by an energy-consuming process, and calcium is rebound or extruded, inducing relaxation. In contrast, evoked contractions and β_2 adrenergic agonists induce increase in the cytosol concentration of cyclic-AMP. A suggested mechanism is that β-agonists activate plasma membrane adenylate cyclase, which increases the rate of conversion of ATP to cyclic-AMP. Cyclic-AMP activates cytosol protein kinases, which phosphorylate the lipoproteins of cell membranes and induce Ca^{2+} binding in nonionic form to produce relaxation.

Cyclic-AMP is converted to AMP by specific phosphodiesterases that are inhibited by Ca^{2+}. Thus, β-adrenergic agonists relax myocytes via the adenylate cyclase cyclic-AMP mechanism. Excessive contractility with elevated cytosol Ca^{2+} from evoked contractions alters the equilibrium of ATP→cyclic-AMP→AMP system by blocking the conversion of cyclic-AMP→AMP and raising the cytosol level of cyclic-AMP. The latter mechanism may provide a degree of protection to the uterus from overstimulation.

Two opposing hypotheses for the organization of myocytes in contractions have been suggested. The traditional model assumes a macroscopic site of pacemaker activity and a spreading wave of contractility throughout the organ, recruiting muscle cells to the event. Pathophysiologic events are described in terms of excessive numbers or inappropriate sites of pacemaker function or resistance to conduction and contraction by special geographic areas (placental site). The alternate hypothesis suggests all contractions are the manifestations of subcellular phenomena. When a sufficient number of myocytes become highly excitable and develop depolarization rhythms characteristic of clinical labor or menses, effective contractions are present. One of the singular characteristics of the contractions of normal spontaneous labor or menses is the variability of the waveform, amplitude, and beat-to-beat period within the experience of a single woman's labor or menses, suggesting that this system is dependent on the physiologic behavior of the population of myocytes participating in each contraction rather than on well-defined organizational systems. The absence of any integrating nerve structures similar to those of the heart or intestine reinforces the belief that clinical contractile phenomena are controlled by the physiologic state of the myocytes and the influences on them provided by naturally occurring ecbolic or relaxing agents.

Control Systems for Prelabor and Labor Activity

As each new naturally occurring ecbolic or uterine relaxing agent has been identified, a hypothesis for the control of the onset of labor or a role for the agent in pathophysiologic phenomena has been constructed. While no hypothesis has been supported by sufficient evidence to achieve acceptance, these researches have provided a new level of understanding of uterine clinical pharmacology and a useful background of knowledge for clinical practice.

Estrogens are important because they maintain membrane excitability, and, in association with stretch, they control the quantity of actomyocin and thus the working potential of the uterus. The uterus appears to be less active under the pharmacologic influence of progesterone because of reduced numbers of burst discharges and less conductance. Both sex steroids require intracellular protein receptors for effective action, and both experience metabolic conversion to steroids of lesser biologic activity. Until the changes at the cellular level during the transition of the uterus from prelabor to labor are known, the role of estrogen and progesterone in the onset of labor will remain undefined. Certainly, labor is not associated with abrupt changes in the production or circulating levels of either sex hormone.

Oxytocin is secreted in increasing quantities by the neurohypophysis during labor, and the placenta produces an inactivating enzyme, cysteineaminopeptidase (oxytocinase), which appears in maternal plasma in increasing concentrations during pregnancy. Membrane receptors for oxytocin have been identified in the myocyte, but changes in receptor content during pregnancy have not been investigated. Plasma levels of oxytocin are not significantly different prior to or early in labor but do rise in late labor.

Prostaglandins (PG) E and F are produced by the decidua and possibly by the myocytes. Spontaneous contractions of myometrial biopsies from the pregnant uterus are inhibited by PG synthetase inhibitors, but the muscle tissue continues to contract when stimulated by oxytocin or α-adrenergic agonists. Amnion contains large quantities of arachidonic acid esters. It is suggested that destabilization of amnion cell lysozymes releases hydrolases, which, in turn, result in the availability of large quantities of free arachidonic acid to the decidua and chorion, where it is converted by PG synthetase to active PG. It is hypothesized that chorion or decidual PG synthetase is subtrate limited, and the availability of free arachidonic acid results in active PG production by the chorion and/or decidua. PGs then soak into the myometrium and initiate labor. Attempts to study maternal plasma PG metabolite concentrations before and after the onset of labor have not identified acute increases in association with the onset of labor with certainty.

Adrenergic α-agonists stimulate contractility; β-agonists stimulate relaxation. Adrenergic receptor blocking agents obtund the effects of α- and β-agonists in the myometrium as they do elsewhere. Fluorescent staining of the myometrium for catecholamines during pregnancy reveals marked reduction in comparison to the nonpregnant uterus, presumably the result of loss of receptor sites. It is assumed that this result aids in the maintenance of uterine relaxation and blunts the impact of ecbolic adrenergic agonists or sympathetic discharges upon the myometrium.

The Phase of Expulsion of the Fetus

After the cervix is completely dilated (second stage of labor), the forces provided by the contractions of the uterus, augmented by maternal voluntary expulsive efforts, expel the fetus. Most women delivering their first child require about 20 contractions over a period of 60 minutes; multiparas usually require about 10 contractions over a period of less than 30 minutes.

The main forces of resistance to delivery in the second stage of labor are the inferior fascia of the urogenital diaphragm and the levator ani muscle (Fig. 23–6A, B). The urogenital diaphragm consists of a fascial

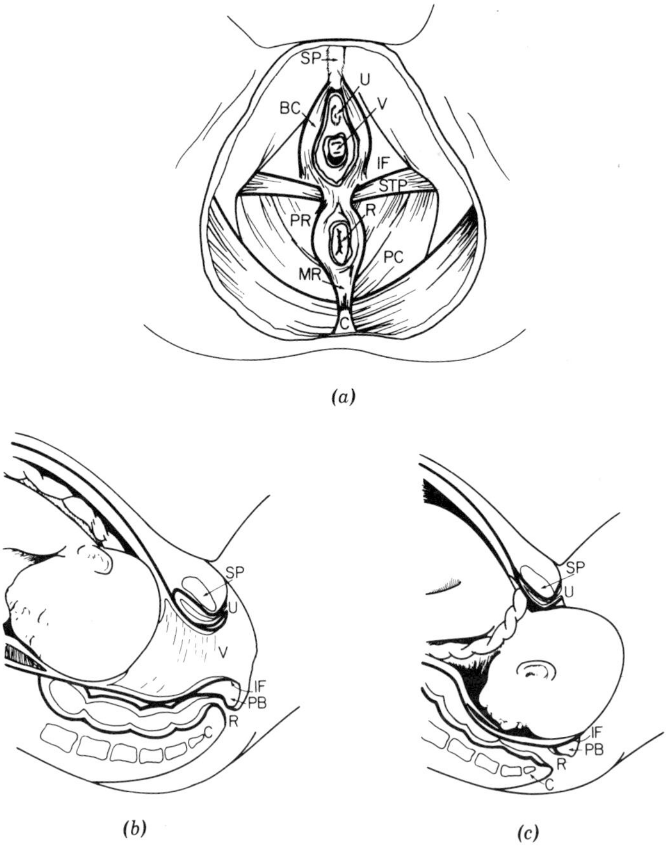

Figure 23–6. Diagrammatic views of the major soft tissue resistances to the second stage of labor: (*a*) inferior view of pelvic floor; (*b*) lateral view prior to second stage; (*c*) lateral view late in second stage of labor. See text for details. SP = symphysis pubis, U = urethra, BC = bulbocavernosus muscle, V = vagina, IF = inferior fascia of urogenital diaphragm, STP = superficial transversus perinei muscle, R = rectum, PR = puborectalis muscle, PC = pubococcygeus muscle, PB = perineal body, MR = median raphe, C = coccyx.

sheet which fills the anterior portion of the outlet of the bony pelvic girdle. It is attached to the medial and anterior margins of the pubic arch and terminates posteriorly at the level of the anus (i.e., at the level of the tuber ischii). The urethra and vagina perforate this fascia and are securely fastened to it. The hymeneal ring identifies the level of the fascia in the pelvic outlet on clinical examination. The posterior portion of the pelvic outlet, posterior to the anus and anterior to the tip of the sacrum, is filled by the sling of the pubococcygeal and puborectalis portions of the levator ani muscle. The perineal body (may all forgive the "anatomist" who first used the term "body") is a confusing term used to designate the anatomic fact that the vagina and rectum are separated by a greater volume of structures within 5–6 cm of their outlets than they are farther up (superiorly and posteriorly) the canals of both structures. The perineum is comprised of muscles that are divided into deep (superior) and superficial (inferior) sections by the fascia of the urogenital diaphragm. Inferiorly are the bulbocavernosus, ischiocavernosus, and superficial transversus perinei muscles. Superiorly is the deep transversus perinei muscle, which is attached to the lower (posterior) portion of the pubic rami and to each other at the median raphe. A midline perineotomy (episiotomy), performed to enlarge the outlet of the birth canal, severs the median raphe and cuts the perineal skin and the epithelium and fibromuscular structure of the vaginal wall, separates the right from the left superficial and deep transversus perinei and bulbocavernosus muscles, and splits the inferior fascia of the urogenital diaphragm from the hymeneal ring to the edge of the anal sphincter.

As the fetal head descends on the pelvic floor, the structures described above are pushed inferiorly. Because each structure is firmly attached to bone at some point, stretching of the muscles and fascia is required if delivery is to be accomplished. As the perineum is pushed more and more inferiorly, the levator muscle is also pushed posteriorly, and the inferior fascia of the urogenital diaphragm is pushed anteriorly. The orifices of the urethra and vagina now point more anteriorly than inferiorly, and the orifice of the anus points more posteriorly than inferiorly. The orifices of the vagina and anus both are pushed open in the process of stretching the pelvic floor. When the fetal head has advanced a sufficient distance inferiorly so that the fetal neck comes under the arch of the pubis, the resistance of the pelvic floor forces the head anteriorly (delivery by extension of the fetal head), and no further movement of the pelvic floor in an inferior direction is required. Stretch of the superficial and deep muscles of the urogenital diaphragm continues until the fetal head escapes from the mother (Fig. 23–6C).

Potential traumas that arise from this stage of the birth process are both immediate and remote. Laceration of the perineal body, causing separation of the sphincter ani with extension of the laceration up the rectal wall, results in the formation of a cloaca and fecal incontinence. This trauma results from a laceration of the perineum, which extends in a posterior direction. Laceration of the perineal body may extend superiorly along either posterior sulcus of the vagina, splitting the vaginal wall on either or both sides and extending into the levator muscle. This laceration is associated with severe hemorrhage. As the inferior fascia of the urogenital diaphragm is pushed inferiorly and anteriorly, it pulls the urethra, which is firmly fixed to it, away from its normal position behind the symphysis pubis. The urethra is maintained in its normal anatomic position by a set of lateral ligaments that are attached to the posterior aspect of the symphysis pubis. As the urethra is pulled inferiorly, these supports can be torn with loss of support for the urethra (urethrocele), which may result in urinary stress incontinence. (See Chapter 12.)

The Phase of Expulsion of the Placenta

After the fetus has been expelled, the uterus contracts and reduces its interior volume and area, shearing the placenta from the uterine wall at the decidual layer. With another contraction or two, the placenta is forced into the vagina and can be expelled by the woman if she is not entirely supine. This third stage of labor usually requires only a few minutes, certainly less than 15. During the time that the placenta is being sheared from the uterine wall, maternal blood escapes from the spiral arterioles, and an average of 600 ml is lost through the vagina over the first 24 hours of the puerperium. Losses in excess of 1 liter result in hypovolemia of clinical significance.

ABNORMAL (DYSFUNCTIONAL) LABOR

Early diagnosis of abnormal labor is important in the preservation of fetal health. A large study conducted in the 1950s demonstrated that abnormal labor results in children with two to four times the rate (the variance depends on the cause) of neurologic damage than those born after normal labor. Hopefully, current interventive techniques will improve these results. More than one cause of abnormal labor may exist concurrently and complicate diagnostic analysis. Complete investigation is mandatory for selection of proper treatment. Abnormal labor may result from the following categories of abnormalities:

1. Abnormal contractility
2. Abnormal cervical compliance

3. Obstruction of the birth canal
4. Obstruction of fetal origin
5. Obstruction of placental origin
6. Inadequate voluntary effort

Abnormal Contractility

Abnormal contractility is the commonest cause of dysfunctional labor in well-nourished, healthy maternity populations. Exceptions can always be identified, but when contractions occur less than twice per 10 minutes and average less than 25 mm Hg in amplitude, the progress of labor, as evidenced by the rate of dilatation of the cervix, is usually abnormal. No proven etiologies have been identified, but a variety of clinical conditions contribute. Among the suspects are supine position, fear and anxiety, and premature use of regional anesthesia. The observation of hypocontractility in some patients with obstructed labor complicates etiologic diagnosis.

Hypercontractility is observed in association with pregnancy-induced hypertension and abruptio placentae and is manifested by more than five contractions in 10 minutes with or without elevated resting pressure. Hypercontractility usually results in effective labor progress if the cervix is compliant, but it induces fetal asphyxia.

After the placenta is sheared from the uterine wall and delivered into the vagina, cessation of blood flow through the spiral arterioles is dependent on continued contraction of the myocytes until thrombosis occurs. Ineffective contractility of the uterus after separation of the placenta results in profuse uterine hemorrhage.

Abnormal Cervical Compliance

In about 3% of patients with abnormal labor, other causes of abnormal labor are not present, and the cervix remains firm and poorly effaced. Failure to progress in labor is attributed to underveloped cervical compliance. We have no means to assess this diagnosis objectively but suspect it by exclusion of other causes. This diagnosis assumes our knowledge is complete, and if the problem is not the result of other pathologic conditions, it must be the cervical compliance that is pathologic. Until we know the mechanisms by which cervical compliance is achieved, this pathophysiology will always be of questionable validity.

Obstruction of the Birth Canal

The bony pelvis may be too small in one or more dimensions, usually a manifestation of nutritional deprivation during childhood, or it may be asymmetric, the result of traumatic fracture or abnormal gaits during childhood when the pelvis is growing. Tumors of the pelvic reproductive organs (myoma, ovarian cysts), congenital malformations (pelvic kidney, double uterus), inspissated feces, or a full bladder may prevent entry of the fetus into the pelvis.

Obstruction of Fetal Origin

The fetus may be generally macrosomic, as observed in untreated or poorly managed diabetic women or grossly obese women, or a fetal part may be pathologically enlarged (hydrocephaly or obstruction to fetal urinary outflow). More commonly, the fetus produces obstructed labor by presenting diameters and circumferences to the birth canal that will not permit safe vaginal delivery, such as transverse lie of the fetus within the uterus.

Obstruction of Placental Origin

Occasionally, the trophoblast implants sufficiently low in the uterus to prevent entry of the fetus into the pelvis. If the placenta is sufficiently low (placenta previa), the initiation of severe bleeding usually results in the need for abdominal delivery, and the obstruction does not become manifest. However, if the placenta is implanted on the posterior surface of the lower uterine segment, vaginal bleeding may not be a problem, but the placenta overlying the promontory of the sacrum may prevent entry of the presenting part of the fetus into the birth canal.

Reduced Patient Effort

Events controlling cervical effacement and dilatation are outside the woman's voluntary control (the first stage of labor). However, after dilatation is complete, and the fetal head descends to the pelvic floor, stretching of the levator ani muscles initiates voluntary straining, as if defecating, as an accompaniment to uterine contractions. This effort increases the intrauterine pressure from the usual 40–45 mm Hg to about 160 mm Hg with each contraction. Drugs or anesthetics that obtund or eliminate the woman's ability to sense pressure of the fetal head on the levator muscles or fear and embarassment of untrained, uneducated, and fearful women who think they are defecating in public interfere with this necessary patient effort.

PAIN CONTROL IN LABOR AND THE DELIVERY ROOM

Origin of the Pain

In the first stage of labor, stretch of the collagenous lower uterine segment and cervix is the major cause of pain. This is certain. It has also been suggested that hypoxia of the uterine muscle induced by contractions may contribute to the pain. This is unproven. The cervix does not contain receptors for hot and cold. It may be burned to a char or frozen without patient complaint. However, stretching, as in labor, or crushing, as in grasping the cervix with a tenaculum or biopsy tool, causes pain. The pain of first-stage labor can be as severe as the pain of ureteral colic. The degree of pain varies considerably from woman to woman because of two factors: the efficiency of the labor and the woman's emotional set. If a woman enters labor with her cervix well prepared for labor (see section on labor), blessed with pelvic bones of ideal size and form, a myometrium that contracts strongly and frequently, and an average-sized fetus presenting as a vertex with the occiput transverse or anterior, her labor will be efficient (short) and her pain less. The woman's emotional set is determined by her maturity, knowledge, and desire for maternity. The illegitimately pregnant, immediately postadolescent, wholly ignorant woman who is plagued with an inefficient labor will suffer incredible torture if not assisted. The mature woman who has chosen motherhood, who understands her role in labor and is prepared for it, and who, because of her socioeconomic blessings, has a very efficient labor, can often proceed through the first stage of labor with little discomfort.

In the second stage of labor, the pain sensation is identical to that associated with an intense desire to defecate. The pelvic floor (the levator muscle particularly) is stretched. Reactions vary considerably. If the woman anticipates this sensation, she may not find it painful. If she is unprepared, a hysterical response may ensue. Imagine the discomfort of a woman thinking she is being "torn apart" and stooling in bed or on the delivery table (in public). In a woman having her first baby, the vaginal orifice can dilate easily to about 5 or 6 cm. Beyond that, lacerations of the pelvic floor, pelvic diaphragm, and the periurethral and vaginal areas may occur. For this reason, an episiotomy (perineotomy) is performed and the incision repaired. Anesthesia is required for this.

Physician's Role

The physician's motivations for providing pain relief in obstetric labor are: 1) the milk of human kindness—labor is painful; 2) labor is part of the developing maternal-child relationship, and we would like the mother not to have a negative memory of the experience. If her experience is a good one, the minor unpleasant details of the event fade very quickly. Pain relief is routinely required for episiotomy and repair and pain relief is a necessity for operative delivery. Ideally, pain relief techniques should be thoroughly effective, should not damage the mother, and should not alter normal fetal homeostasis. The statement is, of course, philosophic, and it must be remembered that there are always small costs associated with pain relief.

Pain Transmission

The pain of the first stage of labor rises in the sympathetic fibers that travel through the pericervical ganglia via the pelvic plexus, through the hypogastric plexus, through the lumbar and lower thoracic plexus, and enter the spinal cord at the level of the 11th and 12th dorsal vertebrae. The parasympathetic portion consists of a few fibers that enter the second, third, and fourth sacral segments. The motor pathway to the uterus leaves the spinal cord at the level of the seventh and eighth thoracic vertebrae. The parasympathetic portion consists of a few fibers that leave the second, third, and fourth sacral segments. Block of the spinal nerves up to the level T 10–12 does not interfere with uterine motor activity. However, anesthesia that rises higher than that level may reduce uterine contractility. Any anesthetic agent that causes loss of sensation in the pelvic floor will, of course, eliminate the bearing-down reflex of the second stage.

Techniques of Pain Relief

Psychologic Preparation

Three basic techniques of psychologic preparation for labor are: 1) knowledge, security in the professional environment, and familiarity with surroundings. These emotionally supportive techniques have been known through the ages, in fact, ever since some society decided that women in labor should have a birth attendant. This technique was espoused by Dr. Grantly Dick Read in England, and the technique is known by his name; 2) psychoprophylactic technique. This technique is generally known as the Lamaze technique because of its introduction in Western Europe by this French obstetrician. However, it has been utilized in the Soviet Union for many years and is based essentially on Pavlovian principles, depending largely on conditioning through breathing and other skeletal muscle exercises. The woman and her labor coach are both instructed in the method, and the coach actively participates by assisting and encouraging the woman; 3) positive reinforcement techniques, including hypnosis. All of these techniques can be very helpful to women.

Certainly the informed woman who believes that she is in a health-care environment that wishes to help her, has the talent to help her, and who has friendly and perhaps familiar (husband) faces about her has less fear and anxiety. It should be recognized that hypnosis can have an adverse effect in prepsychotic women. Even the Read and psychoprophylactic techniques can be harmful if there is excessive commitment to these techniques on the part of the woman or physician. Thus, if because of the severity of the woman's pain she is unable to handle her labor without drugs and has been conditioned to believe that this requirement represents a character failure on her part, a serious threat to her fetus, etc., she may have a severe sense of guilt, particularly if the fetus is not born in good health.

Data provided by a San Antonio childbirth training association on 311 women, of whom 180 were having their first, term child, and who received psychoprophylactic training, revealed the following events associated with their labors:

- Ninety required outlet forceps to complete the second stage of labor.
- Seventeen required cesarean section.
- Fifty-seven required epidural, caudal, or saddle-block anesthesia.
- One hundred and sixty-four required pudendal block or local perineal anesthesia.
- Twenty-nine required inhalation anesthesia.

These statistics are not presented as a critique of the psychophylactic method of childbirth but do provide justification for the need of professional attendance if maternal, fetal, and neonatal health are to be optimal. Because the number of women requiring medical assistance were in the majority and could not have been preselected, the argument against home delivery by lay midwives is sound. Furthermore, no data were available from this group of laboring women concerning the number of newborns requiring immediate medical attention.

Drugs

- Analgesics—opiates, usually meperidine
- Sedatives—usually barbiturates
- Hypnotics—usually scopolamine
- Tranquilizers—usually promethazine, diazepam

The major potential harm of opiates is depression of the neonate. Premature and sick term infants are more easily depressed than normal term infants. Intramuscular administration produces longer periods of analgesia than intravenous administration, but fetal depression usually parallels the maternal analgesic effect. There is no evidence that opiates depress the fetus in utero, but they interfere with the initiation of respiration by the neonate.

Scopolamine is an extremely effective drug at producing amnesia. Thus, one can administer smaller doses of analgesic agents and yet not have the mother remember the experience of pain. However, this combination of little or no analgesia and scopolamine produces a patient who is in total delirium and extremely agitated. The woman is wild, unreasonable, often attempts to get out of bed, and needs constant attention. Larger doses of scopolamine have an atropine-like effect on both the maternal and fetal heart rate. Tranquilizers are useful. The effectiveness of promethazine in the usual doses is rather modest; diazepam in the usual clinical doses is effective. It is suspected that if the woman receives multiple doses of diazepam, the neonate may manifest a withdrawal syndrome, characterized by agitation. All these drugs rapidly pass the placenta to the fetus.

Barbiturates should not be used during active labor because there is no known antidote if the fetus should be born in a depressed state. The effects of the barbiturates persist well beyond the immediate delivery room into the neonatal period in the nursery. In modest combinations of opiate analgesics and barbiturates, the opiates are less effective. Scopolamine is not advocated during labor because patients may become wild and combative. Intravenous meperidine in doses of 50 mg, well diluted with either 5% glucose or the patient's own blood, is an effective analgesic and is reasonably short acting. One can anticipate about one to one and one-half hours of good analgesic effect in 85% of patients.

Paracervical Block

With this technique, a local anesthetic agent is injected through the vaginal wall into the paracervical space in an attempt to block pain transmission via the paracervical ganglia and nerves. The results are good in 80–90% of women. The duration of pain relief is in part dependent on the local anesthetic agent used, but it usually lasts for 45–90 minutes. The specific danger of this technique is extremely rapid transmission of anesthetic agents into the fetal circulation. About 30% of fetuses manifest some mild degree of bradycardia and about 5% manifest severe bradycardia. If the infant should be born during this intoxicated state, it may die. The period of bradycardia rarely lasts beyond 25 minutes, and intrauterine fetal death from this technique is extremely rare.

Caudal Anesthesia

This consists of the injection of local anesthesia into the caudal canal via the sacral hiatus (foramen). Any anesthetic solution in the caudal canal will dissect up around the dural sac. Initially, anesthesia will be essentially pelvic but eventually will spread down the legs if repeated solutions are injected into the space. About 80–90% of women get complete pain relief and have no sensation of uterine contractions at all. When some women enter the second stage of labor, they have no desire to push and often require forceps delivery. The sacral hiatus is sufficiently small in 4% to preclude caudal anesthesia. With progressive dilatation of the vessels in the legs, hypotension occasionally occurs in association with caudal anesthesia. After repeated doses, it often becomes difficult to maintain analgesia, even with increasing doses. Caudal is highly recommended for premature labor.

Lumbar Epidural Block

Lumbar epidural block is similar to caudal anesthesia in principle, but the needle is placed through a lumbar interspace, through the ligamentum flavum, into the epidural space. After a test dose (to be certain that the anesthetic agent is not in the subarachnoid space), local anesthetic agent is injected through a catheter that replaced the needle. Use of a catheter for both caudal and lumbar epidural block allows repeat doses to be injected when necessary. Like caudal, lumbar epidural block may induce hypotension (about 5% of the patients). It is a relatively simple and effective method of producing segmental control of the pain of first stage of labor.

Anesthesia required for the second stage of labor is related to the procedure to be done. For a small episiotomy, the perineal area may be infiltrated with local anesthetic. However, this is not as satisfactory as pudendal block because it causes swelling and edema of the local area. The lower half of the vulva and the perineal body are supplied by sensory nerves via the pudendal nerve. These nerve fibers arise in S_2, S_3, and S_4. Sensory fibers in the upper half of the vulva are contained in the ileoinguinal nerve.

Pudendal Block

This consists of the injection of local anesthetic into the pudendal nerve just as it leaves Alcock's canal and circles the ischial spine. It provides good perineal anesthesia but does not provide anesthesia to the upper portion of the labia majora. If combined with a mild amount of meperidine analgesia, both episiotomy and outlet forceps delivery can be performed under pudendal block. It is not adequate anesthesia for midforceps deliveries.

Spinal Anesthesia

The dura in penetrated with the needle placed in an intervertebral space, usually L-4. This is done with the patient in a sitting position and with the use of 10% glucose as the diluent for the local anesthetic agent. The level of anesthesia can be kept at T-10 or below. The pain relief is excellent in greater than 90% of women, but an occasional patient will get an extremely high level with circulatory collapse and respiratory arrest. Postspinal anesthetic headache is a complication in some women.

General Anesthesia

Every general anesthetic agent known has been administered to pregnant women. Nitrous oxide and oxygen is often used in an 80/20 mixture. This is only a weak analgesic mixture that can be used for the woman who needs some assistance during bearing down. It is inadequate for episiotomy. Higher concentrations of N_2O are asphyxial. Cyclopropane is analgesic in 3–5% concentrations and anesthetic in 6–8% concentrations. It is excellent because of its ability to induce anesthesia rapidly. However, 30% of the infants born from patients who have received cyclopropane require resuscitation, and the gas is explosive. Ether is a good anesthetic, but its induction time is very long, and there is a high incidence of fetal narcosis and uterine relaxation. It has few maternal complications, but it is explosive. Halothane provides rapid anesthesia, a low incidence of fetal narcosis, and a high incidence of uterine relaxation. Intravenous thiopental produces rapid induction of unconsciousness but very high fetal levels in a short period of time. Inhalation anesthesia has little or no place in the conduct of a normal delivery but is very useful for complicated deliveries requiring surgical intervention.

Gastric emptying slows with the onset of uterine contractions, and patients in labor have been known to vomit material as late as 24 hours after its ingestion. The most serious complication of inhalation anesthesia is the aspiration of gastric contents (Mendelson's syndrome). The pathology of Mendelson's syndrome arises from the acidity of the aspirate, not from tracheobronchial obstruction by solid food. Having the woman in labor ingest magnesium and aluminum hydroxide is an effective preventive. Gastric washout prior to inhalation anesthesia is even better.

FETAL COST OF LABOR

The maximum opportunity for increasing human longevity and health resides in the care of the fetus and neonate, not in renal dialysis or coronary bypass surgery. Unfortunately, the relative anonymity of the fetus has always interfered with society's wisdom in its choice to invest health dollars.

Birth of an infant poorly prepared for extrauterine life is the largest single cause of perinatal mortality and morbidity. The morbidity is both immediate and remote. For example, 40% of low-birth-weight children have still not caught up to their full-term brethren at the time they enter first grade, and it is doubtful they ever will. In a typical group of nonindigent patients, one can expect to lose 8 fetuses and 8 neonates per 1,000 deliveries of infants weighing 1,000 gm or more. Maternal, obstetric, and fetal disease are major contributors to perinatal mortality. Labor per se in modern care centers ranks rather low as the cause of perinatal loss. Nonetheless, of those mothers who come into term labor with live fetuses, about 1% do not end up with a living child. In such a group of women, if we exclude deaths associated with congenital anomalies, respiratory distress syndrome, and placental complications and consider only mothers and fetuses who enter labor in apparent good health, about 1 in 300 fetuses dies in utero or shortly after delivery of events related to labor. Asphyxia, trauma, infection, and poisoning account for almost all of these deaths.

Asphyxia deaths are related to failure of oxygen transport. On the maternal side, any condition that reduces perfusion pressure in the uterine arteries (hypotension that may accompany spinal or caudal anesthesia or alpha-adrenergic receptor stimulation) may cause intrauterine asphyxia. However, even if the perfusion pressure in the uterine artery is adequate, flow through the intravillous space may be inadequate. The contents of the uterus form a closed hydrostatic system, and pressure in the intervillous space and the amniotic sac are identical both at rest and during a contraction. When the intrauterine pressure exceeds the pressure in the spiral arteriole (the terminal arteries of the maternal blood supply to the intravillous space), flow through the intervillous space ceases (at about 20–25 mm Hg IUP). Spontaneous contractions in normal labor peak at about 50–60 mm Hg IUP, so it is obvious that for the approximate half minute that the IUP is greater than 25 mm Hg, flow through the intervillous space ceases. Fortunately, the quantity of oxygen transferred across the placenta is not flow dependent but concentration dependent. As long as the oxygen concentration in the uterine artery and intervillous space blood is sufficiently higher than that of the umbilical vein, a concentration gradient will be maintained, and oxygen will be transported to the fetal circulation. However, when the intrauterine pressure is raised above 20–25 mm Hg by prolonged uterine contractions (excess oxytocin administration) or by obstetric complications (some cases of polyhydramnios and abruptio placenta), fetal asphyxia eventuates. In addition to the concentration gradient, size of the placental barrier is also important in oxygen transport, for it must be sufficiently thin to permit rapid diffusion and of sufficient area to permit large quantity transport. It is estimated that loss of 70% of the usual functional maternal-fetal interface results in fetal death.

Fetal injury or death as a result of trauma may occur in labors that are totally uncomplicated. Occasionally, one finds obvious gross intracranial hemorrhage typical of trauma without clinical explanation. Most trauma results from the injudicious use of force—breech extraction, forceps, version and extraction, and vacuum extraction. Sometimes the injury can eventuate from faulty technique; in others, it results from poor choice of technique for the pathologic condition present. While there seems to be a thread of deprecation being sewn into medical education in respect to mechanical pathophysiologies in obstetrics, perhaps because most physicians will not practice delivery room obstetrics, it does not follow that these pathologic entities do not occur or that they or their treatments cannot produce fetal death and injury. Most traumatic injuries to the fetus result from the following series of diagnostic failures: to recognize dysfunctional labor early, to apply the clinical and radiologic examinations necessary to make a pathophysiologic diagnosis, and to provide the operative treatment appropriate to the diagnosis with skill.

Infection of the fetus in utero occurs with (95%) or without (5%) rupture of the membranes. It is a very difficult specific diagnosis to make both before and after birth. However, the combination of ruptured membranes and unexplained maternal fever and tachycardia almost invariably suggests chorioamnionitis, and it may be assumed that the fetus is also infected via the mouth, trachea, bronchi, or umbilicus. A mucopurulent discharge can be seen escaping from the cervix on vaginal examination in severe cases. The diagnosis is exceedingly difficult to make even after the fetus is born, and if we insist on a proven diagnosis before treatment, most infected newborns will die. When the triad of ruptured membranes and unexplained maternal fever and tachycardia occurs, antibiotic therapy is in order to protect the mother. Appropriate treatment for the fetus is delivery and vigorous treatment of the newborn with antibiotics. The vast majority of these fetal bacterial infections acquired in utero result from enteric organisms; however, group B streptococci and anaerobic organisms seem to be increasing in importance as pathogens.

Poisoning results from administration of agents used to relieve the pain of labor. When these drugs produce harmful effects, they operate through three pathophysiologic mechanisms:

1. Drug inhibition of the fetal respiratory center (e.g., opiates, barbiturates, ether); this produces no difficulty in utero but interferes with the initiation of respiration.
2. Asphyxial inhibition of the respiratory center induced by maternal hypotension (spinal, caudal), the result of indirect drug action.
3. Direct poisoning by injection of local anesthetic agents into the fetus (paracervical block) or failure of the fetus to metabolize the agent or of the fetal serum proteins to bind the agent.

The commonest drug used, meperidine, causes fetal respiratory depression infrequently. However, effects are additive, so that the combination of large doses of meperidine with inhalation anesthesia, with prematurity, or with clinical conditions inducing fetal asphyxia tends to have a far greater inhibiting effect on the initiation of respiration than meperidine alone.

SUGGESTED READING

Hendricks CH, Brenner WE, Kraus G: The normal cervical dilatation pattern in late pregnancy and labor. *Am J Obstet Gynecol* 106:1065, 1970.

Pritchard JA, MacDonald PC: *Williams Obstetrics,* ed. 15. New York, Appleton-Century-Crofts, 1976.

Speroff L: What initiates labor? *Contemp Obstet Gynecol* 7:113, 1976.

QUESTIONS

1. Draw a diagram of the nonpregnant uterus labeling corpus, cervix, and internal and external os, indicating the shape and volume of the cavity. Identify the percentage of muscle in (a) upper corpus ____% and (b) lower cervix ____%.
2. Draw a diagram of a pregnant uterus at term in labor. Label upper uterine segment, lower uterine segment, external os, internal os. Identify the muscle-collagen ratio in the upper and lower uterine segment.
3. The uterus accommodates to the enlarging conceptional mass by muscle (a) ________ and (b) ________ and (c) ________ of collagen.
4. The major role of the cervix in pregnancy is to serve as a ________.
5. Changes in the shape of the cervix and its dilatation are accomplished by (a) ________ and (b) changes in cervical ________.
6. Is cervical dilatation initiated *prior to* or *at* the onset of labor? ________
7. Uterine activity in pregnancy becomes more pronounced after ________ weeks of gestation.
8. The intrauterine environment is a closed hydrostatic system. It follows, then, that the intrauterine pressure in the intervillous space and the amniotic sac are (a) the same or (b) different.

9. A patient with triplets at term will have the same intrauterine pressure in normal labor as a patient in premature labor.
TRUE FALSE
10. Until labor is established, uterine contractility in late pregnancy is characterized by variability. TRUE FALSE
11. Myometrial oxytocin sensitivity and spontaneous activity change similarly in the last half of pregnancy. TRUE FALSE
12. Uterine contractions average a *peak* pressure of ________.
13. In an uncomplicated patient in late pregnancy, an episode of active uterine contractility sensed by the patient but not resulting in cervical change is called ________.
14. Patients entering labor have accomplished some effacement and dilatation.
    a. nulliparas are shown more effaced than multiparas
    TRUE FALSE
    b. nulliparas are more dilated than multiparas
    TRUE FALSE
15. The work of the first stage of labor is the completion of cervical (a)________ and cervical (b)________.
16. Prior to the onset of labor, the external os of the cervix is at the level of the ________. At full dilatation, it is at the level of the ________.
17. Is the process of dilatation in labor the result of (a) isotonic or (b) isometric contractions of myometrial cells? ________.
18. In a coordinate plot of cervical dilatation (Y) against time (X) would the normal curve approximate (a) a straight line or (b) an exponential function? ________.
19. In normal active prelabor, the plot of cervical dilatation against time tends to produce a straight line curve because some of the force of myometrial activity is used for ________.
20. Dilatation of the cervix from 4 to 10 cm should be accomplished within (a)________ hours in a nullipara and (b)________ hours in a multipara.
21. Delivery should be accomplished within (a)________ hours after admission in multiparas and (b)________ hours in nulliparas.
22. Rupture of the membranes prior to the onset of labor produces "dry labors" that tend to be prolonged and difficult.
TRUE FALSE

23. Rupture of the membranes provides the opportunity for the ingress of bacteria into the formerly sterile amniotic sac.
TRUE FALSE

24. Multiparas require (a)__________ contractions in the second stage or a maximum of about (b)__________.

25. Nulliparas requires (a)__________ contractions in the second stage or a maximum or about (b)__________.

26. Prolongation of the second stage beyond the normal limits produces asphyxia of the fetus. TRUE FALSE

27. The tissues cut in a midline episiotomy are: (a)__________ (b)__________ (c)__________ (d)__________ (e)__________ (f)__________

28. All labor other than normal is _______________.

29. All etiologies of dysfunctional labor can be classified into three large groups, which are:

__________ __________ __________

30. Give one reason why meperidine is the preferred narcotic analgesic in obstetrics.

31. Pain of the first stage arises from cervical (a) STRETCH, (b) COMPRESSION, or (c) DEFORMATION. _______________

32. Pain of the second stage arises from pelvic floor stretching.
TRUE FALSE

33. Block of the paracervical sympathetic ganglia permits painless episiotomy. TRUE FALSE

34. Block of pelvic sensations (e.g., spinal anesthesia) may (a) PROLONG or (b) SHORTEN the second stage of labor; and may (c) INCREASE or (d) DECREASE the frequency of operative delivery. (Circle correct one.)

35. Paracervical block may produce fetal intoxication by local anesthetic agents manifested by (a) FETAL TACHYCARDIA, (b) FETAL BRADYCARDIA, or (c) NO CHANGE IN FETAL HEART RATE. _______________

36. Pudendal block anesthesia provides pain relief to (a) UPPER VULVA, (b) LOWER VULVA, or (c) MEDIAN RAPHE OF THE PERINEAL BODY. _______________

37. Block of spinal nerves up to T-10–12 does not interfere with uterine motor activity. TRUE FALSE

38. IV Meperidine produces a shorter period of analgesia than IM TRUE FALSE
39. Diazepam reduces fetal medullary cardiac center activity. TRUE FALSE
40. Diazepam produces bradycardia in the newborn infant. TRUE FALSE
41. The major concern in the use of opiates in labor is ________ of the neonate.
42. Scopolamine works as an effective analgesic during labor. TRUE FALSE
43. The pain relief mechanism of lumbar epidural block and caudal anesthesia is quite (a) SIMILAR or (b) DISSIMILAR________
44. Mendelson's syndrome (aspiration of gastric contents in association with anesthetic inhibition of the cough reflex) results from the (a) ACIDITY OF GASTRIC CONTENTS, (b) TRACHEOBRONCHIAL OBSTRUCTION, and may be prevented by (c) ANTACIDS or (d) GASTRIC WASHING PRIOR TO ADMINISTRATION OF ANESTHETIC AGENT. ________.
45. The major mechanisms of fetal and neonatal death are (a) ________, (b)________, (c)________, and (d)________.
46. Give three maternal pathophysiologic mechanisms for fetal asphyxia. (a)________ (b)________ (c)________
47. Give four fetal pathophysiologic mechanisms for asphyxia.
    (a)________ (b)________
    (c)________ (d)________
48. Most but not all traumatic injury to the fetus is avoidable. TRUE FALSE
49. Most obstetric trauma to the fetus occurs in association with ________ labor.
50. Most but not all fetal infections occur by the transvaginal route. TRUE FALSE
51. Most fetal poisoning is associated with pain relief. TRUE FALSE
52. List three mechanisms for harm to the fetus or neonate produced by pain-relief therapeutics.
    (a) ________
    (b) ________
    (c) ________
53. What are the two cardinal signs of intrauterine infection acquired

during labor with ruptured membranes and no evidence of any other infectious disease? (a)__________ (b)__________

54. If intrauterine infection occurs, the fetus may be born with the following abnormal conditions:
(a)________________ (b)________________.

ANSWERS

1. (a) 35%, (b) 5%
2. See Figure 23–1.
3. (a) hyperplasia, (b) hypertrophy, (c) replication
4. sphincter
5. (a) uterine contractions, (b) compliance
6. prior to
7. 20
8. the same (a)
9. FALSE
10. TRUE
11. TRUE
12. 50–60 mm Hg
13. False labor
14. (a) TRUE, (b) FALSE
15. (a) effacement, (b) dilatation
16. ischial spine, pelvic inlet
17. (a) isotonic
18. (b) exponential function
19. effacement
20. (a) 4 hours, (b) 3½ hours
21. (a) 7 hours, (b) 11 hours
22. FALSE
23. TRUE
24. (a) 4–10, (b) 30 minutes
25. (a) 20, (b) 60 minutes
26. TRUE
27. skin, superficial fat, superficial transverse perinea muscle, bul-

bocavernosus muscle, inferior fascia of the urogenital diaphragm, deep transversus perinea.

28. dysfunctional
29. the force, the passenger, the passage
30. less respiratory depression
31. (a) STRETCH
32. TRUE
33. FALSE
34. (a) PROLONG, (c) INCREASE
35. (b) FETAL BRADYCARDIA
36. (b) LOWER VULVA, (c) MEDIAN RAPHE OF THE PERINEAL BODY
37. TRUE
38. TRUE
39. TRUE
40. FALSE
41. respiratory depression
42. FALSE
43. (a) SIMILAR
44. (a) ACIDITY OF GASTRIC CONTENTS, (c) ANTACIDS, (d) GASTRIC WASHING PRIOR TO ADMINISTRATION
45. (a) asphyxia, (b) trauma, (c) infection, (d) poisoning
46. (a) failure of perfusion pressure, (b) uterine artery vasoconstriction, (c) excessive contractility
47. (a) cord obstruction, (b) loss of cotyledonary terminal villi circulation, (c) fetal anemia, (d) fetal congenital heart disease
48. TRUE
49. dysfunctional
50. TRUE
51. TRUE
52. (a) inhibition of respiratory center by drugs
    (b) inhibition of respiratory center by asphyxia
    (c) direct poisoning of the fetus
53. (a) maternal fever, (b) maternal tachycardia
54. (a) pneumonia, (b) bacteremia

twenty-four

the puerperium

OBJECTIVES

1. Know the anatomic, physiologic, and endocrinologic changes that take place during the puerperium.
2. Know the processes responsible for normal breast development and preparation for and establishment of lactation.
3. Become familiar with appropriate breast-feeding techniques.
4. Become familiar with fertility risk and fertility control techniques as applied to the puerperium.
5. Understand the importance of maternal-infant bonding.

The puerperium extends from the end of the third stage of labor until the time at which the pelvic organs have returned to the nonpregnant state, traditionally about six weeks postpartum. A series of well-defined anatomic, physiologic, and endocrinologic changes take place during this time.

ANATOMIC CHANGES

The most striking anatomic change is decrease in uterine size with concomitant decrease in abdominal girth. Involution of the uterus is accomplished by decrease in size of individual myometrial fibers. The process is so rapid that the uterus is no longer palpable abdominally within 10 days of delivery.

After expulsion of the placenta, the uterine cavity collapses to its normal flattened state. Thrombosis of vessels in the placental site occurs, followed in several days by necrosis and slough of the superficial layers of tissue in this area. The remaining decidua spongiosa is also sloughed, causing a bloody discharge known as the lochia rubra. Regeneration of the endometrium is accomplished by growth of glands and stroma from the decidua basalis over the entire cavity, including the placental site. Lochial discharge may continue from 10 to 20 days postpartum but gradually becomes serous in character rather than bloody.

The primiparous uterus never returns to the shape and contour characteristic of the never-pregnant (nulligravidous) state, but the uterus of the previously parous woman will return to essentially the same size and contour after each pregnancy. Once the state of grand multiparity (greater than five term pregnancies) is reached, the nonpregnant uterine size is discernibly larger than it is for women of low parity.

Failure of the uterus to return to normal, nonpregnant size within six weeks is defined as subinvolution. This process may be of greater or lesser significance, depending on accompanying symptoms and the woman's previous status. Subinvolution accompanied by irregular vaginal bleeding and uterine tenderness strongly suggests the presence of retained placental fragments and, in rare instances, trophoblastic disease. Curettage should be carried out in such women for both diagnosis and treatment. If no pain or bleeding is present, the subinvolution may represent an extreme of the physiologic norm or the presence of a uterine lesion such as a myoma. Leiomyomas tend to grow during pregnancy under the influence of estrogen. If myomas are suspected, the woman may be observed for several weeks to define further the process.

PHYSIOLOGIC CHANGES

The most striking physiologic changes in the early puerperium are related to circulating blood volume. Increase in blood volume during pregnancy is accomplished by an increase in red cell mass and a greater increase in plasma volume. After delivery, decrease in red cell mass occurs because of blood loss during delivery. Decrease in plasma and extracellular fluid volume is accomplished through a massive diuresis that occurs in the first 12–24 hours postpartum. The diuresis continues over the next few days and is accompanied by a weight loss of 4–5 lb as the excess fluid is mobilized and blood volume returns to normal. The pulse rate becomes significantly lower because decrease in uterine blood flow effectively increases the circulating blood volume.

Because of the mobilization of extracellular fluid and diuresis, cardiac patients must be observed very closely for signs of congestive heart failure. Patients who have had spinal anesthesia or a difficult delivery with possible periurethral injury must be observed postpartum for urinary retention and secondary bladder atony.

Once the immediate puerperium has passed, the major physiologic changes that occur are related to pregnancy-related effects on connective tissue. It has been noted that the connective tissue of the abdominal wall and pelvis becomes somewhat lax during pregnancy to accommodate the growing uterus and to permit easy passage of the fetus through the bony pelvis. These changes are said to result from the secretion of a protein hormone called relaxin by the corpus luteum of pregnancy. The hormone reportedly effects the pattern of collagen binding to permit greater stretching of the connective tissue. Reversal of this process occurs during the puerperium, but time limits have not been established. It has been traditional to limit the exercise of the puerperal woman to avoid undue stress on abdominal and pelvic connective tissue while this reversal is taking place. Firm evidence is lacking to indicate the length of time that excessive exercise should be avoided and whether, in fact, the puerperal woman needs to be limited beyond what she feels she is physically able to do.

LACTATION

Establishment of normal lactation depends on proper growth and development of the breast both before and during pregnancy. Estrogen secretion beginning at puberty stimulates growth of the nipples and ducts. With the onset of ovulation, progesterone production stimulates growth of the lobular-alveolae. During pregnancy, high levels of estradiol and progesterone stimulate further growth and development of the lobular-alveolar complexes. In addition, prolactin, growth hormone, chorionic somatomammotrophin, insulin, thyroxine, and cortisol are necessary for maximal differentiation of breast tissue. The first histologic signs of lactogenesis can be detected in the second trimester. Lactation may occur in any pregnancy terminated after the second trimester. Lactation during pregnancy is apparently prevented by an estrogen-dependent inhibition of lactogenesis at the level of the breast. This finding is paradoxical because the progressive elevation of prolactin levels during pregnancy appears to be an estrogen-dependent phenomenon also. Prolactin levels increase about 10-fold during pregnancy, going from a mean non-pregnant serum level of 20 ng/ml to a level of 200 ng/ml at term.

Lactation generally does not begin until the third postpartum day, when serum estrogen levels have dropped sufficiently to release local inhibition. Prior to that time, the breast secretes a serous fluid called colostrum, which is an important source of immunoglobulins for the neonate. Suckling during this period is important by providing passive immunity to the infant against a number of viruses and bacterial organisms, especially gastrointestinal pathogens. Suckling at this time also serves as a further stimulus to prolactin production.

Lactation involves two interrelated processes, synthesis and secretion of milk by the mammary epithelium and ejection of milk from the alveoli into the ducts (milk letdown). The former process is prolactin dependent and the latter oxytocin dependent. Suckling by the infant or stripping or pumping of the breasts stimulates encapsulated and unencapsulated receptors in the nipples and ducts. The stimuli are relayed via the lemniscal and reticular afferent systems to the hypothalamus where they then have two end results, production of oxytocin by the posterior pituitary and production of the lactogenic complex (prolactin, growth hormone, and ACTH) by the anterior pituitary. The latter results in increased milk production by the mammary epithelium, while the former results in contraction of the mammary myoepithelium with ejection of milk from the alveoli into the larger ducts. As estrogen and progesterone levels drop postpartum, prolactin levels drop, also. Unless suckling is established, lactation will cease.

Maintenance of lactation does not depend on continuous elevation of prolactin. In the nursing mother, prolactin levels gradually return to normal nonpregnant values. However, a transient prolactin spike occurs after each episode of suckling, thus preparing the breasts for the next feeding. The prolactin level reached with each spike approaches term pregnant values but gradually diminishes over the weeks and months following delivery even though nursing continues. In the non-nursing mother, prolactin returns to nonpregnant baseline values over a period of 15–30 days following delivery.

Suppression of lactation can be achieved only if undertaken shortly after delivery. Medications that have been employed include long-acting injectable testosterone-progesterone combinations and oral estrogens in a decreasing dosage schedule. Neither of these has significantly greater success than administration of a placebo. Recently, a nonsteroidal synthetic estrogen, chlorotrianisene, given in high doses, has been shown to be effective in suppressing lactation. In order for any of these medications to be effective, they must be administered prior to the time that lactation has been established. They are not effective in "drying up" the breasts of nursing mothers. If the medication is not effective, and the mother does not wish to nurse, symptoms of engorgement should be treated symptomatically with good breast support, ice packs, and analgesics. Oxytocin has been administered to promote breast emptying in cases of acute engorgement, but the drainage may stimulate or simulate the suckling reflex and perpetuate the problem.

Weaning is carried out most satisfactorily for both mother and infant by dropping feedings over a period of several days to weeks rather than by suddenly discontinuing breast-feeding. Discomfort should be treated in the same fashion as it is for immediate postpartum breast engorgement.

In order for nursing to be a satisfactory experience for both mother and infant, the mother must be instructed in proper breast-feeding techniques. The mother should be in a comfortable position for both herself and the baby. The baby should be positioned so that its hands cannot push the nipple away. To initiate nursing, the mother should stroke the baby's cheek with her nipple. This will stimulate the rooting reflex and encourage the baby to turn toward the nipple and begin sucking. The baby should not be permitted to nurse for too long to avoid excoriation of the nipples. Nursing should begin on alternate breasts with each feeding. The first breast that is suckled with each feeding will empty more than the second because the baby will suck hardest in the first 5–10 minutes of feeding. The baby should probably not nurse for more than 10–15 minutes on each breast. If the nipples become cracked, they should be left open to the air rather than treated with cream, lotion, or astringents. These agents will neither soften nor harden the nipples but will only tend to continue any irritation.

The nursing mother should be made aware that her infant will want to eat more often than a bottle-fed infant. Because curd size of breast milk is smaller than that of cow's milk, gastric emptying time is faster in breast-fed than in bottle-fed infants. It is not uncommon for the breast-fed neonate to require feeding every two hours. For this reason the nursing mother should be made aware that the infant will make great demands on her time and energy. Other activities, such as housekeeping and food preparation, should be kept to a minimum.

Adequate nutrition is of utmost importance to the lactating woman both for herself and her infant. Caloric intake required for adequate breast-feeding is in the range of 3,500 calories a day. Prenatal iron and folic acid supplements should be continued. Fluid intake should also be in the range of 2 liters a day, if possible. The composition of the diet should be similar to that recommended for pregnancy. Medications known to be secreted in breast milk should be avoided.

RESUMPTION OF OVULATION

Although circulating levels of estradiol and progesterone drop precipitately following delivery, gonadotrophins remain suppressed for a variable period of time. It is not uncommon for the patient three to four weeks postpartum to demonstrate vaginal cytology similar to that seen in postmenopausal patients.

In the nonlactating female, gonadotrophins begin to rise two to three weeks postpartum, and ovulation resumes within 6–20 weeks after delivery. It is not uncommon for the first episode of vaginal bleeding to be anovulatory, but this is by no means the rule. The nonpregnant patient should always be considered potentially fertile.

In lactating women, ovulation tends to be suppressed for a longer period of time. Onset of ovulation usually occurs between 3 and 18 months postpartum. However, lactation does not prevent ovulation in a large percentage of women and should not be considered as adequate contraception. Twenty percent of actively nursing women will resume ovulation by the third postpartum month. Inhibition of ovulation in the lactating woman has been attributed to ovarian refractoriness to gonadotrophin stimulation. This refractoriness may be mediated via prolactin or by other processes stimulated by the suckling reflex.

RESUMPTION OF COITAL ACTIVITY

The appropriate time for the resumption of coitus after delivery has been the subject of great controversy. In the past, it was traditional to interdict coitus until after a six-week postpartum examination. There appears to be no valid scientific basis for such a judgment. Therefore, it is necessary to evaluate the potential hazards of coitus and to determine when these hazards no longer exist. The primary hazards of puerperal coitus are hemorrhage, sepsis, pain, and conception. The first three of these problems may arise either from the uterus or an episiotomy site or laceration. Risk of both hemorrhage and sepsis should be minimal after the 10th postpartum day. By this time, the uterus has involuted markedly, the endometrium has begun to reepithelialize, and the cervix has closed. The episiotomy site should be healing well by this time. Pain in an episiotomy site may persist beyond this time. In addition, relative atrophy of the vaginal mucosa may make coitus painful. It seems reasonable to suggest that the parturient may attempt coitus after the third postpartum week but that alternate methods of sexual activity should be encouraged if coitus is too painful. Water-soluble lubricants may be helpful initially. Persistence with coitus in the presence of pain should be avoided in order to prevent the development of response patterns that may be detrimental to satisfactory function once full healing has taken place.

The risk of fertility is always present regardless of duration of time from delivery. Therefore, the patient should be counseled in satisfactory contraceptive techniques prior to dismissal from the hospital.

PUERPERAL CONTRACEPTION

The patient who desires oral contraceptives may begin therapy within the week after delivery. Oral contraceptives are contraindicated in lactating women for the duration because they are secreted in the breast milk and may decrease milk synthesis. The effect of contraceptive steroids on the nursing infant is uncertain, so it seems best to avoid this possible hazard. Oral contraceptive therapy is also contraindicated in patients with thrombophlebitis or a history of thromboembolic disorders. Oral contraceptives may also be contraindicated in patients who required induction of ovulation in order to conceive.

As neither the intrauterine device nor the diaphragm can be adequately fitted during the immediate puerperium, patients who desire these methods of contraception or patients for whom the use of oral contraceptives is contraindicated should be instructed in the use of barrier methods such as foam and condoms. Continuing methods of contraception should then be fully discussed and attended to at the six-weeks postpartum examination.

BONDING

One aspect of the puerperium that has received great attention of late is that of maternal-infant bonding. The term "bonding" refers to the formation of a strong emotional attachment between mother and child that begins within the first moments of the child's life.

The experiences of the first minutes and hours postpartum may be crucial to the formation of that set of behavior patterns that make a woman willing, if not anxious, to perform tasks for the care and nurturing of her child that might be considered distasteful in other circumstances.

Early forced separation of mothers from infants such as prematures interferes with formation of the maternal-infant bond. Several studies have demonstrated a high incidence of child neglect and child beating in cases in which early maternal-infant contact was prevented. Factors that have been considered crucial in the process of bonding include mutual touching, eye contact, fondling, entrainment (movement in response to speech), infant cry, and high-pitched maternal cooing.

Established practices of hospital-based delivery that generally permit limited maternal-infant contact are the greatest threat to the formation of adequate bonding if the delivery process is normal. The current trend toward home deliveries is essentially the result of antibonding hospital practices. If the current high level of infant survival is to be maintained, a more homelike hospital delivery atmosphere must be achieved that offers the advantages of a hospital delivery without the disadvantage of forced maternal-infant separation.

SUGGESTED READING

Kennell JH, Klaus MH: *Maternal-Infant Bonding*, St. Louis, C.V. Mosby Co, 1976.

Speroff L, Glass RH, Kase NG: *Clinical Gynecologic Endocrinology and Infertility* ed. 2. Baltimore, Williams & Wilkins, 1978.

QUESTIONS

1. Involution of the uterus should be complete in the majority of patients by the
   A. 10th postpartum day
   B. 3rd postpartum week
   C. 6th postpartum week
   D. 8th postpartum week

2. Bloody discharge called lochia commonly continues for
   A. 2–3 days postpartum
   B. 1–2 weeks postpartum
   C. 10–20 days postpartum
   D. 4–6 weeks postpartum
3. Return of maternal blood volume to normal is accomplished primarily by:
   A. blood loss at delivery
   B. massive postpartum diuresis
   C. lochial discharge
   D. expulsion of fetus and placenta with loss of placental volume and amniotic fluid
4. Name the eight hormones required for appropriate breast development and preparation for lactation.
5. Which of the following hormones is primarily responsible for lactogenesis in response to suckling?
   A. oxytocin
   B. chorionic somatomammotrophin
   C. estradiol
   D. progesterone
   E. prolactin
6. Which of the following hormones is primarily responsible for milk letdown?
   A. oxytocin
   B. chorionic somatomammotrophin
   C. estradiol
   D. progesterone
   E. prolactin
7. Which factor accounts for the need for more frequent feeding in breast-fed infants?
   A. inability to get as much milk from the breast as can be obtained from a bottle
   B. small curd size of breast milk with rapid gastric emptying
   C. desire of the infant for close contact with the mother
   D. necessity for keeping the breasts empty to prevent engorgement
8. Resumption of ovulation in the nonnursing mother commonly occurs
   A. 2–3 weeks postpartum
   B. 4–6 weeks postpartum
   C. 6–20 weeks postpartum
   D. 3–18 months postpartum

9. Resumption of ovulation in the lactating mother commonly occurs
   A. 2–3 weeks postpartum
   B. 4–6 weeks postpartum
   C. 6–20 weeks postpartum
   D. 3–18 months postpartum
10. Name four factors important in the formation of maternal-infant bonding.

ANSWERS

1. C
2. C
3. B
4. estradiol, progesterone, cortisol, thyroxine, growth hormone, prolactin, chorionic somatomammotrophin, insulin
5. E
6. A
7. B
8. C
9. D
10. eye contact, touch, cry, entrainment

index